David.

Wavelets

Wavelets

Tools for Science & Technology

Stéphane Jaffard
Université Paris XII
Institut Universitaire de France

Yves Meyer
Ecole Normale Supérieure de Cachan
Académie des Sciences

Robert D. Ryan
Paris, France

siam.
Society for Industrial and Applied Mathematics
Philadelphia

Copyright ©2001 by the Society for Industrial and Applied Mathematics.

10 9 8 7 6 5 4 3 2 1

All rights reserved. Printed in the United States of America. No part of this book may be reproduced, stored, or transmitted in any manner without the written permission of the publisher. For information, write to the Society for Industrial and Applied Mathematics, 3600 University City Science Center, Philadelphia, PA 19104-2688.

Library of Congress Cataloging-in-Publication Data

Jaffard, Stéphane, 1962-
   Wavelets : tools for science & technology / Stéphane Jaffard, Yves Meyer, Robert D. Ryan.
     p. cm.
      "This new book began as a one-chapter revision of Wavelets: algorithms & applications, (SIAM, 1993), which is based on lectures Yves Meyer delivered at the Spanish Institute in Madrid in February 1991" --Pref.
     Includes bibliographical references and indexes.
     ISBN 0-89871-448-6
      1. Wavelets (Mathematics) I. Meyer, Yves. II. Ryan, Robert D. (Robert Dean) 1933- III. Title.

QA403.3 .J34 2001
515'.2433--dc21

00-051607

siam is a registered trademark.

Contents

Preface to Revised Edition ix

Preface from the First Edition xiii

Chapter 1. Signals and Wavelets 1
  1.1 What is a signal? 1
  1.2 The language and goals of signal and image processing 2
  1.3 Stationary signals, transient signals, and adaptive coding 6
  1.4 Grossmann–Morlet time-scale wavelets 8
  1.5 Time-frequency wavelets from Gabor to Malvar and Wilson 9
  1.6 Optimal algorithms in signal processing 10
  1.7 Optimal representation according to Marr 12
  1.8 Terminology 13
  1.9 Reader's guide 13

Chapter 2. Wavelets from a Historical Perspective 15
  2.1 Introduction 15
  2.2 From Fourier (1807) to Haar (1909), frequency analysis becomes
       scale analysis 16
  2.3 New directions of the 1930s: Paul Lévy and Brownian motion 20
  2.4 New directions of the 1930s: Littlewood and Paley 21
  2.5 New directions of the 1930s: The Franklin system 23
  2.6 New directions of the 1930s: The wavelets of Lusin 25
  2.7 Atomic decompositions from 1960 to 1980 26
  2.8 Strömberg's wavelets 28
  2.9 A first synthesis: Wavelet analysis 29
  2.10 The advent of signal processing 31
  2.11 Conclusions 32

Chapter 3. Quadrature Mirror Filters 35
  3.1 Introduction 35
  3.2 Subband coding: The case of ideal filters 36
  3.3 Quadrature mirror filters 37
  3.4 Trend and fluctuation 40
  3.5 The time-scale algorithm of Mallat and the time-frequency
       algorithm of Galand 40
  3.6 Trends and fluctuations with orthonormal wavelet bases 42

3.7	Convergence to wavelets	43
3.8	The wavelets of Daubechies	46
3.9	Conclusions	46

Chapter 4. Pyramid Algorithms for Numerical Image Processing — 49
4.1	Introduction	49
4.2	The pyramid algorithms of Burt and Adelson	50
4.3	Examples of pyramid algorithms	54
4.4	Pyramid algorithms and image compression	55
4.5	Pyramid algorithms and multiresolution analysis	57
4.6	The orthogonal pyramids and wavelets	58
4.7	Biorthogonal wavelets	63

Chapter 5. Time-Frequency Analysis for Signal Processing — 67
5.1	Introduction	67
5.2	The collections Ω of time-frequency atoms	69
5.3	Mallat's matching pursuit algorithm	71
5.4	Best-basis search	72
5.5	The Wigner–Ville transform	72
5.6	Properties of the Wigner–Ville transform	74
5.7	The Wigner–Ville transform and pseudodifferential calculus	76
5.8	Return to the definition of time-frequency atoms	79
5.9	The Wigner–Ville transform and instantaneous frequency	79
5.10	The Wigner–Ville transform of asymptotic signals	81
5.11	Instantaneous frequency and the matching pursuit algorithm	83
5.12	Matching pursuit and the Wigner–Ville transform	84
5.13	Several spectral lines	85
5.14	Conclusions	86
5.15	Historical remarks	86

Chapter 6. Time-Frequency Algorithms Using Malvar–Wilson Wavelets — 89
6.1	Introduction	89
6.2	Malvar–Wilson wavelets: A historical perspective	90
6.3	Windows with variable lengths	92
6.4	Malvar–Wilson wavelets and time-scale wavelets	94
6.5	Adaptive segmentation and the split-and-merge algorithm	95
6.6	The entropy of a vector with respect to an orthonormal basis	96
6.7	The algorithm for finding the optimal Malvar–Wilson basis	97
6.8	An example where this algorithm works	99
6.9	The discrete case	99
6.10	Modulated Malvar–Wilson bases	100
6.11	Examples	102
6.12	Conclusions	104

Chapter 7. Time-Frequency Analysis and Wavelet Packets — 105
7.1	Heuristic considerations	105
7.2	The definition of basic wavelet packets	108
7.3	General wavelet packets	111
7.4	Splitting algorithms	112

7.5	Conclusions	114

Chapter 8. Computer Vision and Human Vision — 117
8.1	Marr's program	117
8.2	The theory of zero-crossings	120
8.3	A counterexample to Marr's conjecture	121
8.4	Mallat's conjecture	122
8.5	The two-dimensional version of Mallat's algorithm	124
8.6	Conclusions	125

Chapter 9. Wavelets and Turbulence — 127
9.1	Introduction	127
9.2	The statistical theory of turbulence and Fourier analysis	128
9.3	Multifractal probability measures and turbulent flows	130
9.4	Multifractal modeling of the velocity field	131
9.5	Coherent structures	137
9.6	Couder's experiments	139
9.7	Marie Farge's numerical experiments	140
9.8	Modeling and detecting chirps in turbulent flows	141
9.9	Wavelets, paraproducts, and Navier–Stokes equations	145
9.10	Hausdorff measure and dimension	147

Chapter 10. Wavelets and Multifractal Functions — 149
10.1	Introduction	149
10.2	The Weierstrass function	150
10.3	Regular points in an irregular background	152
10.4	The Riemann function	157
	10.4.1 Hölder regularity at irrationals	158
	10.4.2 Riemann's function near $x_0 = 1$	163
10.5	Conclusions and comments	164

Chapter 11. Data Compression and Restoration of Noisy Images — 167
11.1	Introduction	167
11.2	Nonlinear approximation and sparse wavelet expansions	168
11.3	Denoising	177
11.4	Modeling images	181
11.5	Ridgelets	184
11.6	Conclusions	185

Chapter 12. Wavelets and Astronomy — 187
12.1	The Hubble Space Telescope and deconvolving its images	187
	12.1.1 The model	187
	12.1.2 Discovering and fixing the problem	188
	12.1.3 IDEA	189
12.2	Data compression	194
	12.2.1 *ht_compress*	194
	12.2.2 Smooth restoration	196
	12.2.3 Comments	197
12.3	The hierarchical organization of the universe	197
	12.3.1 A fractal universe	200

12.4 Conclusions 201

Appendix A. Filter Fundamentals — 203
A.1 The $l^2(\mathbb{Z})$ theory and definitions 203
A.2 The general two-channel filter bank 205

Appendix B. Wavelet Transforms — 209
B.1 The L^2 theory 209
B.2 Inversion formulas 210
    B.2.1 L^2 inversion 211
    B.2.2 Inversion with the Lusin wavelet 213
B.3 Generalizations 215

Appendix C. A Counterexample — 219
C.1 Introduction 219
C.2 The function θ 220
C.3 Representations of $f_0 * \theta_\rho$ and its derivatives 221
C.4 Hunting the zeros of $(f_0 * \theta_\rho)''$ 223
C.5 The functions R, $R * \theta_\rho$, $(R * \theta_\rho)'$, and $(R * \theta_\rho)''$ 225
C.6 $(R * \theta_\rho)''$ and $(R * \theta_\rho)'$ vanish at the zeros of $(f_0 * \theta_\rho)''$ 225
C.7 The behavior of $(R * \theta_\rho)''/(f_0 * \theta_\rho)''$ 226
C.8 Remarks 227
C.9 A case of perfect reconstruction 229

Appendix D. Hölder Spaces and Besov Spaces — 233
D.1 Hölder spaces 233
D.2 Besov spaces 234
D.3 Examples 235

Bibliography — 237

Author Index — 249

Subject Index — 253

Preface to Revised Edition

Wavelet analysis is a branch of applied mathematics that has produced a collection of tools designed to process certain signals and images. This new book is devoted to describing some of these tools, their applications, and their history.

We will trace several of the technical roots of wavelet analysis, going back to the 1930s and before. These are examples of where the mathematical techniques that we now codify as wavelet analysis first appeared. They are for the most part concerned with the internal structure of mathematics itself. We judge that the applied point of view began after World War II and was embedded in a more general philosophical context exemplified by an ambitious program called The Institute for the Unity of Science. This "institute without walls" was a vision, a vision that was shared by such prominent scientists as John von Neumann, Claude Shannon, and Norbert Wiener. It was the time when Claude Shannon discovered the laws that govern the coding and transmission of signals and images. It was the time when Norbert Wiener and John von Neumann unveiled the relationships between mathematical logic, electronics, and neurophysiology. This led to the design of the first computers. It was the time when Dennis Gabor proposed that speech signals should be decomposed into a series of time-frequency atoms he named "logons." It was the time when Eugene P. Wigner and Léon Brillouin introduced the time-frequency plane.

These pioneering scientists opened new avenues in science, and one of these avenues is called time-frequency analysis. Time-frequency analysis, which is based on Gabor wavelets, will be one of the main topics of this book. Gabor wavelets were improved by Kenneth Wilson, Henrique Malvar, and finally by Ingrid Daubechies, Stéphane Jaffard, and Jean-Lin Journé.

In contrast with this established line of research, time-scale analysis has had a harder time. Indeed, time-frequency analysis yields the musical score, the notes with their frequencies and durations of the music we hear. Time-scale analysis focuses on the transients, the attack of the trumpet, which lasts a few milliseconds, and similar nonstationary signals. While time-frequency analysis was born in the 1940s, time-scale analysis emerged in the late 1970s in completely distinct areas such as image processing (E. H. Adelson and P. J. Burt), neurophysiology (David Marr), quantum field theory (Roland Seneor, Jacques Magnen, Guy Battle, Paul Federbush, James Glimm, and Arthur Jaffe), and in geophysics (Jean Morlet). The outstanding collaboration between Alex Grossmann and Jean Morlet gave birth to a new vision that emerged in the 1980s, and the message was the following: While stationary or quasi-stationary signals are adequately decomposed into a series of time-frequency atoms or Gabor-like wavelets, signals with strong transients are better analyzed with the time-scale wavelets developed by Grossmann and Morlet.

A spectacular example where time-frequency analysis and time-scale analysis have been able to compete is the new JPEG-2000 compression standard for still images. This new standard is based on time-scale wavelets. The old JPEG standard was based on an algorithm called the discrete cosine transform, which is a kind of windowed Fourier transform. This algorithm belongs to the time-frequency group. (Here one ought to say "space-frequency," since an image is a two-dimensional signal.) In the case of JPEG-2000, and in similar compression problems, time-scale wavelets have been preferred over time-frequency wavelets. This success story was not available when the original book first appeared.

This new book began as a one-chapter revision of *Wavelets: Algorithms & Applications* (SIAM, 1993), which is based on lectures Yves Meyer delivered at the Spanish Institute in Madrid in February 1991. While Yves Meyer and Robert Ryan were working on the translation and revision of the new chapter, which ultimately became Chapter 11 of the current book, it became clear, based on the many developments in both the theory and applications since 1993, that an extensive revision of the original book was needed. Since Stéphane Jaffard already had suggested a number of changes and additions, particularly in the sections involving the analysis of multifractal functions, where he is a recognized expert, he was invited to join the project. The result of our collaboration is an almost completely new book, and thus we have given it a new title. Although we have retained the core of the first four chapters, many parts of these chapters have been rewritten and expanded, particularly Chapters 1 and 2. Appendix A has been added as an introduction to some basic filter concepts and hence as a complement to Chapter 3. Chapter 5 has been completely rewritten; it contains new material on chirps that was not known when the first edition was published. Chapters 6 and 7 have been slightly expanded, but they generally follow the original texts. Rather than expanding Chapter 8, we have added Appendix C, which is devoted to a complete discussion of a counterexample to a conjecture of Stéphane Mallat on zero-crossings. This counterexample was outlined in the first edition, but this is the first time the details have been published.

Chapters 9 and 10, although based on the first edition, are considerably expanded and hence essentially new. Chapter 9 (formerly Chapter 10) tells a much more complete and up-to-date story about the use of wavelets for the study of turbulence. Chapter 10 (based on the former Chapter 9) contains a complete analysis of the Weierstrass and Riemann functions, plus a general discussion about the use of wavelets to analyze multifractal functions. Appendix B complements Chapter 10 by providing key results (with proofs) about some wavelet transforms and their inverses. The treatment here is perhaps slightly different from other developments of this now-classical theory.

Chapter 11 is the original motivation for this new book, and we consider it the centerpiece. Here we discuss the intriguing interaction between wavelets and nonlinear analysis and the applications of this line of research to image compression and denoising. Since this chapter involves the concepts that may not be familiar to some readers, we have added Appendix D to introduce Hölder and Besov spaces, plus results on their characterizations in terms of wavelet coefficients.

The original edition contained two pages about the then-emerging use of wavelets in astronomy. It was written at a time when the applications of wavelets to astronomy were received with skepticism. Wavelets are today recognized as an essential tool in astronomy. This story has been expanded in Chapter 12, where we have written a detailed analysis of how wavelets are used in two specific algorithms. We also discuss the use of wavelets to understand the hierarchical structure of the universe and its evolution. This is embedded in a historical context going back to the eighteenth century.

The bibliography has been considerably expanded to include research papers from each of the applications discussed, as well as many books and papers of general or historical interest. We have not listed any of the many websites that exist. Instead, we encourage the reader to visit the "official" wavelet site, www.wavelet.org, which is edited by Wim Sweldens with support from Lucent Technologies. Here one will find lists of regularly updated references, a calendar of events, links to homepages of researchers, and links to sites from which wavelet software can be downloaded.

Given the scope of the applications in this book, it is clear that we are not experts in each, and thus we have relied on the help of others. We wish to thank specifically several individuals for their time, patience, and thoughtful comments: Richard Baraniuk, Guy Battle, Albert Bijaoui, Yves Bobichon, Albert Cohen, Joseph L. Gerver, Hamid Krim, John Rayner, Sylvie Roques, Marc Tajchman, Bruno Torrésani, and Eva Wesfreid.

Stéphane Jaffard
Yves Meyer
Robert D. Ryan

Preface from the First Edition

The "theory of wavelets" stands at the intersection of the frontiers of mathematics, scientific computing, and signal processing. Its goal is to provide a coherent set of concepts, methods, and algorithms that are adapted to a variety of nonstationary signals and that are also suitable for numerical signal processing.

This book results from a series of lectures that Mr. Miguel Artola Gallego, Director of the Spanish Institute, invited me to give on wavelets and their applications. I have tried to fulfill, in the following pages, the objective the Spanish Institute set for me: to present to a scientific audience coming from different disciplines, the prospects that wavelets offer for signal and image processing.

A description of the different algorithms used today under the name "wavelets" (Chapters 2–7) will be followed by an analysis of several applications of these methods: to numerical image processing (Chapter 8), to fractals (Chapter 9), to turbulence (Chapter 10), and to astronomy (Chapter 11). This will take me out of my domain; as a result, the last two chapters are merely resumes of the original articles on which they are based.

I wish to thank the Spanish Institute for its generous hospitality as well as its Director for his warm welcome. Additionally, I note the excellent organization by Mr. Perdo Corpas.

My thanks go also to my Spanish friends and colleagues who took the time to attend these lectures.

CHAPTER 1

Signals and Wavelets

The purpose of this chapter is to give the reader a fairly clear idea about the scientific content of the book. All of the themes that will be developed in this study, using the necessary mathematical formalism, already appear in this overture. It is written with a concern for simplicity and clarity and avoids as much as possible the use of formulas and symbols.

Signal and image processing ultimately involve a collection of numerical techniques, or algorithms. But like all other scientific disciplines, signal and image processing assume certain preliminary scientific conventions. We have sought in this first chapter to describe the intellectual architecture underlying the algorithmic constructions that will be presented in other parts of the book.

1.1 What is a signal?

Signal processing has become an essential and ubiquitous part of contemporary scientific and technological activity, and the signals that need to be processed appear in most sectors of modern life. Signal processing is used in telecommunications (telephone and television), in the transmission and analysis of satellite images, and in medical imaging (echography, tomography, and nuclear magnetic resonance), all of which involve the analysis, storage or transmission, and synthesis of complex time series. Signal processing occurs in most late-model automobiles, typically for some monitoring or control function. The record of a stock price is a signal, and so is a record of temperature readings that permit the analysis of climatic variations and the study of global warming.

Does there exist a definition of a signal that is appropriate for the field of scientific activity called signal processing? We will not be mathematically precise on this point; instead, we provide a working definition. A needlessly broad definition of signal could include the sequence of letters, spaces, and punctuation marks appearing in Montaigne's *Essays*, but the tools we present do not apply to such a signal. We note, however, that the structuralist analysis done by Roland Barthes on literary texts shares some interesting similarities with multiresolution analysis (Chapter 4). The point of contact is the notion of scale. Barthes used the idea of scale in his analysis of literary texts, where different scales are represented, for example, by book, chapter, paragraph, sentence, and word. We will see that the definition of multiresolution analysis is built on the concept of scale.

The signals we study will always be sequences of numbers and not sequences of letters, words, or phrases. These numbers often come from measurements, which are typically made using some recording device. We think of these signals as being functions of time like music and speech or, in some cases, as functions of position. For example, by properly associating numbers with the four bases of a DNA molecule, one obtains a signal that can be analyzed by the methods we describe in Chapter 9. Here we are thinking of one-dimensional signals, functions of a single time or space variable.

It is equally important to consider two-dimensional signals, which we call images. Here again, image processing is done on the numerical representation of the image. For a black and white image, the numerical representation is created by covering the image with a sufficiently fine grid and by assigning a numerical gray scale, denoted by $f(x,y)$, to each grid point (x,y). The value of $f(x,y)$ is an average of the gray scales of the image in a neighborhood of (x,y). The image thus becomes a large matrix, and image processing is done on this matrix.

These arrays can be enormous, and as soon as one deals with a sequence of images, such as in television, the volume of numerical data that must be processed becomes immense. Is it possible to reduce this volume by discovering hidden laws, or correlations, that exist between the different pieces of numerical information representing the image? This question leads us naturally to consider some of the goals of the scientific discipline called signal processing.

1.2 The language and goals of signal and image processing

The subjects we are going to study appear in the scientific landscape where parts of mathematical physics, mathematics, and signal processing intersect, and consequently they share language from these disciplines. This can be confusing, so it is useful to explain some of the terms that we will be using. In so doing, we will introduce the signal processing tasks that appear throughout the rest of the book.

"Analysis" has the same meaning in science that it has in ordinary language. The standard dictionary definition of "analyze" is to separate the whole (of either a physical substance or abstract idea) into its essential parts to examine the relationships between these parts as well as their relationship to the whole. The concept of analysis provides a program of work based on this hypothesis: Behind the apparent complexity of the world there is a hidden order that is accessible through analysis. The complexity is due to the mixture, to the combination of simple entities. The objective of analysis is to discover the nature of these constituents and how they relate to one another. This program is one of the pillars of modern science.

In chemistry, this approach led to the preparation of pure substances and to the discovery of molecules and atoms, and it continues today in particle physics. The synthesis of urea by Friedrich Wöhler in 1828 was proceeded by, and based on, its analysis. Analysis often has the same meaning in mathematics. Take, for example, Fourier analysis and assume that the complex object to be studied is a continuous, 2π-periodic function of a real variable. One tries to decompose the function into its structural elements. These are the simplest of the 2π-periodic functions, namely, the sines and cosines. The analysis furnishes the Fourier coefficients. The analysis is validated by a synthesis, and here the synthesis is additive. It amounts to representing the analyzed function by its Fourier series. The synthesis is successful, however, only after the rules for combining the components are established. In our example, this amounts to finding a summation process that ensures the convergence of the Fourier series furnished by the analysis.

In contrast to chemistry, where the constituent parts are well defined, Fourier analysis is not the only way to study the properties of continuous, 2π-periodic functions. For example, by reinterpreting work by G. H. Hardy on a series attributed to Bernhard Riemann, Matthias Holschneider and Philippe Tchamitchian have shown that wavelet analysis is more sensitive and efficient than Fourier analysis for studying the differentiability of the Riemann function at a given point.

In Fourier analysis, the structural elements are unique; they are sines and cosines. However, in wavelet analysis we will encounter many kinds of wavelets and other objects, such as wavelet packets. Unlike Fourier analysis, wavelet analysis favors no particular set of analyzing functions. There are many analyses, and we are led to the concept of a "box of tools" containing different analytic methods. Each of these methods provides a different way to view complexity. The choice of analytic method is justified by the goal of the analysis.

These remarks apply particularly to signal processing. To analyze a signal means, in this book, to look for the constituent elements. These constituent elements are the elementary signals, the simplest signals into which the given signal can be decomposed. But an analysis makes sense only if it enables one to understand the properties of the object being analyzed and to understand its complexity. We will return to this aspect of signal analysis in section 1.3, where we introduce atomic decompositions.

The term "coding" comes from information theory and signal processing, where it, like "analysis," has many uses. "Transform coding" is a general term that refers to taking a linear transform of a signal or image. Fourier analysis is a form of transform coding, as are the algorithms discussed in Chapters 4 and 5. Note that "coding" and "analysis" do not always refer to linear processes. The coding by zero-crossings discussed in Chapter 8 is nonlinear. However, in each case, coding involves methods to transform the recorded numerical signal into another representation that is—depending on the nature of the signals studied—more convenient for some task or further processing. Decoding is simply the inverse of coding, and it means the same thing here as synthesis, or reconstruction.

"Transmission" and "storage" have their ordinary meanings, but in the context of signal processing, these terms can involve layers of complexities. Every transmission channel, whether an old telegraph line or a modern satellite link, has a definite bandwidth and a computable cost for its use. Similarly, every storage medium has performance limitations and a price tag. The costs of information storage and transmission account for much of the economic motivation behind signal processing: The goals are to provide transmission and storage at a given level of performance for the lowest cost. Transmission and storage are often interrelated, in the sense that what is stored must be accessed and transmitted. These ideas will be illustrated later with examples, including the storage of fingerprints by the Federal Bureau of Investigation (FBI) and the storage and transmission of astronomical images (Chapter 12).

The constraints placed on transmission and storage require that information be compressed. For example, it is too slow and too expensive to transmit raw images over the Internet. Before being transmitted, images are compressed using one of several schemes such as Joint Photographic Experts Group (JPEG) and Graphic Interchange Format (GIF). Very roughly, this is how the compression we will discuss works: A digital signal is analyzed, or coded. Either by design or luck, many of the coefficients that come from the coding are either zero or close to zero, and the other coefficients contain the "important information" or "significant features" of the signal. The small coefficients are set equal to zero, and the others are "quantized" and transmitted. These are received at the end of the channel and are used to decode, or synthesize, the signal.

It is important to note that information typically is lost when small coefficients are set equal to zero and when the other coefficients are quantized. The trick, however, is to do the compression is such a way that the lost information is not noticed. If all of this is done cleverly, the reconstructed signal is, for the purposes at hand, as good as the original. A one-dimensional example is the digital telephone: The compression and transmission must be compatible with the 64 Kbit/second standard, which limits without recourse the quantity of information that can be transmitted in one second. At the same time, the quality must be such that the person at the receiver can recognize the voice at the other end.

The compression we have just described should not be confused with another kind of compression that is well known to Internet users, namely, the compression of applications files. Here there must be absolutely no loss of information, and the decompressed file must be bit-for-bit the same as the original. This kind of compression is an example of what is called *entropy coding*, which is another use of "coding." Most, but not all, uses of "coding" refer to either transform coding or entropy coding.

Quantization is an unavoidable (and undesirable) part of this process. Theoretically, the coefficients given by a coding algorithm are arbitrary real or complex numbers, but practically, processors have finite precision, and they produce rational numbers whose dyadic expansions have a fixed length. The desired quality of the restored image, the channel capacity, and the cost dictate the length of the dyadic numbers that will be transmitted. Mapping the coefficients from the coding algorithm into a finite number of "bins" is called quantization or, more precisely, scalar quantization. A more sophisticated process called vector quantization maps vectors of coefficients into "bins" in $\mathbb{R}^n$ (n-dimensional Euclidean space). We will not be discussing quantization, but we wish to emphasize how important quantization is to the overall efficiency of the process. Quantization is an art, and the way it is done can "make or break" an algorithm.

In most of the cases to be discussed, the analysis and synthesis (coding and decoding, or reconstruction) are theoretically invertible processes: There is no loss of information, and one obtains perfect reconstruction of the original signal. Quantization, however, is not an invertible process and, unfortunately, it introduces systematic errors known as quantization noise. It is desired that the algorithms used for coding—taking into account the nature of the signals—reduce the effects of quantization noise. One of the advantages of quadrature mirror filters is that they "trap" this quantization noise inside well-defined frequency channels. These filters will be studied in Chapter 3.

There is another aspect of the coding-transmission-decoding process that needs to be mentioned: Having quantized the coefficients into bins, it is customary to code the bins before transmission. This coding is entropy coding, and as indicated above, it is completely reversible. The idea is to transmit the information as efficiently as possible, using the statistical structure of the information to be transmitted. Perhaps the best-known example of entropy coding is the Morse Code, which codes the most frequently used letters with the simplest sequences of dots and dashes. The total efficiency of a compression scheme depends on the analysis, quantization, and entropy coding and how they work together.

In addition to transmission and storage, there is a collection of signal processing tasks called diagnostics. Roughly speaking, this is like asking and answering a question about a signal. For example, Does a given sample of speech belong to one of several speakers? Or, Is an underwater acoustic signal coming from a submarine or a ship? For the most part, this book does not deal with diagnostics; however, a few comments are indicated.

A diagnostic often depends on extracting a small number of significant parameters from a signal whose complexity and size are overwhelming. Some scientists believe that diagnostics would be easier if the signal or image has been correctly analyzed and compressed. From this point of view, analysis and the diagnostic are naturally related to data compression, and clearly, if this compression is done inappropriately, it can falsify the diagnostic. In the first edition of this book, we took the position that proper compression was relevant, or even necessary, for a given diagnostic task. Our position has changed, based mainly on a series of lectures by David Mumford delivered at the Institut Henri Poincaré in the fall of 1998.[1] We now feel that most diagnostic tasks are related to statistical modeling of a given collection of signals or images. Statistical modeling is an important field of research that is based on a fascinating set of tools. However, a discussion of statistical modeling lies well beyond the scope of this book.

Finally, we mention restoration. Signal restoration is analogous to the restoration of old paintings. It amounts to ridding the signal of artifacts and errors, which we call noise, and to enhancing certain aspects of the signal that have undergone attenuation, deterioration, or degradation. We will discuss an application of wavelets to signal restoration in Chapter 11.

So what are the goals of signal and image processing? Experts in signal processing are asked to develop, for a given class of signals, algorithms that perform certain tasks or operations. These algorithms should lead to the construction of microprocessors, like those that exist in cell phones and automobiles, that execute these tasks automatically. Some of the important tasks have been described above: coding, diagnostics, quantization and compression, transmission or storage, decoding, and restoration.

We will use several examples to illustrate the nature of these operations and the difficulties they present. It will become clear that no "universal algorithm" is appropriate for the extreme diversity of the situations encountered. Thus, a large part of this work is devoted to describing coding or analysis algorithms that can be adapted to particular classes of signals that one needs to process.

Our first example illustrates restoration and diagnostic. One is interested in splitting a signal into the sum of two terms: The first term contains the information one wishes to recover, and the second term is the noise one wishes to erase. The problem is the study of climatic variations and global warming. This problem was discussed in detail by Professor Jacques-Louis Lions at the Spanish Institute in 1990 [174]. In this example, one has fairly precise temperature measurements from different points in the northern hemisphere that were taken over the last two centuries, and one tries to discover if industrial activity has caused global warming. The extreme difficulty of the problem stems from the existence of significant natural temperature fluctuations. Moreover, these fluctuations and the corresponding climatic changes have always existed, as we have learned from paleoclimatology [250]. Thus, to have access to the "artificial" heating of the planet resulting from human activity and to develop a diagnostic, it is essential to analyze, and then to "erase," these natural fluctuations, which play the role of noise.

A more surprising example appears in neurophysiology. The optic nerve's capacity to transmit visual information is clearly less than the volume of information

---

[1] The ideas presented in these lectures can be found in [215] and [216].

collected by all the retinal cells. Thus, there must be low-level processing of information before it transits the optic nerve. David Marr developed a theory to understand the purpose and performance of this low-level processing, which is a type of coding and compression [198]. We present this theory in Chapter 8.

The problems encountered in archiving data—as well as problems of transmission and reconstruction—are illustrated by the FBI's task of storing the American population's fingerprints. Over 200 million fingerprint records must be stored, and the use of inked impressions on paper cards is no longer practical. The FBI began digitizing fingerprint records some years ago as part of a modernization program, but due to the massive amount of data (10 megabytes per record) it was decided that some form of compression was needed. In addition to efficient storage, it was also important to access the fingerprint files quickly and to transmit them electronically throughout the world. The goal was to be able to reconstruct the received image on a laptop computer, and the quality of the reconstructed image had to be such that the end user, whether a fingerprint expert or an automated fingerprint feature extractor, would have no difficulty interpreting the image. It was decided that coding and compression offered the only solution. Different image-compression algorithms were tested, and a wavelet-based algorithm, a variant of one described in Chapter 6, gave the best results, where "best" involved the speed of the algorithm as well as the compression ratio and the quality of the reconstructed image. This established the standard for fingerprint compression and reconstruction that is used today. (For further details, see Christopher Brislawn's paper [41].)

We have just described and illustrated some of the more important goals of signal and image processing that focus on compression, transmission and storage, and the attendant algorithms for coding and reconstruction. It is important to note, however, that there are many other significant problems in signal processing that will not be discussed. In particular, there is a vast area of signal processing based on probability and statistics that is beyond the scope of our work. As mentioned above, statistical modeling is crucial for high-level signal processing tasks like feature or pattern analysis and diagnostics. This is not to say that wavelets do not or will not play a role is this expanded arena; it is rather that here we limit ourselves, for the most part, to a deterministic theory. The few exceptions include some notes on Brownian motion and the appearance of noise in some of the examples.

Before leaving this section, we believe it is important to reiterate a theme hinted at above: For the most part, we will be discussing coding algorithms and the role wavelets play in these algorithms. These techniques are clearly important in today's technology, but they are only a part of the overall process. The quality of the total process depends on blending analysis, quantization, entropy coding, transmission, and decoding—all of which are interdependent—and ultimately, on implementing these processes in hardware.

1.3 Stationary signals, transient signals, and adaptive coding

We have just defined a set of tasks, or operations, to be performed on signals or images. These tasks form a coherent collection. The purpose of this book is to describe a group of coding algorithms that have been shown, during the last few years, to be particularly effective for compression and for analyzing certain signals that are not stationary. We also will describe several "meta-algorithms" that allow one to choose the coding algorithm best suited to a given signal. To approach this problem of choosing an adaptive algorithm, we briefly classify signals by distinguishing stationary signals, quasi-stationary signals, and transient signals.

A signal is stationary if its properties are statistically invariant over time. A well-known stationary signal is white noise, which in its sampled form appears as a series of independent drawings. A stationary signal can exhibit unexpected events, but we know in advance the probabilities of these events. These are the statistically predictable unknowns.

The ideal tool for studying stationary signals is the Fourier transform. In other words, stationary signals decompose canonically into linear combinations of "waves," that is, into sines and cosines. In the same way, some interesting classes of signals that are not stationary decompose more naturally into linear combinations of wavelets. These heuristics should not be taken too literally, since the full class of signals that are not stationary is too large to be processed by a single methodology. The study of nonstationary signals, where transient events appear that cannot be predicted, even statistically with knowledge of the past, necessitates techniques different from Fourier analysis. These techniques, which are specific to the nonstationary character of the signal, include wavelets of the time-frequency type and wavelets of the time-scale type. Time-frequency wavelets are suited, most specifically, to the analysis of quasi-stationary signals, while time-scale wavelets are adapted to signals exhibiting complicated geometrical features. Examples are edges in images and fractal or multifractal signals.

Before defining time-frequency wavelets and time-scale wavelets, we will indicate their common points. They belong to a more general class of algorithms that are encountered in mathematics and in speech processing. Mathematicians speak of *atomic decompositions*, while speech specialists speak of *decompositions in time-frequency atoms*. The scientific reality is the same in both cases.

As we have already mentioned, an atomic decomposition consists in extracting the simple constituents that make up a complicated mixture. Contrary to what happens in chemistry, the "atoms" that are discovered in a signal have no physical reality; they will depend on the point of view adopted for the analysis. These "atoms" will be time-frequency atoms when we study quasi-stationary signals, but they could, in other situations, be replaced by time-scale wavelets, which also are called Grossmann–Morlet wavelets.

These "atoms" or "wavelets" have no more physical existence than a specific number system used to do some numerical computation. Each number system has an internal coherence, but no scientific law asserts that multiplication must necessarily be done in base 10 rather than base 2. On the other hand, we feel the number system used by the Romans is excluded for practical reasons, since it is not particularly suitable for multiplication.

Having different algorithms that allow us to code a signal by decomposing it into time-frequency atoms is a somewhat similar situation. The decision to use one or the other of these algorithms will be made by considering their "performance." How well they perform must be judged in terms of one of the anticipated goals of signal processing. An algorithm that is optimal for compression can be disastrous for analysis: A standard L^2 energy criterion for the compression could cause details that are important for the analysis to be systematically neglected.

These thoughts will be developed and clarified in sections 1.6 and 1.7. At this point, however, we need to be more specific and define wavelets, which we do in the next two sections.

1.4 Grossmann–Morlet time-scale wavelets

Time-scale analysis—which should be called space-scale in the image case, and which is closely related to multiresolution analysis—involves using a vast range of scales. This notion of scale, which appropriately reminds us of cartography, implies that the signal (or image) is replaced, at a given scale, by the best possible approximation that can be drawn at that scale. By "traveling" from the large scales toward the fine scales, one "zooms in" and arrives at more and more precise representations of the given signal.

The analysis is then done by calculating the change from one scale to the next. This produces the details that allow one, by correcting a rather crude approximation, to move toward a better quality representation. This algorithmic scheme is called multiresolution analysis and is developed in Chapters 3 and 4. Multiresolution analysis is equivalent to an atomic decomposition where the atoms are wavelets.

We define these wavelets by starting with a function ψ of the real variable t. This function is called a *mother wavelet* if it is well localized and oscillating. (It resembles a wave because it oscillates, and it is a wavelet because it is localized.) The localization condition is expressed in the usual way by saying that the function decreases rapidly to zero as $|t|$ tends to infinity. The second condition suggests that ψ vibrates like a wave. Mathematically, we require that the integral of ψ be zero and that the other first m moments of ψ also vanish. This is expressed by the relations

$$\int_{-\infty}^{\infty} t^n \psi(t)\, dt = 0 \quad \text{for} \quad n = 0, 1, \ldots, m-1. \tag{1.1}$$

The mother wavelet ψ generates the other wavelets of the family $\psi_{(a,b)}$, $a > 0$, $b \in \mathbb{R}$, by change of scale and translation in time. (The scale of ψ is conventionally one, and that of $\psi_{(a,b)}$ is $a > 0$; the function ψ is conventionally centered around zero, and $\psi_{(a,b)}$ is then centered around b.) Thus we have

$$\psi_{(a,b)}(t) = \frac{1}{\sqrt{a}} \psi\left(\frac{t-b}{a}\right), \quad a > 0, \quad b \in \mathbb{R}. \tag{1.2}$$

Alex Grossmann and Jean Morlet showed in the early 1980s that this collection can be used as if it were an orthonormal basis when ψ is real-valued [133]. This means that any signal of finite energy can be represented as a linear combination of wavelets $\psi_{(a,b)}$ and that the coefficients of this combination are, up to a normalizing factor, the scalar products $\int_{-\infty}^{\infty} f(t)\overline{\psi}_{(a,b)}(t)\, dt$. These scalar products measure, in some sense, the fluctuations of the signal f around the point b, at the scale given by $a > 0$.

It required uncommon scientific intuition to assert, as Grossmann and Morlet did, that this new method of time-scale analysis was suitable for the analysis and synthesis of transient signals. Signal processing experts were at first annoyed by the intrusion of these two poachers on their preserve and made fun of their claims. This polemic had a short life, and in fact, the argument should never have arisen because the methods of time-scale or multiresolution analysis had existed for five or six years under various disguises: in signal analysis under the name of quadrature mirror filters and in image analysis under the name of pyramid algorithms.

The first to report on this was Stéphane Mallat. He constructed a guide that allowed the same signal analysis method to be recognized under very different presentations, including wavelets, pyramid algorithms, quadrature mirror filters, and Littlewood–Paley analysis. Mallat's brilliant observations led to the mathematical definition of multiresolution analysis, which provides a theoretical umbrella for our subject.

Ingrid Daubechies discovered orthonormal wavelet bases having preselected regularity and compact support [71] (see also [73]). The only previously known case was the Haar system (1909), which is not regular. Thus almost 80 years separated Alfred Haar's work and its natural extension by Daubechies. On the other hand, the wavelets invented by Daubechies—or more precisely the biorthogonal versions developed slightly later—have taken less than 10 years to enter the mainstream of technology. The construction of Daubechies wavelets will be discussed in Chapter 3 and biorthogonal wavelets will be discussed in Chapter 4. The relevance of having smooth wavelets will be explained in Chapter 2.

1.5 Time-frequency wavelets from Gabor to Malvar and Wilson

Dennis Gabor, in 1946, was the first to introduce time-frequency wavelets [124], and the functions he used are called Gabor wavelets. He had the idea to divide a wave—whose mathematical representation is $\cos(\omega t + \varphi)$—into segments and to use one of these segments as the analyzing function. This was a piece of a wave, or a wavelet, which had a beginning and an end.

To use a musical analogy, a wave corresponds to a note (A 440, for example) that has been emitted since the origin of time and continues indefinitely, without attenuation, until the end of time. A wavelet then corresponds to the same A 440 that is struck at a certain moment, say, on a piano, and is later muffled by the pedal. In other words, a Gabor wavelet has (at least) three pieces of information: a beginning, an end, and a specific frequency in between.

Difficulties appeared when it was necessary to decompose a signal using Gabor wavelets. As long as one does only continuous decompositions (using all frequencies and all time), Gabor wavelets can be used as if they formed an orthonormal basis, in the same sense described above for the Grossmann–Morlet wavelets. There are problems, however, with a direct discrete version of the Gabor decomposition. In the late 1940s, a number of investigators, including Léon Brillouin, Dennis Gabor, and John von Neumann, felt that the system $e^{2\pi i k x} g(x - l)$, $k, l \in \mathbb{Z}$, where g is the Gaussian $g(x) = \pi^{-1/4} e^{-x^2/2}$, could be used as a basis to decompose any function in $L^2(\mathbb{R})$ (see [40], for example). Two physicists, Roger Balian (1981 in [17]) and Francis Low (1985 in [177]), proved independently that this is not the case. Furthermore, the Balian–Low theorem shows that the particular choice of g is not the problem and that the result cannot be true for any smooth, well-localized function.

It is only recently, by abandoning Gabor's approach, that two scientists working in different fields and in different parts of the world—Henrique Malvar in signal processing in Brasilia and Kenneth Wilson in physics at Cornell University—have discovered time-frequency wavelets having good algorithmic qualities. These special time-frequency wavelets, which we call Malvar–Wilson wavelets, are particularly well suited for coding speech and music.

The decomposition of a signal in an orthonormal basis of Malvar–Wilson wavelets imitates writing music using a musical score. But this comparison is misleading because a piece of music can be written in only one way, whereas there exists a non-denumerable infinity of orthonormal bases of Malvar–Wilson wavelets. Choosing one of these is equivalent to segmenting the given signal and then doing a traditional Fourier analysis on the delimited pieces. What is the best way to choose this segmentation? This question leads us naturally to the next section.

1.6 Optimal algorithms in signal processing

Which wavelet to choose? This question has often been posed at meetings held since 1985 on wavelets and their applications. But this question needs to be sharpened. What freedom of choice is at our disposal? What are the objectives of the choices we make? Can we make better use of the choices offered to us by considering the anticipated goals? These are several of the questions we will try to answer.

The goal we have in mind is aptly illustrated by a remark Benoît Mandelbrot made in an interview on the French radio program *France Culture*. He noted that "the world around us is very complicated" and that "the tools at our disposal to describe it are very weak." It is notable that Mandelbrot used the word "describe" and not "explain" or "interpret." We are going to follow him in this, ostensibly, very modest approach. This is our answer to the problem about the objectives of the choices: Wavelets, whether they are of the time-scale or time-frequency type, will be chosen to describe as well as possible the reality around us. This description may lead to scientific understanding and the formulation of scientific laws, but once formulated, the wavelets themselves disappear. We have no reason to believe that there are scientific laws that are written in terms of wavelets.

Thus our task is to optimize the description. This means that we must make the best use of the resources allocated to us—for example, the number of available bits—to obtain the most precise possible description. To resolve this problem, we must first indicate how the quality of the description will be judged. Most often, the criteria used are mathematical and do not have much to do with the user's point of view. For example, in image processing, most calculations for judging the quality of the description use the quadratic mean value of gray levels. It is clear, however, that our eye is much more sensitive and selective than this quadratic measure. Thus, in the last analysis, we should submit the performance of an "optimal algorithm" to the users, since the average approximation criterion that leads to this algorithm will often be inadequate.

The case of speech (telephonic communication) or music is similar. The systematic research that optimizes the reception quality is based on an L^2 criterion that is mathematically convenient, but it is surely not the criterion used by the human ear.

Ideally, we should have a two-stage program: the first based on mathematical criteria and the second based on user satisfaction. For the most part, the only stage we describe is the "objective search" for an optimal algorithm, even though its optimality is defined in terms of a debatable energy criterion. The search for mathematically tractable criteria that capture the performance of the human eye or ear continues as an open problem at the interface between mathematics and physiology. Progress has been made in this area, and we will discuss in Chapter 11 some new criteria that seem to be closer to the user's point of view (at least for image processing) than the classical energy criteria. For example, in image synthesis, these criteria favor the reconstruction of sharp edges, which the eye is very quick to discern.

Rather than formulate ad hoc algorithms for each signal or each class of signals, we will construct, once and for all, a vast collection called a library of algorithms.

We also will construct a meta-algorithm whose function will be to find the particular algorithm in the library that best serves the given signal, given the criterion for the quality of the description.

The number of signals recorded on $2^{10} = 1024$ points that take only the two values zero and one is 2^{1024}. It would be absurd to store all of these possible signals in our library. We will use a very large "library" to describe the signals, but we exclude this "library of Babel," which would contain all the books, or all the signals in our case. But as everyone knows, the search for a specific book in the library of Babel is an insurmountable task. The "ideal library" must be sufficiently rich to suit all transient signals, but the "books" must be easily accessible.

While a single algorithm, Fourier analysis, is appropriate for all stationary signals, the transient signals are so rich and complex that a single analysis method, whether of time-scale or time-frequency, cannot serve them all.

If we stay in the relatively narrow environment of Grossmann–Morlet wavelets, also called time-scale algorithms, we have only two ways to adapt the algorithm to the signal being studied: We can choose one or another analyzing wavelet, and we can use either the continuous or the discrete version of the wavelets. For example, we can require the analyzing wavelet ψ to be an analytic signal, which means that its Fourier transform $\hat{\psi}(\omega)$ is zero for negative frequencies. In this case, all the wavelets $\psi_{(a,b)}$, $a > 0$, $b \in \mathbb{R}$, generated by ψ also will have this property, and their linear combinations given by the algorithm will be the analytic signal F associated with the real signal f. (For information about analytic signals see sections 2.6 and 2.7 or [222].)

Similarly, we can follow Daubechies and for a given $r \geq 1$, choose for ψ a real-valued function in the class C^r with compact support such that the collection $2^{j/2}\psi(2^j x - k)$, $j, k \in \mathbb{Z}$, is an orthonormal basis for $L^2(\mathbb{R})$. In this discrete version of the algorithm, $a = 2^{-j}$ and $b = k2^{-j}$, $j, k \in \mathbb{Z}$.

In spite of this, the choices that can be made from the set of time-scale wavelets remain limited. The search for optimal algorithms leads us on some remarkable algorithmic adventures, where time-scale wavelets and time-frequency wavelets are in competition, and where they also are compared with intermediate algorithms that mix the two extreme forms of analysis.

These considerations are developed in Chapters 6 and 7 and the question asked some years ago—Which wavelet to choose?—seems no longer relevant. The choices that we can and must consider no longer involve only the analyzing instrument, which is the wavelet. They also involve the methodology employed, which can be a time-scale algorithm, a time-frequency algorithm, or an intermediate algorithm.

Today, the competing algorithms, time-scale and time-frequency, are included in a whole universe of intermediate algorithms. An entropy criterion permits us to choose the algorithm that optimizes the description of the given signal within the given bit allocation.

Each algorithm is presented in terms of a particular orthogonal basis. We can compare searching for the optimal algorithm to searching for the best point of view, or best perspective, to look at a statue in a museum. Each point of view reveals certain parts of the statue and obscures others. We change our point of view to find the best one by going around the statue. In effect, we make a rotation; we change the orthonormal basis of reference to find the optimal basis. These reflections lead us quite naturally to the scientific thoughts of David Marr.

1.7 Optimal representation according to Marr

David Marr was fascinated by the complex relations that exist between the choice of a representation of a signal and the nature of the operations or transformations that such a representation permits. He wrote [198, pp. 20–21]:

> A *representation* is a formal system for making explicit certain entities or types of information, together with a specification of how the system does this. And I shall call the result of using a representation to describe a given entity a *description* of the entity in that representation.
>
> For example, the Arabic, Roman and binary numerical systems are all formal systems for representing numbers. The Arabic representation consists of a string of symbols drawn from the set (0, 1, 2, 3, 4, 5, 6, 7, 8, 9), and the rule for constructing the description of a particular integer n is that one decomposes n into a sum of multiples of powers of 10. . . .
>
> A musical score provides a way of representing a symphony; the alphabet allows the construction of a written representation of words; and so forth. . . .
>
> A representation, therefore, is not a foreign idea at all—we all use representations all the time. However, the notion that one can capture some aspect of reality by making a description of it using a symbol and that to do so can be useful seems to me a fascinating and powerful idea. But even the simple examples we have discussed introduce some rather general and important issues that arise whenever one chooses to use one particular representation. For example, if one chooses the Arabic numerical representation, it is easy to discover whether a number is a power of 10 but difficult to discover whether it is a power of 2. If one chooses the binary representation, the situation is reversed. Thus, there is a trade-off; *any particular representation makes certain information explicit at the expense of information that is pushed into the background and may be quite hard to recover.*[2]
>
> This issue is important, because how information is presented can greatly affect how easy it is to do different things with it. This is evident even from our numbers example: It is easy to add, to subtract, and even to multiply if the Arabic or binary representations are used, but it is not at all easy to do these things—especially multiplication—with Roman numerals. This is a key reason why the Roman culture failed to develop mathematics in the way the earlier Arabic cultures had.

There is an essential difference between Marr's considerations and the algorithms that we develop in the first six chapters. The difference is that the choice of the best representation, according to Marr, is tied to an objective goal. For the problem posed by vision, one goal is to extract the contours, recognize the edges of objects, delimit them, and understand their three-dimensional organization. In contrast, the algorithms we present in this book are aimed only at reducing the amount of data. They were not designed to extract patterns or solve important scientific issues; sometimes they do. One can argue that compression is a necessary first step toward feature extraction and, conversely, that obtaining the "important features"

---

[2] These last italics are ours.

of an image is indeed a form of compression. We, however, strongly believe that pattern recognition is not related to the kind of compression being discussed here. This position is based on our understanding of work by David Mumford.

What we have said so far concerns the use of wavelets for signal and image processing, and indeed this is the major theme of our book. There is, however, a slightly different point of view that focuses on wavelets techniques (analysis and synthesis) as tools within mathematics. This aspect will appear in Chapter 10 where we illustrate the power of wavelet techniques by analyzing two examples of fractal functions: the Weierstrass function and the Riemann function.

1.8 Terminology

The elementary constituents used for signal analysis and synthesis will be called, depending on the circumstances, wavelets, time-frequency atoms, or wavelet packets. The wavelets used will be either Grossmann–Morlet wavelets of the form

$$\psi_{(a,b)}(t) = \frac{1}{\sqrt{a}}\psi\left(\frac{t-b}{a}\right), \quad a > 0, \quad b \in \mathbb{R}, \tag{1.3}$$

the orthonormal wavelet bases that have the form

$$\psi_{j,k}(t) = 2^{j/2}\psi(2^j t - k), \qquad j,k \in \mathbb{Z}, \tag{1.4}$$

or the local Fourier bases of the form

$$w_{k,l}(t) = \omega(t-l)\cos\left[\pi\left(k+\frac{1}{2}\right)(t-l)\right], \quad k \in \mathbb{N}, \quad l \in \mathbb{Z}. \tag{1.5}$$

In the first two cases, we will speak of time-scale algorithms; in the last case, we will speak of time-frequency algorithms. Later we will mix the two points of view and subject the local Fourier bases to dyadic dilations. One thus encounters generalized time-frequency atoms. We will see in Chapter 10 (and in Appendix D) that the orthonormal wavelet bases of the form (1.4) have special properties that are not found in other decomposition algorithms.

We will use only two very large "libraries." The first consists of orthonormal bases whose elements are wavelet packets. In the second, the wavelet packets are replaced by the generalized time-frequency atoms that we have just described.

1.9 Reader's guide

In Chapters 2 through 7, we present the time-scale algorithms (Chapters 3 and 4) and time-frequency algorithms (Chapters 5, 6, and 7). Chapter 2 has a special status. We have tried to retrace some of the paths that led from Fourier analysis (Fourier, 1807) to wavelet analysis (Calderón, 1960, and Strömberg, 1981) and to the core of contemporary mathematics.

Quadrature mirror filters are studied in Chapter 3 in the context of problems posed by the digital telephone. For this revised edition, we have added an appendix that contains elementary information about filters, and thus complements Chapter 3.

The pyramid algorithms described in Chapter 4 concern numerical image processing. They use precisely the quadrature mirror filters of Chapter 3, and they lead either to orthogonal wavelets or to biorthogonal wavelets.

In Chapters 5 through 7, we will study time-frequency algorithms. The Wigner–Ville transform enables the signal to be "displayed in the time-frequency plane."

After indicating the main properties of the Wigner–Ville transform, we show that it leads to an algorithm that allows us to decompose a signal into new time-frequency atoms named "chirplets," which are a kind of frequency-modulated Gabor wavelet. Two other algorithms that provide access to these "atomic decompositions" are presented in Chapter 6 (local Fourier bases) and Chapter 7 (wavelet packets).

The first seven chapters form a coherent unit. This is not the case for the last five chapters; each of them treats a special application of wavelets and time-scale methods. Chapter 8 deals with the possibility of coding an image using the zero-crossings of its wavelet transform. In Chapter 9 we discuss turbulence and some of the recent contributions wavelet analysis has made to this still-unsolved problem. This chapter also serves as an introduction to multifractal analysis; indeed, this subject was initially introduced as a tool for studying turbulence. Multifractal analysis is continued in Chapter 10, where we show how wavelet analysis can be use to determine the Hölder exponents, as a function of position, of a multifractal function. Chapter 10 contains analyses of the Weierstrass and Riemann functions. In Chapter 11 we describe the use of wavelets for denoising signals and images. This chapter also provides a quick look at the connections between wavelets, nonlinear approximation, and Besov spaces—a mixture of seemingly unrelated techniques that is producing surprising and promising results. Chapter 12 is devoted to describing some the wavelet-based techniques that are being used in astronomy.

Four appendices contain complementary material. Appendix A provides a brief introduction to the language and theory of filters. Classical results on the continuous wavelet transform and its inversion are presented in Appendix B. The results in Appendix B apply in particular to the inversion formula used in Chapter 10 for the analysis of Riemann's function. Appendix C contains a presentation of a counterexample to a conjecture about zero-crossings; thus it is properly an appendix to Chapter 8. Although this counterexample has been known, this is the first time that a complete account has been published. Appendix D contains the definitions and a few basic results about Hölder spaces and Besov spaces. These spaces are used elsewhere in the book, particularly in Chapters 9 and 11.

CHAPTER 2

Wavelets from a Historical Perspective

2.1 Introduction

Time-frequency wavelets, which began with work by Dennis Gabor and by John von Neumann in the late 1940s, have a relatively long history in signal processing. Many of the fundamental contributions were subsequently achieved by physicists, and here we are thinking of work by Francis Low, Roger Balian, and Kenneth Wilson. Time-frequency wavelets have been widely used in speech processing, as will be shown in Chapter 6. Mathematicians did not pay much attention to this field. In contrast, the use of time-scale wavelets for signal and image processing is relatively recent, dating from the 1980s. However, in looking back over the history of mathematics, we will uncover at least seven different origins of wavelet analysis. Most of this work was done around the 1930s, and at that time the separate efforts did not appear to be parts of a coherent theory. Only today do we see how this work fits into the history of the theory of wavelets.

We feel that it is important to describe these sources in some detail. Each of them corresponds to a specific point of view and a particular technique, which only now are we able to view from a common scientific perspective. What's more, these specific techniques were rediscovered several years ago by physicists and mathematicians working on wavelets. Matthias Holschneider used, without knowing it, Lusin's technique (1930) to analyze Riemann's function (sections 2.6 and 10.4). Grossmann and Morlet rediscovered Alberto Calderón's identity (1960) 20 years later. And to spare no one, Yves Meyer was not the first to construct a regular, well-localized orthonormal wavelet basis having the algorithmic structure of Haar's system (1909): J.-O. Strömberg had done the same thing five years earlier [245].

Does this mean that everything had already been written? Not at all. More significantly, by rediscovering a number of known results, the "modern" wavelet investigators gave them new life and authority. Our debt to Grossmann and Morlet is not so much for having rediscovered Calderón's identity as it is for having used it to analyze nonstationary signals. This early application of wavelets to signal processing certainly encountered resistance, and Calderón himself found this use of his work incongruous.

The recent history of wavelets has been characterized by another phenomenon that we find scientifically important and sociologically interesting. From the beginning in the early 1980s, the "wavelet group" has consisted of researchers from several quite different disciplines, having different cultures and problems. The exchanges within this group created a dynamic environment that we believe accounts for the rapid advances seen on at least two fronts: the synthesis and structuring of previous and new knowledge to produce a coherent theory of wavelets, and the rapid adoption of wavelet techniques in diverse disciplines outside mathematics. Not surprisingly, the most active interface has been between mathematics and signal processing, and it can be fairly said that most applications in other fields have been through signal or image processing. But we wish to emphasize that the "flow" has been in both directions, and that mathematics has greatly profited by input from the other sciences. The most spectacular example (which is described in section 2.10) is the construction by Ingrid Daubechies of her celebrated orthonormal bases. As will be explained, this construction benefited from work in signal processing. Another example is Stéphane Jaffard's work on the analysis of multifractal functions, which was influenced by work on turbulence by Alain Arneodo and his team.

The history of wavelets is reminiscent of the recent history of fractals. Fractal objects—long before the name was coined—appeared in mathematics more than a century ago. Georg Cantor's triadic set is a prime example. In the mid nineteenth century, nobody would have suspected that fractals could be used to model natural phenomena in physics and chemistry as initiated by Mandelbrot. The success of this modeling led some scientists to speak about "a theory of fractals." This theory has been questioned with skepticism by certain mathematicians. Nevertheless, one must acknowledge that Mandelbrot's scientific vision has been a source of inspiration in contemporary mathematics. At a more artificial level, both fractals and wavelets involve "scale," and so they enjoy a natural relationship, as we hope to show in Chapters 9 and 10.

Our immediate objective in this chapter is to describe the links between signal processing and the different mathematical efforts that developed outside the "theory of wavelets"; a larger objective is to show how wavelet-based techniques in signal processing have been applied to disciplines outside mathematics.

2.2 From Fourier (1807) to Haar (1909), frequency analysis becomes scale analysis

We first go back to the origins, that is, to Joseph Fourier. As is well known, Fourier asserted in 1807 that any 2π-periodic function f could be represented by the sum

$$a_0 + \sum_{k=1}^{\infty}(a_k \cos kx + b_k \sin kx),$$

which is called its *Fourier series*. The coefficients a_0, a_k, and b_k ($k \geq 1$) are given by

$$a_0 = \frac{1}{2\pi} \int_0^{2\pi} f(x)\, dx$$

and by

$$a_k = \frac{1}{\pi} \int_0^{2\pi} f(x) \cos kx\, dx, \qquad b_k = \frac{1}{\pi} \int_0^{2\pi} f(x) \sin kx\, dx.$$

When Fourier announced his surprising results, neither the notion of function nor that of integral had yet received a precise definition. We can say conservatively that the mathematical justification of Fourier's statement played an essential role in the evolution of the ideas mathematicians have had about these concepts.[3]

---

[3]We recommend *Fourier Series and Wavelets* by J.-P. Kahane and P. G. Lemarié-Rieusset [164] for the history of Fourier series.

Before Fourier's work, entire series were used to represent and manipulate functions, and thus the most general functions that could be constructed were endowed with very special properties. Furthermore, these properties were unconsciously associated with the notion of function itself. By passing from a representation of the form

$$a_0 + a_1 x + a_2 x^2 + a_3 x^3 + \cdots$$

to one of the form

$$a_0 + (a_1 \cos x + b_1 \sin x) + (a_2 \cos 2x + b_2 \sin 2x) + \cdots,$$

Fourier discovered, without knowing it, a new functional universe.

In 1873, Paul Du Bois-Reymond constructed a continuous, 2π-periodic function of the real variable x whose Fourier series diverged at a given point. If Fourier's assertion were true, it could not be so in the naïve sense imagined by Fourier. At that time, three new avenues were opened to mathematicians, and all three have led to important results:

(1) They could modify the notion of function and find one that is adapted, in a certain sense, to Fourier series.

(2) They could modify the definition of convergence of Fourier series.

(3) They could find other orthogonal systems for which the divergence phenomenon discovered by Du Bois-Reymond in the case of the trigonometric system cannot happen.

The functional concept best suited to Fourier series was created by Henri Lebesgue. It involves the space $L^2[0, 2\pi]$ of functions that are square integrable on the interval $[0, 2\pi]$. The sequence

$$\frac{1}{\sqrt{2\pi}}, \ \frac{1}{\sqrt{\pi}} \cos x, \ \frac{1}{\sqrt{\pi}} \sin x, \ \frac{1}{\sqrt{\pi}} \cos 2x, \ \frac{1}{\sqrt{\pi}} \sin 2x, \ \ldots \quad (2.1)$$

is an orthonormal basis for this space. Furthermore, the coefficients of the decomposition in this orthonormal basis form a square-summable series, and this expresses the conservation of energy: The quadratic mean value of the developed function f is (up to a normalization factor) the sum of the squares of the coefficients. Finally, the Fourier series of f converges to f in the sense of the quadratic mean.

The second way that was followed to avoid the difficulty raised by Du Bois-Reymond was to modify the notion of convergence. If the partial sums s_n are replaced by the Cesàro sums $\sigma_n = (s_0 + \cdots + s_{n-1})/n$, then everything falls into place: The Fourier series of a continuous function f converges uniformly to f.

The third route leads to wavelets. This was followed by Haar, who asked himself this question in his thesis: Does there exist another orthonormal system $h_0, h_1, \ldots, h_n, \ldots$ of functions defined on $[0,1]$ such that for any continuous function f defined on $[0,1]$, the series

$$\langle f, h_0 \rangle h_0(x) + \langle f, h_1 \rangle h_1(x) + \cdots + \langle f, h_n \rangle h_n(x) + \cdots$$

converges to $f(x)$ uniformly on $[0,1]$? Here we have written

$$\langle u, v \rangle = \int_0^1 u(x) \overline{v}(x) \, dx,$$

where $\bar{v}(x)$ is the complex conjugate of $v(x)$, and we have chosen the interval [0,1] for convenience.

As we will see, this problem has an infinite number of solutions. In 1909, Haar discovered the simplest solution and at the same time opened one of the routes leading to wavelets [134].

Haar begins with the function h such that $h(x) = 1$ for $x \in [0, \frac{1}{2})$, $h(x) = -1$ for $x \in [\frac{1}{2}, 1)$, and $h(x) = 0$ for $x \notin [0, 1)$. For $n \geq 1$, he writes $n = 2^j + k$, $j \geq 0$, $0 \leq k < 2^j$, and defines $h_n(x) = 2^{j/2} h(2^j x - k)$. The support of h_n is exactly the dyadic interval $I_n = [k 2^{-j}, (k+1) 2^{-j})$, which is included in [0,1) when $0 \leq k < 2^j$. To complete the set, define $h_0(x) = 1$ on [0,1). Then the sequence $h_0, h_1, \ldots, h_n, \ldots$ is an orthonormal basis (also called a Hilbert basis) for $L^2[0,1]$.

The uniform approximation of $f(x)$ by the sequence

$$S_n(f)(x) = \langle f, h_0 \rangle h_0(x) + \cdots + \langle f, h_n \rangle h_n(x)$$

is nothing more than the classical approximation of a continuous function by step functions whose values are the mean values of $f(x)$ on the appropriate dyadic intervals.

We can criticize the Haar construction on a couple of points. On one hand, the atoms h_n used to construct the continuous function f are not themselves continuous functions, and thus there is a lack of coherence. But there is a more serious criticism. Suppose that instead of being continuous on the interval [0,1], f is a C^1 function, which means that f is continuous and has a continuous derivative. Then the approximation of f by step functions would be completely inappropriate. In this case, a suitable approximation would be the one created from the graph of f by inscribing polygonal lines.

These two defects of the Haar system and the idea of approximating the graph of $f(x)$ with inscribed polygonal lines led Faber and Schauder to replace the functions h_n of the Haar system by their primitives. This research began in 1910 and continued until 1920.

Define the "triangle function" Δ by $\Delta(x) = 0$ if $x \notin [0,1]$, $\Delta(x) = 2x$ if $0 \leq x \leq \frac{1}{2}$, and $\Delta(x) = 2(1-x)$ if $\frac{1}{2} \leq x \leq 1$. Faber [100] and Schauder [234] considered the sequence Δ_n, $n \geq 1$, defined by

$$\Delta_n(x) = \Delta(2^j x - k) \quad \text{for} \quad n = 2^j + k, \quad j \geq 0, \quad 0 \leq k < 2^j.$$

The support of Δ_n is the dyadic interval $I_n = [k 2^{-j}, (k+1) 2^{-j}]$, and on this interval, Δ_n is the primitive of h_n multiplied by $2 \cdot 2^{j/2}$.

For $n = 0$, we set $\Delta_0(x) = x$, and we add the function $\Delta_{-1}(x) = 1$ to complete the set of functions. Then the sequence $\Delta_{-1}, \Delta_0, \ldots, \Delta_n, \ldots$ is a *Schauder basis* for the Banach space E of continuous functions on [0,1]. This means that every continuous function f on [0,1] can be written as

$$f(x) = a + bx + \sum_{n=1}^{\infty} \alpha_n \Delta_n(x) \tag{2.2}$$

and that the series has the following properties: The convergence is uniform on [0,1] and the coefficients are unique.

We note that the Haar system is not a Schauder basis of E because a Schauder basis of a Banach space must be made up of vectors of that space, and the functions h_n are not continuous.

The coefficients in (2.2) can be calculated directly by induction. We have $f(0) = a$ and $f(1) = a + b$, which gives a and b. This allows us to consider a function $f(x) - a - bx$, which is zero at $x = 0$ and $x = 1$ (and which we again denote by f). Once this reduction is made, we have $f(\frac{1}{2}) = \alpha_1$, which allows us to consider a function equal to zero at $x = 0$, $x = \frac{1}{2}$, and $x = 1$. The calculation continues with $f(\frac{1}{4}) = \alpha_2$ and $f(\frac{3}{4}) = \alpha_3$, and so on. If we do not wish to "peel" f this way, the coefficients α_n can be computed directly by the formula

$$\alpha_n = f\left(\left(k + \frac{1}{2}\right)2^{-j}\right) - \frac{1}{2}[f(k2^{-j}) + f((k+1)2^{-j})], \qquad (2.3)$$

where $n = 2^j + k$, $j \geq 0$, and $0 \leq k < 2^j$.

We can give a further interpretation to (2.2). If, instead of being continuous, f were in C^1, then we could differentiate (2.2) term by term and obtain the expansion of f' in the Haar basis. If f is in C^1, the series (2.2) converges uniformly to f and the series differentiated term by term converges uniformly to f'. Does this mean that the functions Δ_n, $n \geq 0$, with the added function 1, constitute a Schauder basis for the Banach space $C^1[0,1]$? As before, this is not the case because the functions Δ_n do not belong to the space in question.

Following Hölder, we define the space $C^h[0,1]$, for $0 < h < 1$, by the relation $|f(x) - f(y)| \leq C|x - y|^h$ for some constant $C > 0$ and for all $x, y \in [0,1]$. Then it is clear from (2.3) that $|\alpha_n| \leq C2^{-(j+1)h}$ if f belongs to C^h. Since $2^j \leq n < 2^{j+1}$, we can also write $|\alpha_n| \leq Cn^{-h}$, $n \geq 1$. The converse, although much less evident, is nevertheless true when $0 < h < 1$. It is not true if $h = 1$.

Physicists are interested in the *Hölder spaces* C^h because they occur naturally in the study of fractal structures. In fact, physicists wish to know more. They are interested in functions f whose *Hölder exponents* $h(x_0)$ vary from one point to another. This pointwise definition is slightly different: We say that f satisfies a *Hölder condition* of exponent h, $0 < h < 1$, at a point x_0 if

$$|f(x) - f(x_0)| \leq C|x - x_0|^h. \qquad (2.4)$$

More generally, if $m < h < m + 1$, $m \in \mathbb{N}$, then this definition should read

$$|f(x) - P_m(x - x_0)| \leq C|x - x_0|^h, \qquad (2.5)$$

where P_m is a polynomial of degree m. Then the *Hölder exponent* of f at x_0 is denoted by $h(x_0)$ and defined to be the supremum of the h that satisfy (2.5).

Contemporary science deals with numerous physical phenomena having multifractal structures. By this we mean that the Hölder exponents of the function representing the structure vary from point to point in a particularly erratic way. To be more precise, we consider the set of points E_α where the Hölder exponent $h(x_0)$ takes the value α. If these E_α are fractal sets, we say that f is multifractal. In this case, scientists are interested in determining the Hausdorff dimension $d(\alpha)$ of E_α as a function of α. (Hausdorff dimension is defined in section 9.10.)

An example from mathematics of a multifractal object is the celebrated "nondifferentiable" function attributed to Bernhard Riemann, which is defined by $\sum_{n=1}^{\infty} \sin(\pi n^2 x)/\pi n^2$. This example illustrates the point that the Fourier series of a function provides no directly accessible information about the function's multifractal structure. By using the wavelets of Lusin (which we present in section 2.6), Holschneider and Tchamitchian obtained a new proof of Gerver's theorem

[127, 128], which states that Riemann's function is differentiable at certain rational multiples of π [144]. More recently, Stéphane Jaffard has provided a complete analysis of the multifractal nature of Riemann's function using wavelet techniques [153]. We describe this work in Chapter 10.

A second example is the signal coming from fully developed turbulence. The multifractal structure of this signal has been studied by Alain Arneodo and his collaborators. We present this example in Chapter 9.

Conceivably, the pointwise Hölder exponents could be computed by going back to the definition. However, the example of the Riemann function shows that such an approach is too crude to yield practical results. Furthermore, for applications outside mathematics, this approach offers no way to take into consideration the inevitable noise that alters a signal. The Schauder basis presents the same difficulties because the calculation of the coefficients α_n (according to (2.3)) calls directly on explicit values of the signal.

Today, we are fortunate to have much more subtle ways to attack this problem. Specifically, the pointwise Hölder exponents are now determined using wavelet analysis. The wavelet coefficients replace those given by formula (2.3). They are less sensitive to noise because they measure, at different scales, the average fluctuations of the signal. These methods will be described in Chapters 9 and 10.

2.3 New directions of the 1930s: Paul Lévy and Brownian motion

Brownian motion is a random process. We will limit our discussion to the one-dimensional case. We thus write $X(t, \omega)$ for the Brownian motion: t denotes time, ω belongs to a probability space Ω, and $X(t, \omega)$ is regarded as a real-valued function of time depending on the parameter ω.

To obtain a realization of Brownian motion, we choose a particular orthonormal basis $Z_i(t)$, $i \in I$, for the usual Hilbert space $L^2(\mathbb{R})$. Then we know that the derivative (in the sense of distributions) $\frac{d}{dt}X(t, \omega)$ is written as

$$\frac{d}{dt}X(t, \omega) = \sum_{i \in I} g_i(\omega) Z_i(t),$$

where the $g_i(\omega)$, $i \in I$, are independent, identically distributed Gaussian random variables with zero mean.

The problem is to choose the best possible representation of Brownian motion. As in all signal processing problems, it is certainly advisable to have in mind what we wish to study. If we wish to examine the spectral properties of Brownian motion, we are led to select the Fourier representation. The real line is cut into intervals $[2l\pi, 2(l+1)\pi]$, $l \in \mathbb{Z}$, and the trigonometric system is used on each of the intervals. In its real form, this is the trigonometric system (2.1).

However, if we wish to highlight the local regularity of Brownian motion, Fourier analysis is inadequate. On the other hand, the analysis using the Schauder basis immediately reveals the Hölder regularity C^α, $\alpha < \frac{1}{2}$, of the Brownian motion trajectories.

We start with the Haar basis for $L^2(\mathbb{R})$ composed of the functions $h_n(t - l)$, $n \geq 0$, $l \in \mathbb{Z}$, and expand the white noise $\frac{d}{dt}X(t, \omega)$ in this orthonormal basis. By taking primitives, we obtain the development of Brownian motion in the Schauder basis. To simplify matters, we restrict the discussion to Brownian motion on the interval [0,1]. For this case, $l = 0$, and

$$X(t,\omega) = g_0(\omega)t + \frac{1}{2}\sum_{n=1}^{\infty} 2^{-j/2} g_n(\omega)\Delta_n(t),$$

where the $g_n(\omega)$ are independent, identically distributed Gaussian random variables with mean zero and variance one. This expansion often is called the "midpoint displacement construction." This refers to the specific geometry of the "error term" $\alpha(j,k)\Delta(2^j x - k)$ in the Schauder basis expansion. Adding this term amounts to moving the midpoint of the preceding (piecewise affine) approximation of $f(x)$. This midpoint displacement is precisely $\alpha(j,k)$. In the case of Brownian motion, these displacements are $2^{-j/2} g_n(\omega)$.

To verify that the function $X(t,\omega)$ belongs to the Hölder space C^α for almost all $\omega \in \Omega$, it is sufficient to show that $2^{-j/2}|g_n(\omega)| \leq C(\omega) 2^{-j\alpha}$. If, for almost all $\omega \in \Omega$, one had $\sup_{n \geq 0} |g_n(\omega)| < \infty$, then the trajectories of the Brownian motion would almost surely belong to the space $C^{1/2}$. But this is not the case, and instead we have $\sup_{n \geq 2}(|g_n(\omega)|/\sqrt{\log n}) < \infty$ for almost all $\omega \in \Omega$. Then the criterion for Hölder regularity gives

$$|X(t+h,\omega) - X(t,\omega)| \leq C(\omega)\sqrt{h \log \frac{1}{h}},$$

where $C(\omega) < \infty$ for almost all $\omega \in \Omega$.

We see from this theorem of Paul Lévy how a representation in a particular basis can provide easy access to certain aspects of a problem. In this case, the Schauder basis provides quick access to local regularity properties of Brownian trajectories. As we were told by Gérard Kerkyacharian, this elegant proof was not given by Lévy, although the tools we are using were available to him. Zbigniew Ciesielski [53] was the first to relate the midpoint displacement construction of Brownian motion to its global regularity.

Fabrice Sellan [2] has extended this analysis to the case of fractional Brownian motion, as it was proposed by Mandelbrot and J. W. van Ness to model certain noise (see also [111]). He has found wavelets that, when suitably normalized, do for fractional Brownian motion what the Schauder basis did for ordinary Brownian motion. The coefficients in Sellan's basis are uncorrelated Gaussians. This representation allows one to simulate precisely and efficiently the long-range correlations found in fractional Brownian motion. (For more about this, see the note at the end of Chapter 4.) Albert Benassi, Stéphane Jaffard, and Daniel Roux have generalized these ideas to the "elliptic Gaussian fields" [31]. This work demonstrates that multiresolution methods are well adapted to the analysis and synthesis of some Gaussian processes.

2.4 New directions of the 1930s: Littlewood and Paley

We have shown with the example of Brownian motion how the Schauder basis provides direct and easy access to local regularity properties. On the other hand, the analysis of these properties using the trigonometric system is considerably more involved.

Similar difficulties are encountered when we try to localize the energy of a function. To be more precise, we first focus on 2π-periodic functions f and their Fourier series expansions. The integral $\frac{1}{2\pi}\int_0^{2\pi}|f(x)|^2\,dx$, which is the mean value of the energy, is given directly by the sum of the squares of the Fourier coefficients. However, it is often important to know if the energy is concentrated around a few points or if it is distributed over the whole interval $[0,2\pi]$. This determination can be made by calculating $\frac{1}{2\pi}\int_0^{2\pi}|f(x)|^4\,dx$ or, more generally, $\frac{1}{2\pi}\int_0^{2\pi}|f(x)|^p\,dx$ for $2 < p < \infty$. When the energy is concentrated around a few points, this integral will be much larger than the mean value of the energy, while it will be the same order of magnitude when the energy is evenly distributed. We write $\|f\|_p = (\frac{1}{2\pi}\int_0^{2\pi}|f(x)|^p\,dx)^{1/p}$ and, for obvious reasons of homogeneity, we compare the norms $\|f\|_p$ to determine if the energy is concentrated or dispersed. But if p is different from 2, we can neither calculate nor even estimate these norms $\|f\|_p$ by examining the Fourier coefficients of f. The information needed for this calculation is hidden in the Fourier series of f; to reveal it, it is necessary to subject the series to manipulations that were discovered by Littlewood and Paley as long ago as 1930.

Littlewood and Paley define the *dyadic blocks* $\Delta_j f$ by

$$\Delta_j f(x) = \sum_{2^j \leq k < 2^{j+1}} (a_k \cos kx + b_k \sin kx),$$

where $a_0 + \sum_{k=1}^{\infty}(a_k \cos kx + b_x \sin kx)$ denotes the Fourier series of f. Then

$$f(x) = a_0 + \sum_{j=0}^{\infty} \Delta_j f(x),$$

and the fundamental result of Littlewood and Paley is that there exists for each p, $1 < p < \infty$, two constants $C_p \geq c_p > 0$ such that

$$c_p\|f\|_p \leq \left\|\left(|a_0|^2 + \sum_{j=0}^{\infty}|\Delta_j f(x)|^2\right)^{1/2}\right\|_p \leq C_p\|f\|_p. \tag{2.6}$$

If $p = 2$, $C_p = c_p = 1$, and there is equality in (2.6).

Up to this point, wavelets have not yet appeared. The path that leads from the work of Littlewood and Paley to wavelet analysis passes through the research done by Antoni Zygmund's group at the University of Chicago. Zygmund and the mathematicians around him sought to extend to n-dimensional Euclidean space the results obtained in the one-dimensional periodic case by Littlewood and Paley.

It was at this point that a "mother wavelet" ψ appeared. It is an infinitely differentiable, rapidly decreasing function, defined on the Euclidean space $\mathbb{R}^n$, whose Fourier transform $\hat{\psi}$ satisfies the following four conditions, where α is chosen by hypothesis in the interval $(0, \frac{1}{3}]$:

(1) $\hat{\psi}(\xi) = 1$ if $1 + \alpha \leq |\xi| \leq 2 - 2\alpha$.

(2) $\hat{\psi}(\xi) = 0$ if $|\xi| \leq 1 - \alpha$ or $|\xi| \geq 2 + 2\alpha$.

(3) $\hat{\psi}(\xi)$ is infinitely differentiable on $\mathbb{R}^n$.

(4) $\sum_{-\infty}^{\infty}|\hat{\psi}(2^{-j}\xi)|^2 = 1$ for all $\xi \neq 0$.

Condition (4) is not as complicated as it appears. It is sufficient to verify it for $1 - \alpha \leq |\xi| \leq 2 - 2\alpha$, and then only two cases arise: If $1 - \alpha \leq |\xi| \leq 1 + \alpha$, condition (4) reduces to $|\hat{\psi}(\xi)|^2 + |\hat{\psi}(2\xi)|^2 = 1$, while if $1 + \alpha \leq |\xi| \leq 2 - 2\alpha$, it is automatically satisfied since one term is equal to one and all the others are zero.

Condition (4) implies that the analysis of Littlewood–Paley–Stein (whose definition will be given in a moment) conserves L^2 energy. In the one-dimensional case, this same condition is satisfied by every mother wavelet ψ having the property that $2^{j/2}\psi(2^j x - k)$, $j, k \in \mathbb{Z}$, is an orthonormal basis for $L^2(\mathbb{R})$. It also anticipates similar conditions shared by the quadrature mirror filters (Chapter 3) and the Malvar–Wilson wavelets (Chapter 6).

The theory for $\mathbb{R}^n$ proceeds by setting $\psi_j(x) = 2^{nj}\psi(2^j x)$ and replacing the dyadic blocks of Littlewood and Paley with the convolutions $\Delta_j(f) = f * \psi_j$. The *Littlewood–Paley–Stein function* is defined by

$$g(x) = \left(\sum_{-\infty}^{\infty} |\Delta_j(f)(x)|^2 \right)^{1/2}.$$

If f belongs to $L^2(\mathbb{R}^n)$, the same is true for g, and $\|f\|_2 = \|g\|_2$ (the conservation of energy).

If $1 < p < \infty$, there exist two constants $C_p \geq c_p > 0$ such that for all functions f belonging to $L^p(\mathbb{R}^n)$,

$$c_p \|g\|_p \leq \|f\|_p \leq C_p \|g\|_p, \tag{2.7}$$

where

$$\|f\|_p = \left(\int_{\mathbb{R}^n} |f(x)|^p \, dx \right)^{1/p}.$$

The Littlewood–Paley–Stein function g provides a method for analyzing f in which a major role is played by the ability to vary arbitrarily the scales used in the analysis; by the same token, the notion of frequency plays a minor role. The dilations of size 2^j are present in the definition of the operators Δ_j. Nevertheless, conditions (1) and (2) endow these operators with a frequency content. *The sequence of operators Δ_j, $j \in \mathbb{Z}$, constitutes a bank of band-pass filters, oriented on frequency intervals covering approximately one octave.* Littlewood–Paley techniques have been extensively developed by Stein and his collaborators. We refer to [242], [243], and [114] for detailed descriptions of their applications in analysis.

Thanks to the work of Marr and Mallat (which we describe in Chapter 8), the Littlewood–Paley analysis provides an effective algorithm for numerical image processing.

2.5 New directions of the 1930s: The Franklin system

In 1927, Philip Franklin, who was a professor at the Massachusetts Institute of Technology, had the idea to create an orthonormal basis from the Schauder basis by using the Gram–Schmidt process. This produces a sequence $(f_n)_{n \geq -1}$ beginning with $f_{-1}(x) = 1$, $f_0(x) = 2\sqrt{3}(x - \frac{1}{2}), \ldots$, which is an orthonormal basis for $L^2[0,1]$. This sequence $(f_n)_{n \geq -1}$ is called the Franklin system and satisfies

$$\int_0^1 f_n(x) \, dx = \int_0^1 x f_n(x) \, dx = 0 \quad \text{for} \quad n \geq 1.$$

The Franklin system has advantages over both the Schauder basis and the Haar basis. It can be used to decompose any function f in $L^2[0,1]$, which the Schauder basis does not allow, and it can be used to characterize the spaces C^α, $0 < \alpha < 1$, by the relation $|\langle f, f_n\rangle| \leq Cn^{-1/2-\alpha}$, which the Haar basis does not allow. Thus the Franklin system works as well in relatively regular situations as it does in relatively irregular situations.

The weakness of the Franklin basis is that it no longer has a simple algorithmic structure. The functions of the Franklin basis, unlike those of the Haar basis or those of the Schauder basis, are not derived from a fixed function ψ by integer translations and dyadic dilations. This defect caused the Franklin system to be unattractive for applications.

Zbigniew Ciesielski revived the forgotten Franklin system in 1963 by showing that it is localized [54, 55]. There exists an exponent $\gamma > 0$ and a constant $C > 0$ such that

$$|f_n(x)| \leq C 2^{j/2} \exp(-\gamma|2^j x - k|)$$

if $0 \leq x \leq 1$, $n = 2^j + k$, $0 \leq k < 2^j$, and

$$\left|\frac{d}{dx} f_n(x)\right| \leq C 2^{3j/2} \exp(-\gamma|2^j x - k|).$$

Thus, on a mathematical level, everything works as if $f_n(x) = 2^{j/2}\psi(2^j x - k)$, where ψ is a Lipschitz function having exponential decay.

Today we are aware of a much closer relationship between the Franklin system and wavelets (see [150]). Asymptotically the functions of the Franklin system become arbitrarily close to the orthonormal wavelet basis discovered by Strömberg in 1980. In fact, for $n = 2^j + k$, $0 \leq k < 2^j$,

$$f_n(x) = 2^{j/2}\psi(2^j x - k) + r_n(x)$$

where, for a certain constant C,

$$\|r_n(x)\|_2 \leq C(2-\sqrt{3})^{d(n)}, \qquad d(n) = \inf(k, 2^j - k). \tag{2.8}$$

The function ψ, which was discovered in 1980 by Strömberg, is completely explicit. It has the following three properties:

(1) ψ is continuous on the whole real line, it is linear on the intervals

$$[1,2], [2,3], \ldots, [l, l+1], \ldots,$$

and it is linear on the intervals

$$\left[\frac{1}{2}, 1\right], \left[0, \frac{1}{2}\right], \left[-\frac{1}{2}, 0\right], \ldots, \left[-\frac{l+1}{2}, -\frac{l}{2}\right], \ldots.$$

(2) $|\psi(x)| \leq C(2-\sqrt{3})^{|x|}$.

(3) $2^{j/2}\psi(2^j x - k)$, $j, k \in \mathbb{Z}$, is an orthonormal basis for $L^2(\mathbb{R})$.

Note that $(2-\sqrt{3}) < 1$; hence condition (2) means that ψ decreases rapidly at infinity, and (2.8) means that $\|r_n(x)\|_2$ is small when $d(n)$ is large.

2.6 New directions of the 1930s: The wavelets of Lusin

This section is in the right place historically, but scientifically it should come after the next section, since Lusin's work is an example of continuous wavelet expansions.

The interpretation of Lusin's work in terms of the theory of wavelets would probably astonish its author. But it is certainly the best reading, the one that gives the greatest beauty to Lusin's work. We begin by introducing the object of Lusin's study, namely, the Hardy spaces $H^p(\mathbb{R})$, where $1 \leq p \leq \infty$. Let P denote the open, upper half-plane defined by $z = x + iy$ and $y > 0$. A function $f(x+iy)$ belongs to $H^p(\mathbb{R})$ if it is holomorphic in the half-plane P and if

$$\sup_{y>0} \left(\int_{-\infty}^{\infty} |f(x+iy)|^p \, dx \right)^{1/p} < \infty. \tag{2.9}$$

When this condition is satisfied, the upper bound, taken over $y > 0$, is also the limit as y tends to zero. Furthermore, $f(x+iy)$ converges to a function denoted by $f(x)$ when y tends to zero, where convergence is in the sense of the L^p norm. The space $H^p(\mathbb{R})$ can thus be identified with a closed subspace of $L^p(\mathbb{R})$, which explains the notation.

Hardy spaces play a fundamental role in signal processing. One associates with a real-valued signal f, defined for all $t \in \mathbb{R}$ of finite energy, the analytic signal F for which f is the real part. By hypothesis, the energy of f is $\int_{-\infty}^{\infty} |f(t)|^2 \, dt < \infty$, and we require that F have finite energy as well. This implies that F belongs to the Hardy space $H^2(\mathbb{R})$. Then $F(t) = f(t) + ig(t)$, and the function g is the Hilbert transform of f. For further information about analytic signals, the reader may refer to [222]. One may also consult the remarkable exposition by Jean Ville [254].

Read in the light of the theory of wavelets, Lusin's work concerns the analysis and synthesis of functions in $H^p(\mathbb{R})$ using "atoms" or "basis elements," which are the elementary functions of $H^p(\mathbb{R})$. In fact, these are the functions $(z - \overline{\zeta})^{-2}$, where the parameter $\zeta = u + iv$ belongs to P. In Lusin's work, the Hardy space $H^p(\mathbb{R})$ was used as a tool to provide a better understanding of the $L^p(\mathbb{R})$ space. More specifically, singular operators were shown to be bounded on $L^p(\mathbb{R})$ by first studying their action on $H^p(\mathbb{R})$. In the latter case, such an operator is understood through its action on the building blocks $(z - \overline{\zeta})^{-2}$, $\zeta \in P$.

Thus one wishes to obtain effective and robust representations of the functions f in $H^p(\mathbb{R})$ of the form

$$f(z) = \iint_P (z - \overline{\zeta})^{-2} \alpha(\zeta) \, du \, dv, \tag{2.10}$$

where $\zeta = u + iv$ and where $\alpha(\zeta)$ plays the role of the coefficients. These coefficients should be simple to calculate, and their order of magnitude should provide an estimate of the norm of f in $H^p(\mathbb{R})$. Furthermore, we are interested in relating the decomposition of f given by (2.10) to a wavelet decomposition as defined in the next section.

The synthesis is obtained by the following rule. We start with an *arbitrary* measurable function $\alpha(\zeta)$, subject only to the following condition introduced by Lusin: The quadratic functional A defined by

$$A(x) = \left(\iint_{\Gamma(x)} |\alpha(u+iv)|^2 v^{-2} \, du \, dv \right)^{1/2},$$

where $\Gamma(x) = \{(u,v) \in \mathbb{R}^2 \mid v > |u-x|\}$, must be such that $\int_{-\infty}^{\infty}(A(x))^p\, dx$ is finite. Note that this condition involves only the modulus of the coefficients $\alpha(\zeta)$.

If the integral $\int_{-\infty}^{\infty}(A(x))^p\, dx$ is finite, then necessarily

$$f(x) = \iint_P (x-\bar\zeta)^{-2}\alpha(\zeta)\, du\, dv$$

belongs to $H^p(\mathbb{R})$, and if $1 \le p < \infty$,

$$\|f\|_p \le C(p)\left(\int_{-\infty}^{\infty}(A(x))^p\, dx\right)^{1/p}. \tag{2.11}$$

The left member of (2.11) is the norm of f in $H^p(\mathbb{R})$, as defined by (2.9). The estimate given by the right member of (2.11) is sometimes very crude. If, for example, $f(x) = (x+i)^{-2}$ and if one makes the natural choice of the Dirac measure at the point i for $\alpha(\zeta)$, then the second member of (2.11) is infinite. This paradox arises because the representation (2.10) is not unique.

To obtain a unique decomposition, which we call the natural decomposition, we restrict the choice to $\alpha(\zeta) = \frac{2i}{\pi}vf'(u+iv)$. When we do this, the two norms $\|f\|_p$ and $\|A\|_p$ become equivalent if $1 \le p < \infty$. Today this natural choice of coefficients has an interesting explanation. This interpretation, which depends on the contemporary formalism of wavelet theory, is given in the following section.

2.7 Atomic decompositions from 1960 to 1980

Guido Weiss, in collaboration with Ronald Coifman, was the first to interpret, as we have just done, Lusin's theory in terms of atoms and atomic decompositions. The atoms are the simplest elements of a function space, and the objective of the theory is to find, for the usual function spaces, the atoms and the "assembly rules" that allow one to reconstruct all the elements of the function space using these atoms.

In the case of the holomorphic Hardy spaces of the last section, the atoms were the functions $(z-\bar\zeta)^{-2}$, $\zeta \in P$, and the assembly rules were given by the condition on Lusin's function A.

For the spaces $L^p[0,2\pi]$, $1 < p < \infty$, the atoms cannot be the functions $\cos kx$ and $\sin kx$, $k \ge 1$, because this choice does not lead to assembly rules that are sufficiently simple and explicit to be useful in practice. Marcinkiewicz showed in 1938 that the simplest atomic decomposition for the spaces $L^p[0,1]$, $1 < p < \infty$, is given by the Haar system. The Franklin basis would have served as well, and from the scientific perspective given by wavelet theory, the Franklin basis and Littlewood–Paley analysis are naturally related.

One of the approaches to atomic decompositions is given by Calderón's identity. To explain Calderón's identity, we start with a function ψ belonging to $L^2(\mathbb{R}^n)$. (Later in this history, Grossmann and Morlet called this function an analyzing wavelet.) Its Fourier transform $\hat\psi(\xi)$ is subject to the condition that

$$\int_0^\infty |\hat\psi(t\xi)|^2 \frac{dt}{t} = 1 \tag{2.12}$$

for almost all $\xi \in \mathbb{R}^n$. If ψ belongs to $L^1(\mathbb{R}^n)$, condition (2.12) implies that

$$\int_{\mathbb{R}^n} \psi(x)\, dx = 0.$$

We write $\tilde{\psi}(x) = \overline{\psi}(-x)$, $\psi_t(x) = t^{-n}\psi(\frac{x}{t})$, and $\tilde{\psi}_t(x) = t^{-n}\tilde{\psi}(\frac{x}{t})$. Let Q_t denote the operator defined as convolution with ψ_t; its adjoint Q_t^* is the operator defined as convolution with $\tilde{\psi}_t$.

Calderón's identity is a decomposition of the identity operator, written symbolically as

$$I = \int_0^\infty Q_t Q_t^* \frac{dt}{t}.$$

This means that for all $f \in L^2(\mathbb{R})$,

$$f = \int_0^\infty Q_t[Q_t^*(f)] \frac{dt}{t},$$

where the limit of this improper integral is to be taken in the sense of $L^2(\mathbb{R})$.

Grossmann and Morlet rediscovered this identity in 1980, 20 years after the work of Calderón. However, with this rediscovery, they gave it a different interpretation by relating it to the coherent states of quantum mechanics [133]. They defined wavelets (generated from the analyzing wavelet ψ) by

$$\psi_{(a,b)}(x) = a^{-n/2}\psi\left(\frac{x-b}{a}\right), \quad a > 0, \quad b \in \mathbb{R}^n.$$

In the analysis and synthesis of an arbitrary function f belonging to $L^2(\mathbb{R}^n)$, these wavelets $\psi_{(a,b)}$ are going to play the role of an orthonormal basis. The wavelet coefficients $W(a,b)$ are defined by

$$W(a,b) = \langle f, \psi_{(a,b)} \rangle, \tag{2.13}$$

where $\langle u, v \rangle = \int u(x)\overline{v}(x)\,dx$. The function f is analyzed by (2.13). The synthesis of f is given by

$$f(x) = \int_0^\infty \int_{\mathbb{R}^n} W(a,b)\psi_{(a,b)}(x)\,db \frac{da}{a^{n+1}}. \tag{2.14}$$

This is a linear combination of the original wavelets using the coefficients given by the analysis.

We return to the specific case of the Hardy spaces $H^p(\mathbb{R})$ for $1 \leq p < \infty$. The analyzing wavelet $\psi(x) = \frac{1}{\pi}(x+i)^{-2}$ chosen by Lusin is the restriction to the real axis of the function $\frac{1}{\pi}(z+i)^{-2}$; it is holomorphic in P and belongs to all of the Hardy spaces. The Fourier transform of ψ is $\hat{\psi}(\xi) = -2\xi e^{-\xi}$ for $\xi \geq 0$ and $\hat{\psi}(\xi) = 0$ if $\xi \leq 0$. Condition (2.12) is not satisfied; however, we have

$$\int_0^\infty |\hat{\psi}(t\xi)|^2 \frac{dt}{t} = \begin{cases} 1 & \text{if } \xi > 0, \\ 0 & \text{if } \xi \leq 0. \end{cases} \tag{2.15}$$

Condition (2.15) implies that the wavelets $\psi_{(a,b)}$ generate $H^2(\mathbb{R})$ instead of $L^2(\mathbb{R})$ when $a > 0$, $b \in \mathbb{R}$.

The wavelet coefficients of a function f belonging to the Hardy space $H^2(\mathbb{R})$ are then

$$W(a,b) = \langle f, \psi_{(a,b)} \rangle = \frac{1}{\pi} \int_{-\infty}^\infty f(x) \frac{a\sqrt{a}}{(x-b-ia)^2}\,dx.$$

By Cauchy's formula, this is equal to $2ia\sqrt{a}f'(b+ia)$, since f is holomorphic in P. Thus the representation (2.14) of a function in the Hardy space $H^2(\mathbb{R})$ coincides with the natural representation that we defined in the preceding section.

2.8 Strömberg's wavelets

The real version of the holomorphic Hardy space $H^1(\mathbb{R})$ is denoted by $\mathcal{H}^1(\mathbb{R})$. It is composed of the real-valued functions $u(x)$ for which there exists a real-valued function $v(x)$ such that $u(x)+iv(x)$ belongs to $H^1(\mathbb{R})$. In other words, $u(x)$ belongs to $\mathcal{H}^1(\mathbb{R})$ if and only if u and its Hilbert transform $\tilde{u}$ belong to $L^1(\mathbb{R})$.

Research on atomic decompositions for the functions in the Hardy space $\mathcal{H}^1(\mathbb{R})$ takes two completely different approaches: One involves the atomic decomposition of Coifman and Weiss, and the other concerns the search for unconditional bases for the space $\mathcal{H}^1$. Here is an outline of these theories.

Coifman and Weiss showed that any function f of $\mathcal{H}^1$ can be written as

$$f(x) = \sum_{k=0}^{\infty} \lambda_k a_k(x), \qquad (2.16)$$

where the coefficients λ_k are such that $\sum_{k=0}^{\infty} |\lambda_k| < \infty$ and where the functions a_k are atoms of $\mathcal{H}^1$. The conditions for a function to be an atom are the following: For each a_k, there exists an interval I_k such that $a_k(x) = 0$ outside of I_k, $|a_k(x)| \leq 1/|I_k|$ ($|I_k|$ is the length of I_k), and $\int_{I_k} a_k(x)\,dx = 0$. These three conditions imply that the norms of the a_k in $\mathcal{H}^1$ are bounded by a fixed constant. The price to pay for this extraordinary decomposition is that it is not given by a linear algorithm, and this naturally raises the problem of finding one.

Finding an unconditional basis means constructing, once and for all, a sequence of functions b_k of $\mathcal{H}^1$ that are linearly independent, in a very strong sense, and such that any function f of $\mathcal{H}^1$ can be decomposed in the form

$$f(x) = \sum_{k=0}^{\infty} \beta_k b_k(x),$$

where the scalars β_k are defined explicitly by the formulas

$$\beta_k = \int f(x) g_k(x)\,dx.$$

Here the g_k are specific functions in the dual of $\mathcal{H}^1$; that is, they are BMO functions.

The strong independence property is this: There exists a constant C such that if two sequences of coefficients β_k and λ_k satisfy $|\beta_k| \leq |\lambda_k|$ for all k, then

$$\left\| \sum_{k=0}^{\infty} \beta_k b_k(x) \right\| \leq C \left\| \sum_{k=0}^{\infty} \lambda_k b_k(x) \right\|,$$

where $\|\cdot\|$ is the norm of the function space $\mathcal{H}^1$.

Wojtaszczyk proved that the Franklin system $\{f_n\}_{n\in\mathbb{N}}$ without the function $f(x) = 1$ is an unconditional basis for the subspace of $\mathcal{H}^1(\mathbb{R})$ composed of functions that vanish outside the interval $[0,1]$ [262]. Strömberg showed that the orthonormal wavelet basis $2^{j/2}\psi(2^j x - k)$, $j,k \in \mathbb{Z}$, defined in section 2.5 is, in fact, an unconditional basis for the space $\mathcal{H}^1(\mathbb{R})$ [245].

Does there exist a relation between these two types of atomic decompositions? We first point out the main difference: The decomposition (2.16) of a function is not unique, and in some sense the atoms a_k must be fitted to the function f. Thus the decomposition algorithm is not linear. On the other hand, one way to construct the atoms for (2.16) is to start with the expansion of f in an orthonormal basis of compactly supported wavelets and to group the wavelets to form the atoms. These groups of wavelets are a little like the dyadic blocks of Littlewood and Paley; however, in this case, they are defined by considering the moduli of the coefficients $\alpha_{j,k}$ of this series. The interested reader is referred to [203]. (The construction of wavelets with compact support will be developed in Chapter 3.)

2.9 A first synthesis: Wavelet analysis

Thanks to the historical perspective that we enjoy today, we can relate the Haar system, the Littlewood–Paley decomposition (1930), the version of Franklin's basis given by Strömberg (1981), and Calderón's identity (1960) to one another.

This first synthesis will be followed by a more inclusive synthesis that encompasses the techniques of numerical signal and image processing. This second synthesis will lead to Daubechies's orthonormal bases.

This first synthesis is based on the definition of the "wavelet" and on the concept of "wavelet analysis." We will see that the success of this synthesis depends on a certain lack of specificity in the original definition. When wavelets were first defined, mathematicians had not created a general formalism covering all of the examples we presented above. A physicist and a geophysicist, Grossmann and Morlet, provided a definition and a way of thinking based on physical intuition that was flexible enough to cover all these cases. Starting with the Grossmann–Morlet definition, we will present two other definitions and indicate how they are related.

The first definition of a wavelet, which is due to Grossmann and Morlet, is quite broad. *A wavelet is a function ψ in $L^2(\mathbb{R})$ whose Fourier transform $\hat{\psi}(\xi)$ satisfies the condition $\int_0^\infty |\hat{\psi}(t\xi)|^2 \frac{dt}{t} = 1$ almost everywhere.*

The second definition of a wavelet is adapted to the Littlewood–Paley–Stein theory. *A wavelet is a function ψ in $L^2(\mathbb{R}^n)$ whose Fourier transform $\hat{\psi}(\xi)$ satisfies the condition $\sum_{-\infty}^\infty |\hat{\psi}(2^{-j}\xi)|^2 = 1$ almost everywhere.* If ψ is a wavelet in this sense, then $\sqrt{\log 2}\,\psi$ satisfies the Grossmann–Morlet condition.

The third definition refers to the work of Haar and Strömberg. *A wavelet is a function ψ in $L^2(\mathbb{R})$ such that $2^{j/2}\psi(2^j x - k)$, $j, k \in \mathbb{Z}$, is an orthonormal basis for $L^2(\mathbb{R})$.* Such a wavelet ψ necessarily satisfies the second condition.

This shows that in going from the first to the third definition we are adding more conditions and thus narrowing the choice of functions that will be wavelets. What we gain is a more economical (less redundant) representation of the analyzed function. In the general Grossmann–Morlet theory—which is identical to Calderón's theory—the wavelet analysis of a function f yields a function $W(a, b)$ of $n + 1$ variables $a > 0$ and $b \in \mathbb{R}^n$. This function is defined by (2.13): $W(a, b) = \langle f, \psi_{(a,b)} \rangle$, where $\psi_{(a,b)}(x) = a^{-n/2}\psi(\frac{x-b}{a})$, $a > 0$, $b \in \mathbb{R}^n$.

In the Littlewood–Paley theory, a is replaced by 2^{-j}, while b is denoted by x. Thus, if Γ is the multiplicative group $\{2^{-j}, j \in \mathbb{Z}\}$, then the Littlewood–Paley analysis is obtained by restricting the Grossmann–Morlet analysis to $\Gamma \times \mathbb{R}^n$.

In the Franklin–Strömberg theory, a is replaced by 2^{-j} and b is replaced by $k2^{-j}$, where $j, k \in \mathbb{Z}$. In other words, the analysis of f in the Franklin–Strömberg basis is obtained by restricting the Littlewood–Paley analysis to the "hyperbolic lattice" S in $(0, \infty) \times \mathbb{R}$ consisting of the points $(2^{-j}, k2^{-j})$, $j, k \in \mathbb{Z}$. The logical relations between these wavelets analyses are easy to verify.

We start with the Grossmann–Morlet analysis, which is equivalent to the Calderón identity. This is written $I = \int_0^\infty Q_t Q_t^* \frac{dt}{t}$, where $Q_t(f) = f * \psi_t$. This becomes $I = \sum_{-\infty}^{\infty} \Delta_j \Delta_j^*$ in the Littlewood–Paley theory. Indeed, if $t = 2^{-j}$, then one has $Q_t(f) = \Delta_j(f)$. Replacing t by 2^{-j} and the integral $\int_0^\infty u(t)\frac{dt}{t}$ by the sum $\sum_{-\infty}^{\infty} u(2^{-j})$ is completely classic.

To relate Littlewood–Paley analysis to the analysis that is obtained using the orthogonal wavelets of Franklin and Strömberg, we write $\psi_j(x) = 2^j \psi(2^j x)$ and $\tilde\psi_j(x) = 2^j \tilde\psi(2^j x)$, where $\tilde\psi(x) = \overline{\psi}(-x)$. We let $\Delta_j(f)$ denote the convolution product $f * \psi_j$ and Δ_j^* denote the adjoint of the operator $\Delta_j : L^2(\mathbb{R}) \to L^2(\mathbb{R})$. Then $\Delta_j^*(f) = f * \tilde\psi_j$. The coefficients $\alpha(j,k)$ of the decomposition of f in Strömberg's orthonormal basis are then given by

$$\begin{aligned}\alpha(j,k) &= 2^{j/2} \int f(x)\overline{\psi}(2^j x - k)\, dx \\ &= 2^{-j/2}(f * \tilde\psi_j)(k 2^{-j}) \\ &= 2^{-j/2}\Delta_j^* f(k 2^{-j}).\end{aligned}$$

Thus we see that the wavelet coefficients are obtained by sampling the dyadic blocks $\Delta_j^*(f)$ on the grid $2^{-j}\mathbb{Z}$. This sampling is consistent with Shannon's theorem.

In all three cases, wavelet analysis is followed by a synthesis that reconstructs f from its wavelet transform. In the case of Grossmann–Morlet wavelets, this synthesis is given by the identity (2.14), which we rewrite here:

$$f(x) = \int_0^\infty \int_{\mathbb{R}^n} W(a,b)\psi_{(a,b)}(x)\, db \frac{da}{a^{n+1}}. \tag{2.17}$$

In the case of the Littlewood–Paley analysis, the integral $\int_0^\infty u(a)\frac{da}{a}$ is replaced, as we have already mentioned, by the sum $\sum_{-\infty}^{\infty} u(2^{-j})$ and (2.17) becomes

$$f(x) = \sum_{-\infty}^{\infty} \int_{\mathbb{R}^n} (\Delta_j^* f)(b)\psi_j(x - b)\, db. \tag{2.18}$$

Finally, in the case of Strömberg's orthogonal wavelets in one dimension, the last integral becomes a sum, and (2.18) becomes

$$f(x) = \sum_{-\infty}^{\infty} \sum_{-\infty}^{\infty} \alpha(j,k)\psi_{j,k}(x).$$

The preceding arguments may seem less than exciting, since the hypotheses on ψ are designed specifically for the analysis of the space L^2 of square-summable functions. This is the setting in which Grossmann and Morlet wrote their theoretical work. But this is evidently a sort of regression, for we have just shown that across a century of mathematical history, wavelet analysis was created specifically to analyze function spaces other than L^2. Fourier analysis serves admirably for the analysis of L^2.

If we want wavelets to be useful for the analysis of other function spaces, it is necessary to impose conditions on the wavelets in addition to those we have already given. Until now we have required only that the analysis preserves energy or, equivalently, that the synthesis gives an exact reconstruction (although this equivalence is not immediately obvious). These new conditions concern the regularity of the wavelet ψ, the decay at infinity of ψ, and the number of vanishing moments of ψ. For example, we can require that ψ belongs to the Schwartz class and that all of its moments vanish. Or, in the case of Daubechies's wavelets, we can require that ψ has m continuous derivatives, that it has compact support, and that its first $r + 1$ moments vanish.

The properties of the Strömberg wavelet are intermediate: It has exponential decay, as does its first derivative, and

$$\int_{-\infty}^{\infty} \psi(x)dx = \int_{-\infty}^{\infty} x\psi(x)dx = 0.$$

These new wavelets are particularly useful. For example, the Daubechies wavelets just mentioned can be used to analyze the functions in $L^{s,p}$, the space of functions in L^p whose derivatives of order $s < \inf(r, m)$ are also in L^p.

2.10 The advent of signal processing

If history had stopped with this first synthesis, the Daubechies orthonormal bases, which improve the rudimentary Haar basis, would never have been discovered. A new start was made in 1985 by Stéphane Mallat when he was still a graduate student. Mallat discovered the similarities between the following objects:

(a) the quadrature mirror filters, which were invented by Croisier, Esteban, and Galand for the digital telephone;

(b) the pyramid algorithms of Burt and Adelson, which are used in the context of numerical image processing;

(c) the orthonormal wavelet bases discovered by Strömberg and Meyer.

The relations between these concepts will be explained in the next two chapters.

By using the relation between wavelets and quadrature mirror filters, Daubechies was able to complete Haar's work. For each integer r, Daubechies constructs an orthonormal basis for $L^2(\mathbb{R})$ of the form $2^{j/2}\psi_r(2^j x - k)$, $j, k \in \mathbb{Z}$, having the following properties:

(a) The support of ψ_r is the interval $[0, 2r + 1]$.

(b) $0 = \int_{-\infty}^{\infty} x^n \psi_r(x)\,dx = 0$ for $0 \le n \le r$.

(c) ψ_r has γr continuous derivatives, where the constant γ is about $1/5$.

When $r = 0$, this reduces to the Haar system.

Daubechies's wavelets provide a much more effective analysis and synthesis than that obtained with the Haar system. If the function being analyzed has m continuous derivatives, where $0 \le m \le r + 1$, then the coefficients $\alpha(j, k)$ from its decomposition in the Daubechies basis will be of the order of magnitude $2^{-(m+1/2)j}$, while it would be of the order $2^{-3j/2}$ with the Haar system. This means that as soon as the analyzed function is regular, the coefficients one keeps (those exceeding the machine precision) will be much fewer than in the case of the Haar system. Thus one speaks of signal compression. Furthermore, this property has a purely local aspect because Daubechies's wavelets have compact support.

Synthesis using Daubechies's wavelets also gives better results than the Haar system. In the latter case, a regular function is approximated by functions that have strong discontinuities. This produces an annoying "blocking effect" when images are compressed using Haar wavelets, as the reader can verify by referring to the images of Jean Baptiste Joseph Fourier in Figure 2.1. These remarkable qualities of Daubechies's bases explain their undisputed success.

FIG. 2.1. *Jean Baptiste Joseph Fourier (1768–1830): The image on the right was produced by analyzing the original image on the left using Haar wavelets. It was reconstructed from the 600 largest wavelet coefficients and shows the characteristic blocking effect that is the signature of Haar compression. Courtesy of Académie des Sciences–Paris and Jean-Loup Charmat.*

2.11 Conclusions

The status of wavelet analysis within mathematics is unique. Indeed, mathematicians have been working on various forms of wavelet decompositions for a fairly long time. Their goal was to provide direct and easy access to various function spaces. But during this period, which stretches from 1909 to 1980, from Haar to Strömberg, there was very little scientific interchange between mathematicians (of the "Chicago School"), physicists, and experts in signal processing. Not knowing about the mathematical developments and faced with the pressure of specific needs within their disciplines, the last two groups were led to rediscover wavelets.

For example, Marr did not know about Calderón's work on wavelets (dating from 1960) when he announced the hypothesis that we analyze in detail in Chapter 8. Similarly, G. Battle and P. Federbush were not aware of Strömberg's basis when they needed it to do renormalization computations in quantum field theory [108] (see also [27] and [29]). As it was stressed by Battle [28, p. 87], "The physics community was intuitively aware of wavelets years before anything better than the Haar basis was mathematically known to exist. This cultural knowledge dates back to a paper by Wilson [261] on the renormalization group." In the numerous fields of science and technology where wavelets appeared at the end of the 1970s, they were handcrafted by the scientists and engineers themselves. Their use has never resulted from proselytism by mathematicians.

Battle's comment raises another point: We have given a brief historical review of some of the mathematical origins of what is now known as the theory of time-scale wavelets. This historical perspective is, however, incomplete for two reasons. First, we focused the discussion on mathematics. We are sure that diligent detectives could write a similar story about the appearance of wavelet techniques in physics—

and perhaps in other fields of science. Indeed, as we have mentioned, David Marr built his own wavelets in an image processing context, while several groups of physicists working in quantum field theory designed ad hoc orthonormal wavelet bases (see, for example, [108], [130], and [261]).

Second, we focused on time-scale wavelets. Furthermore, we note that time-frequency wavelets were, for a rather long time, neglected by mathematicians. They were, however, popular in signal processing. Indeed, computing the scalar product between a given signal f and a Gabor wavelet $g(t-\tau)e^{i\omega t}$ amounts to performing a windowed Fourier analysis. Moreover, Gabor wavelets were immediately welcomed in physics and signal processing, while time-scale wavelets had a harder time. Finally, Gabor wavelets can be described as an orbit under the action of the Weyl–Heisenberg group, which is playing a key role in quantum mechanics. This discussion will be postponed until Chapters 5 and 6, where the interaction between quantum mechanics and time-frequency analysis will be studied in more detail.

Today the boundaries between mathematics and signal and image processing have faded, and mathematics has benefited from the rediscovery of wavelets by experts from other disciplines. The detour through signal and image processing was the most direct path leading from the Haar basis to Daubechies's wavelets.

CHAPTER 3

Quadrature Mirror Filters

3.1 Introduction

In his thesis, "Codage en sous-bandes: théorie et applications à la compression numérique du signal de parole," Claude Galand carefully described the quadrature mirror filters (which he invented in collaboration with Esteban and Croisier [68]) and their anticipated applications [125]. He also posed some very important problems that would lead to the discovery of wavelet packets (Chapter 7) and Malvar–Wilson wavelets (Chapter 6).

Galand's work was motivated by the possibility of improving the digital telephone, a technology that involves transmitting speech signals as sequences of 0's and 1's. However, as Galand remarked, these techniques extend far beyond digital speech, since facsimile, video, databases, and many other forms of information travel over telephone lines. At present, the bit allocation used for telephone transmission is the well-known 64 kilobits per second. Galand sought, by using coding methods tailored to speech signals, to transmit speech well below this standard. To validate the method he proposed, Galand compared it with two other techniques for coding sampled speech: predictive coding and transform coding.

Linear prediction coding amounts to looking for the correlations between successive values of the sampled signal. These correlations are likely to occur on intervals of the order of 20 to 30 milliseconds. This leads one to cut the sampled signal $x(n)$ into blocks defined by $1 \leq n \leq N$, $N+1 \leq n \leq 2N$, etc., and then to seek, for each block, coefficients a_k, $1 \leq k \leq p$, that minimize the quadratic mean $\frac{1}{N}\sum_1^N |e(n)|^2$ of the prediction errors defined by

$$e(n) = x(n) - \sum_{k=1}^p a_k x(n-k).$$

In general, p is much smaller than N. To transmit a block $x(n)$, it suffices to transmit the first p values $x(1), \ldots, x(p)$, the p coefficients $a_1, \ldots, a_p$, and the prediction errors $e(n)$. The method is efficient if most of the prediction errors are near zero. When they fall below a certain threshold, they are not transmitted, and significant compression can result.

A form of transform coding consists of cutting the sampled signal into successive blocks of length N, as we have just done, and then using a unitary transformation A to transform each block (denoted by X) into another block (denoted by Y). The block Y is then quantized, with the hope that, for a suitable linear transformation, the Y blocks will have a simpler structure than the X blocks. Subband coding will be presented in the next section.

For a stationary Gaussian signal, the theoretical limits of the minimal distortion that can be obtained by the three methods are the same. However, as Galand showed, this assumes, in the case of subband coding, that the width of the frequency channels tends to zero and that their number tends to infinity. In Galand's work, these frequency channels are obtained through a treelike arrangement of quadrature mirror filters. This construction leads precisely to wavelet packets, which we discuss in detail in Chapter 7. Today we know that wavelet packets based on filters with finite length do not enjoy the frequency localization that Galand had hoped for.

In the cases of linear prediction coding and transform coding, the theoretical limits of the minimal distortion are calculated as the lengths of the blocks tend to infinity, while conserving the stationarity hypothesis.

If the three types of coding yield asymptotically the same quality of compression, why introduce subband coding? Galand saw two advantages: the simplicity of the algorithm and the possibility that subband coding would reduce the unpleasant effects of quantization noise as perceived at the receiver. By quantizing inside each subband, the signal would tend to mask the quantization noise, and it would be less apparent.

The same argument has been repeated by Adelson, Hingorani, and Simoncelli for numerical image processing [3]. The use of pyramid algorithms and wavelets allows aspects of the human visual system to be taken into account so that the signal masks the noise. The perceptual quality of the reconstructed image is improved even though the theoretical compression calculations do not distinguish this method from the others. It should be observed, however, that these theoretical compression calculations are based on a very specific hypothesis that is clearly not fulfilled in the case of images, which are not well modeled by a Gaussian stationary processes (see, for example, [95]).

Readers not familiar with the theory of filters may wish to look at Appendix A; it provides an elementary introduction to the language and notation that is used in this chapter.

3.2 Subband coding: The case of ideal filters

To illustrate the ideas, we follow Galand and begin with a simplistic example. For a fixed $m \geq 2$, let I denote an interval of length $\frac{2\pi}{m}$ within $[0,2\pi]$, and let l_I^2 denote the Hilbert space of sequence $(c_k)_{k \in \mathbb{Z}}$ satisfying $\sum_{-\infty}^{\infty} |c_k|^2 < \infty$ and such that $f(\theta) = \sum_{-\infty}^{\infty} c_k e^{ik\theta}$ is zero outside the interval I. This subspace l_I^2 will be called a frequency channel.

If $(c_k)_{k \in \mathbb{Z}}$ is a sequence belonging to l_I^2, the subsequence $(c_{km})_{k \in \mathbb{Z}}$ provides an optimal, compact representation of the original. In fact, for $\theta \in I$,

$$\frac{1}{m} f(\theta) = \frac{1}{m}\left[f(\theta) + f\left(\theta + \frac{2\pi}{m}\right) + \cdots + f\left(\theta + \frac{2\pi(m-1)}{m}\right)\right] = \sum c_{km} e^{ikm\theta},$$

since $f(\theta) = 0$ for $\theta \notin I$. Thus we have $\sum_{-\infty}^{\infty} |c_{km}|^2 = \frac{1}{m} \sum_{-\infty}^{\infty} |c_k|^2$, and this relation expresses the redundancy contained in the original sequence (c_k), which is strongly correlated. This means that the original sequence contains m times the numerical data needed to reconstruct f on I, knowing that f vanishes outside of I. This observation is a form of Shannon's theorem, and the critical sampling c_{km} is done at the Shannon–Nyquist rate.

The ideal subband coding scheme consists of first filtering the incoming signal into m frequency channels associated with the intervals $[\frac{2\pi l}{m}, \frac{2\pi(l+1)}{m}]$, $0 \leq l \leq m-1$, and then *subsampling* the corresponding outputs, retaining only one point in m. This operation, which consists of restricting a sequence defined on $\mathbb{Z}$ to $m\mathbb{Z}$, is called *decimation* and is denoted by $m \downarrow 1$. This ideal subband coding scheme is illustrated in Figure 3.1.

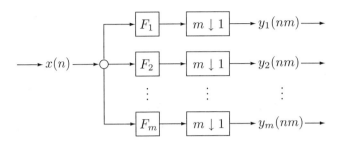

FIG. 3.1. *Subband coding scheme.*

The scheme for reconstructing the original signal is the dual of the analysis scheme. We began by extending the sequences $y_1(nm), \ldots, y_m(nm)$ by inserting 0's at all integers that are not multiples of m. Next we filter this "absurd decision" by using the adjoint filters $F_1^*, \ldots, F_m^*$. The output returns the original signal $(x(n))$. The reconstruction is illustrated in Figure 3.2.

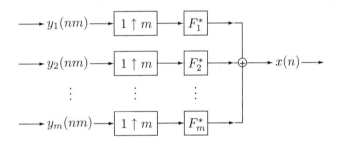

FIG. 3.2. *Reconstruction.*

One can, as Galand did, hope for the best and try to replace the index functions of the intervals $[\frac{2\pi l}{m}, \frac{2\pi(l+1)}{m}]$ with more regular functions of the form $w(mx - 2\pi l)$, $0 \leq l \leq m-1$. If the $w(x) = w_m(x)$ were a finite trigonometric sum, then the filters $F_1, \ldots, F_m$ would have finite length, which is essential for applications. But the Balian–Low theorem (Chapter 6) tells us that such a function w cannot be constructed if we demand that it be regular and well localized (uniformly in m). Consequently, it is not possible to realize the ideal subband coding scheme just described if we require that the filters $F_1, \ldots, F_m$ have finite length and, at the same time, provide good frequency definition.

3.3 Quadrature mirror filters

Faced with the impossibility of realizing subband coding using m bands covering the frequency space regularly and having finite-length filters—whose length must be Cm, for some $C > 0$, as required by the Heisenberg uncertainty principle—Galand limited himself to the case $m = 2$. He then had the idea to effect a finer frequency tiling by suitably iterating the two-band process. We will see in Chapter 7 that this arborescent scheme leads directly to wavelet packets, but we will also see that these wavelet packets do not have the desired spectral properties. Subband coding using two frequency channels works perfectly. We are going to describe it in detail.

The input signals are arbitrary sequences $(x(n))_{n\in\mathbb{Z}}$ with finite energy:

$$\sum_{-\infty}^{\infty} |x(n)|^2 < \infty,$$

which means that $x \in l^2(\mathbb{Z})$. In the context of the digital telephone, we assume that the original speech signals have been sampled to give the signals $(x(n))_{n\in\mathbb{Z}}$.

Let D denote the decimation operator $D : l^2(\mathbb{Z}) \to l^2(2\mathbb{Z})$ that consists of retaining only the terms with even index in a sequence $(x(n))_{n\in\mathbb{Z}}$. (D is also denoted by $2 \downarrow 1$.) The adjoint operator

$$E = D^* : l^2(2\mathbb{Z}) \to l^2(\mathbb{Z})$$

is the crudest possible extension operator. It consists, starting with a sequence $(x(2n))_{n\in\mathbb{Z}}$, in constructing the sequence defined on $\mathbb{Z}$ obtained by inserting 0's at the odd indices. Thus we get the sequence

$$(\ldots, 0, x(-4), 0, x(-2), 0, x(0), 0, x(2), 0, x(4), 0, \ldots).$$

To simplify the notation we write X in place of $(x(n))_{n\in\mathbb{Z}}$. These input signals X are first filtered using two filters F_0 and F_1. Later, we will require that F_0 be a low-pass filter (in a sense that will be made precise), and, consequently, F_1 will be a high-pass filter. However, there is no distinction between the two filters at this point.

The outputs $X_0 = F_0(X)$ and $X_1 = F_1(X)$ are two signals $(x_0(n))_{n\in\mathbb{Z}}$ and $(x_1(n))_{n\in\mathbb{Z}}$ with finite energy.

X_0 and X_1 are subsampled with the decimation operator $D = 2 \downarrow 1$. Then we have $Y_0 = D(X_0) = (x_0(2n))_{n\in\mathbb{Z}}$ and $Y_1 = D(X_1) = (x_1(2n))_{n\in\mathbb{Z}}$.

We write

$$\|Y_0\| = \left(\sum_{-\infty}^{\infty} |x_0(2n)|^2\right)^{1/2} \quad \text{and} \quad \|Y_1\| = \left(\sum_{-\infty}^{\infty} |x_1(2n)|^2\right)^{1/2}.$$

The two filters F_0 and F_1 are called *quadrature mirror filters* if, for all signals X of finite energy, one has

$$\|Y_0\|^2 + \|Y_1\|^2 = \|X\|^2. \tag{3.1}$$

Denote the operator $DF_0 : l^2(\mathbb{Z}) \to l^2(2\mathbb{Z})$ by T_0, and similarly let T_1 denote the operator $DF_1 : l^2(\mathbb{Z}) \to l^2(2\mathbb{Z})$. It can be shown that (3.1) is equivalent to

$$I = T_0^* T_0 + T_1^* T_1. \tag{3.2}$$

In (3.2), $I : l^2(\mathbb{Z}) \to l^2(\mathbb{Z})$ is the identity operator. What is much less evident is that the vectors $T_0^* T_0(X)$ and $T_1^* T_1(X)$ are always orthogonal, which is a consequence of the following theorem.

THEOREM 3.1. *Let $F_0(\theta)$ and $F_1(\theta)$ denote the transfer functions of the filters F_0 and F_1. Then the following two properties are equivalent to each other and to the property expressed by (3.1):*

(i) *The matrix* $\frac{1}{\sqrt{2}}\begin{bmatrix} F_0(\theta) & F_1(\theta) \\ F_0(\theta+\pi) & F_1(\theta+\pi) \end{bmatrix}$ *is unitary for almost all* $\theta \in [0, 2\pi]$.

(ii) *The operator* $(T_0, T_1) : l^2(\mathbb{Z}) \to l^2(2\mathbb{Z}) \times l^2(2\mathbb{Z})$ *is an isometric isomorphism.*

Recall that the sequence of Fourier coefficients of the 2π-periodic function $F_0(\theta)$ is the impulse response of the filter F_0, and similarly for $F_1(\theta)$ and F_1.

Condition (3.2) is called the perfect reconstruction property. The input signal X is the sum of two orthogonal signals $T_0^* T_0(X)$ and $T_1^* T_1(X)$, where the signals $T_0(X)$ and $T_1(X)$ were given by the analysis. The operators $T_0^* = F_0^* E$ and $T_1^* = F_1^* E$ are applied to two sequences sampled on the even integers. These are first extended in the crudest way, which is by replacing the missing values with 0's. Next, this seemingly nonsensical step is corrected by passing the sequences through the filters F_0^* and F_1^*, which are the adjoints of F_0 and F_1. The correct result is read at the output (Figure 3.3).

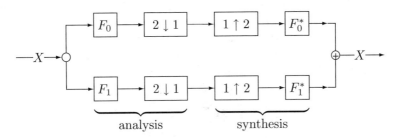

Fig. 3.3. *The complete scheme, analysis and synthesis.*

Condition (ii) means that quadrature mirror filters constitute orthogonal transformations of a particular type, while (i) allows us to construct quadrature mirror filters that have finite impulse response. To see this, we start with a trigonometric polynomial

$$m_0(\theta) = \alpha_0 + \alpha_1 e^{i\theta} + \cdots + \alpha_N e^{iN\theta}$$

such that $|m_0(\theta)|^2 + |m_0(\theta + \pi)|^2 = 1$ for all θ. Next, we write

$$F_0(\theta) = \sqrt{2}\,\overline{m_0}(\theta), \qquad F_1(\theta) = \sqrt{2}\,e^{i\theta} m_0(\theta + \pi).$$

Then it follows directly that (i) is satisfied. The following five examples illustrate the definition of quadrature mirror filters.

The first example is essentially a counterexample because it is never used, for a reason that will become clear later. It consists of bypassing the operators F_0 and F_1 and defining T_0 and T_1 directly. Define $T_0(X)$ to be the restriction of the sequence $X = (x(n))_{n \in \mathbb{Z}}$ to the even integers, and define $T_1(X)$ to be the restriction of this sequence to the odd integers. This is equivalent, in our notation, to taking F_0 equal to the identity I, and to taking F_1 to be the shift operator defined by $(F_1 X)(n) = x(n - 1)$. Condition (3.1) is trivially satisfied, and the unitary matrix in (i) is

$$\frac{1}{\sqrt{2}}\begin{bmatrix} 1 & e^{i\theta} \\ 1 & -e^{i\theta} \end{bmatrix}.$$

The second example is more interesting. The filter operators are defined by

$$(F_0 X)(n) = \frac{1}{\sqrt{2}}(x(n) + x(n+1)),$$

$$(F_1 X)(n) = \frac{1}{\sqrt{2}}(x(n) - x(n+1)).$$

The unitary matrix in (i) is

$$\frac{1}{2}\begin{bmatrix} 1 + e^{-i\theta} & 1 - e^{-i\theta} \\ 1 - e^{-i\theta} & 1 + e^{-i\theta} \end{bmatrix},$$

and the orthonormal wavelet basis associated with this choice will be the Haar system.

The third example recaptures the ideal filters presented at the beginning of the chapter. The 2π-periodic function $m_0(\theta)$ is one on $[0,\pi)$ and zero on $[\pi, 2\pi)$, and $m_1(\theta) = 1 - m_0(\theta)$. As above, define $F_0(\theta) = \sqrt{2}\,\overline{m_0}(\theta)$ and $F_1(\theta) = \sqrt{2}\,\overline{m_1}(\theta)$.

In the fourth example, $m_0(\theta)$ becomes the characteristic function of the interval $[-\frac{\pi}{2}, \frac{\pi}{2})$ when it is restricted to $[-\pi, \pi)$, and $m_1(\theta) = 1 - m_0(\theta)$.

The last example is a smooth modification of the preceding one. With $0 < \alpha < \frac{\pi}{2}$, we ask that $m_0(\theta)$ be 2π-periodic, equal to one on the interval $[-\frac{\pi}{2} + \alpha, \frac{\pi}{2} - \alpha]$, equal to zero on $[\frac{\pi}{2} + \alpha, \frac{3\pi}{2} - \alpha]$, even, and infinitely differentiable. In addition, we impose the condition

$$|m_0(\theta)|^2 + |m_0(\theta + \pi)|^2 = 1. \tag{3.3}$$

Then write $m_1(\theta) = e^{-i\theta}\,\overline{m_0}(\theta + \pi)$, $F_0(\theta) = \sqrt{2}\,\overline{m_0}(\theta)$, $F_1(\theta) = \sqrt{2}\,\overline{m_1}(\theta)$, and we obtain two quadrature mirror filters.

3.4 Trend and fluctuation

Let H denote the Hilbert space $l^2(\mathbb{Z})$ of all sequences $(x(n))_{n \in \mathbb{Z}}$ such that $\sum_{-\infty}^{\infty} |x(n)|^2 < \infty$. Write H_0 and H_1 for the two subspaces $T_0^* T_0(H)$ and $T_1^* T_1(H)$. If F_0 and F_1 are quadrature mirror filters, then by Theorem 3.1, H will be the direct orthogonal sum of H_0 and H_1.

Write $m_0(\theta) = \frac{1}{\sqrt{2}}\,\overline{F_0}(\theta)$ and assume that $m_0(\pi) = 0$ and that this zero has order $q \geq 1$. Then $|m_0(\theta)|^2 = 1 + O(|\theta|^{2q})$ and $m_1(\theta) = O(|\theta|^q)$ as θ tends to zero. Under these conditions, we say that F_0 is a low-pass filter and that F_1 is a high-pass filter, even though this terminology may not always be strictly justified.

When these conditions are satisfied, the trend and the fluctuation around this trend of a signal X are defined by $\tilde{X}_0 = T_0^* T_0(X)$ and $\tilde{X}_1 = T_1^* T_1(X)$, respectively. Note that the trend and fluctuation are defined in terms of a given pair of filters. They are not intrinsic properties of the function X, but they are handy heuristics. The trend $\tilde{X}_0 = T_0^* T_0(X)$ is generally "smoother" than X, in the sense that the low-pass filter F_0 removes high frequencies from X. In fact, one often says that $\tilde{X}_0$ is "twice as smooth as X," which is another useful heuristic. These heuristics are consistent with a theorem by S. Bernstein that relates the smoothness of a function with the size of the support of its Fourier transform.

3.5 The time-scale algorithm of Mallat and the time-frequency algorithm of Galand

It is amazing to reread Galand's thesis in the light of present understanding. Indeed, Galand's goal was to obtain finer and finer frequency resolutions by appropriately iterating the quadrature mirror filters. This is possible, however, only in the case of the ideal filters in our third example, but we cannot use these ideal filters because they have an infinite impulse response. In spite of this criticism, we will return to Galand's point of view in Chapter 7, and it will lead us to wavelet packets.

Thus we see that Galand was looking for time-frequency algorithms. But his fundamental discovery, quadrature mirror filters, was diverted from that end by Mallat, who used quadrature mirror filters to construct time-scale algorithms using a hierarchical scheme.

Mallat considers an increasing sequence $\Gamma_j = 2^{-j}\mathbb{Z}$ of nested grids that go from the "fine grid" Γ_N, $N \geq 1$, to the "coarse grid" Γ_0. The signal to be analyzed has been sampled on the fine grid (we will come back to the sampling technique when studying the convergence problem), and our starting point is thus a sequence $f = f_0$ belonging to $l^2(\Gamma_N)$.

In addition, two quadrature mirror filters F_0 and F_1 are given. (We will see later what conditions they must satisfy.) These same filters will be used throughout the discussion.

We process the signal f by decomposing it into $T_0 f$ and $T_1 f$, which we also call the trend and fluctuation. The trend $T_0 f = DF_0 f$ has been downsampled and "lives" the coarser grid Γ_{N-1}; it represents a new signal that is decomposed again into a trend and fluctuation. The fluctuations are never analyzed in this scheme, and the algorithm follows a "herringbone" pattern illustrated in Figure 3.4. (To be precise, the operators $T_0 = DF_0$ and $T_1 = DF_1$ should have another index k to indicate that they "live" on the grid Γ_{N-k}. We have used a simplified notation to emphasize that the filters are always the same; they are just expanded at each step to fit the coarser grid.)

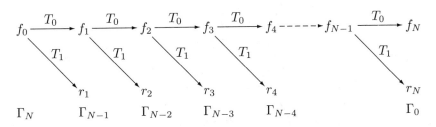

FIG. 3.4. *Mallat's algorithm.*

The input signal $f \in l^2(\Gamma_N)$ is finally represented by the sequence $r_1, \ldots, r_N$ of fluctuations and by the last trend $f_N \in l^2(\Gamma_0)$. The transformation that maps f onto $(r_1, \ldots, r_N, f_N)$ is composed of a sequence of transformations, each of which is invertible because of the perfect reconstruction property of the quadrature mirror filters. Thus f can be computed directly from $(r_1, \ldots, r_N, f_N)$.

The significance of Mallat's algorithm stems from the following observation: For appropriate choice of the filters F_0 and F_1, there are numerous cases where the fluctuations $r_1, \ldots, r_N$ are, at different steps, extremely small. Coding the signal thus comes down to coding the last trend f_N as well as those coefficients of the fluctuations that are above the threshold fixed by the quantization. Notice that the last trend contains 2^{-N} times less data than the input signal. If, in addition, many of the terms in the fluctuations are essentially zero, then the amount of data that must be stored or transmitted can be appreciably less than the data representing

3.6 Trends and fluctuations with orthonormal wavelet bases

We propose to describe the asymptotic behavior of Mallat's algorithm as the number of stages N tends to infinity. To do this, it is first necessary to present the continuous version of this algorithm. This involves orthonormal wavelet bases in the following "complete form," which means that we have a wavelet plus a *multiresolution analysis*. This will be explained here for $L^2(\mathbb{R})$, and it will be discussed again in the next chapter for $L^2(\mathbb{R}^n)$, where we formally define a multiresolution analysis.

We begin with a function φ belonging to $L^2(\mathbb{R})$ that has the following property:

$$\varphi(x-k),\ k \in \mathbb{Z},\ \text{is an orthonormal sequence in } L^2(\mathbb{R}). \tag{3.4}$$

Let V_0 denote the closed linear subspace of $L^2(\mathbb{R})$ generated by this sequence. For the other $j \in \mathbb{Z}$, define the spaces V_j in terms of V_0 by simply changing scale. This means that

$$f(x) \in V_0 \iff f(2^j x) \in V_j. \tag{3.5}$$

The other hypotheses are these: The V_j, $j \in \mathbb{Z}$, form a nested sequence; their intersection $\cap_{-\infty}^{\infty} V_j$ reduces to $\{0\}$; and their union $\cup_{-\infty}^{\infty} V_j$ is dense in $L^2(\mathbb{R})$. We then write $\varphi_{j,k}(x) = 2^{j/2}\varphi(2^j x - k)$, $j, k \in \mathbb{Z}$, and define the trend f_j, at scale 2^{-j}, of a function $f \in L^2(\mathbb{R})$ by

$$f_j(x) = \sum_{k \in \mathbb{Z}} \langle f, \varphi_{j,k} \rangle \varphi_{j,k}(x).$$

The fluctuations—or details in the case of an image—are denoted by d_j and defined by $d_j(x) = f_{j+1}(x) - f_j(x)$. To analyze these details further, we let W_j denote the orthogonal complement of V_j in V_{j+1}, so that $V_{j+1} = V_j \oplus W_j$. Then there exists at least one function ψ belonging to W_0 such that $\psi(x-k)$, $k \in \mathbb{Z}$, is an orthonormal basis of W_0. Such a function ψ, called the mother wavelet, has the following properties:

$$\psi_{j,k}(x) = 2^{j/2}\psi(2^j x - k), \qquad j, k \in \mathbb{Z}, \tag{3.6}$$

is an orthonormal basis for $L^2(\mathbb{R})$, and, more precisely, for all $j \in \mathbb{Z}$, we have

$$d_j(x) = \sum_{k \in \mathbb{Z}} \beta_{j,k} \psi_{j,k}(x). \tag{3.7}$$

The details at a given scale are thus linear combinations of the "elementary fluctuations," which are the wavelets related to that scale.

Given two functions φ and ψ that satisfy the conditions just described (which are called the father wavelet and mother wavelet, respectively), it is possible to define two quadrature mirror filters F_0 and F_1 by using the operators $T_0 = DF_0$ and $T_1 = DF_1$. This is done by relating the approximation of the function space $L^2(\mathbb{R})$ that is given by the nested sequence of subspaces V_j to the approximation of the real line $\mathbb{R}$ that is given by the nested sequence of grids $\Gamma_j = 2^{-j}\mathbb{Z}$.

To do this, we consider that the function $\varphi_{j,k}(x) = 2^{j/2}\varphi(2^j x - k)$ is centered around the point $k2^{-j}$, which would be the case if φ were an even function. We associate the point $k2^{-j}$ with the function $\varphi_{j,k}$. This gives a correspondence between Γ_j and the orthonormal basis $\{\varphi_{j,k},\ k \in \mathbb{Z}\}$ of V_j. At the same time, $l^2(\Gamma_j)$ is identified isometrically with V_j.

To define the operator $T_0 : l^2(\Gamma_{j+1}) \to l^2(\Gamma_j)$, it is sufficient to define its adjoint $T_0^* : l^2(\Gamma_j) \to l^2(\Gamma_{j+1})$. It is constructed by starting with the isometric embedding $V_j \subset V_{j+1}$ and by identifying V_j with $l^2(\Gamma_j)$ and V_{j+1} with $l^2(\Gamma_{j+1})$, as explained above. This adjoint T_0^* is a partial isometry.

The orthonormal basis $\psi_{j,k}$ of W_j allows us to identify W_j with $l^2(\Gamma_j)$ in the same way. The isometric embedding of $W_j \subset V_{j+1}$, interpreted with this identification, becomes the partial isometry

$$T_1^* : l^2(\Gamma_j) \to l^2(\Gamma_{j+1}).$$

(A mapping $T : H_1 \to H_2$ from one Hilbert space to another is called a *partial isometry* if $\|Tx\|_{H_2} = \|x\|_{H_1}$ for all $x \in H_1$, which is equivalent to saying T preserves inner products; that is, $\langle Tx, Ty \rangle_{H_2} = \langle x, y \rangle_{H_1}$ for all $x, y \in H_1$. "Partial" means there is no assumption that the mapping is onto.)

Finally, the couple (φ, ψ) is represented by the couple (T_0, T_1) or, which amounts to the same thing, by the pair (F_0, F_1) of the two quadrature mirror filters. This crucial observation is due to Mallat.

Mallat also posed the converse problem: Given two quadrature mirror filters F_0 and F_1, is it possible to associate with them two functions φ and ψ having properties (3.4), (3.5), and (3.6)? Although the converse is incorrect in general, it is correct in numerous cases, and this led to the construction of Daubechies's wavelets.

Our first and third examples of quadrature mirror filters show that the converse is generally false. There are no functions φ and ψ behind these numerical algorithms.

The second example is related to the Haar system. The function φ is the characteristic function of $[0,1)$, and V_j is composed of step functions that are constant on each interval $[k2^{-j}, (k+1)2^{-j}),\ k \in \mathbb{Z}$.

The fourth example leads to Shannon's wavelets. The function φ is the cardinal sine defined by $\frac{\sin \pi x}{\pi x}$.

Finally, the last example is more interesting because both of the functions φ and ψ belong to the Schwartz class $\mathcal{S}(\mathbb{R})$, which consists of the infinitely differentiable functions that decrease rapidly at infinity.

In the next section, we are going to pass from the analysis of sampled functions to the analysis of functions defined on $\mathbb{R}$ by passing to the limit in the discrete algorithms. To do this, we will give sufficient conditions on the transfer function $F_0(\theta) = \sqrt{2}\,\overline{m_0}(\theta)$ to construct a multiresolution analysis starting with two quadrature mirror filters F_0 and F_1.

3.7 Convergence to wavelets

Before one can restrict a very irregular function f belonging to $L^2(\mathbb{R})$ to a grid $\Gamma = h\mathbb{Z},\ h > 0$, it is often necessary to smooth the function by filtering. This filtering ought to be done according to specific rules. These rules are designed so that in the event f is very regular, f can be reconstructed from the sampled version with good accuracy using interpolation. The proper sampling technique is a direct consequence of Shannon's work.

We filter f by forming the convolution $f * g_h$, where $g_h(x) = h^{-1}g(h^{-1}x)$ and where g is chosen so that it and its Fourier transform $\hat{g}$ satisfy the following three conditions:

(1) $g \in C^r$ and $g, g', \ldots, g^{(r)}$ all decrease rapidly at infinity.

(2) $\int_{-\infty}^{\infty} g(x)\,dx = 1$.

(3) $\hat{g}(2k\pi) = 0$ for $k \in \mathbb{Z}$, $k \neq 0$.

One can then restrict the filtered signal $f * g_h$ to the grid $h\mathbb{Z}$.

We assume that these conditions are satisfied throughout the discussion, and we begin by reconsidering Mallat's "herringbone" algorithm. Start by fixing f in $L^2(\mathbb{R})$ and sample f on the grid Γ_N using the preconditioning filter $f * g_N$, where $g_N(x) = 2^N g(2^N x)$.

We wish to study the asymptotic behavior of Mallat's algorithm as N tends to infinity. The limit we are looking for is defined as follows: Fix the index j of the grid Γ_j. (Starting with Γ_0, we will look at $\Gamma_1, \Gamma_2, \Gamma_3, \ldots$.) Then we seek the (simple) limits of the sequences

$$f_N(k), r_N(k), r_{N-1}(2^{-1}k), \ldots, r_{N-j}(2^{-j}k), \ldots$$

as N tends to infinity, j and k being fixed. Refer to the "herringbone" scheme (section 3.5 and Figure 3.4) for the definitions of $f_N, r_N, r_{N-1}, \ldots$. Here are the results [210], which do not depend on the choice of g as long as it satisfies the conditions stated above. (We note that this last point will come up again in Chapter 4 when presenting the two-dimensional version of this theorem.)

THEOREM 3.2. *Assume that the impulse responses of the quadrature mirror filters F_0 and F_1 decrease rapidly at infinity and that the transfer function $F_0(\theta)$ of F_0 satisfies $F_0(0) = \sqrt{2}$ and $F_0(\theta) \neq 0$ if $-\frac{\pi}{2} \leq \theta \leq \frac{\pi}{2}$. Then Mallat's "herringbone" algorithm, applied to $f * g_N$, as indicated above, converges to the analysis of f in an orthonormal wavelet basis. More precisely,*

$$\lim_{N \to +\infty} f_N(k) = \int_{-\infty}^{\infty} f(x)\overline{\varphi}(x-k)\,dx,$$

$$\lim_{N \to +\infty} r_N(k) = \int_{-\infty}^{\infty} f(x)\overline{\psi}(x-k)\,dx,$$

$$\vdots$$

$$\lim_{N \to +\infty} r_{N-j}(2^{-j}k) = \int_{-\infty}^{\infty} f(x)\overline{\psi}_{j,k}(x-k)\,dx.$$

Observe that $F_0(0) = \sqrt{2}$ means that F_0 is a low-pass filter and that F_1 is a high-pass filter. The functions φ and ψ are, respectively, the father and the mother of the orthonormal wavelet basis, as explained in the preceding section.

We will not prove this theorem, but we will discuss the hypotheses and how they relate to this analysis.

Assuming that we have arrived at an orthonormal wavelet basis, we know that the function φ must satisfy the functional equation

$$\varphi(x) = \sqrt{2} \sum_{k \in \mathbb{Z}} h_k \varphi(2x - k), \tag{3.8}$$

where the sequence $\{h_k\}_{k\in\mathbb{Z}}$ is in $l^2(\mathbb{Z})$. This follows from the inclusion $V_0 \subset V_1$ and the fact that $\sqrt{2}\varphi(2x-k)$, $k \in \mathbb{Z}$, is an orthonormal basis for V_1. Another requirement is that $|\hat{\varphi}(0)| = |\int \varphi(x)\,dx| = 1$. This is equivalent to $\cup_{-\infty}^{\infty} V_j$ being dense in $L^2(\mathbb{R})$, although this is not obvious. (The proof of this and of the other assertions in this section can be found in [60].) We assume (multiplying by a constant with modulus one if necessary) that $\int \varphi(x)\,dx = 1$.

Taking the Fourier transform of both sides of (3.8) shows that

$$\hat{\varphi}(\xi) = \Big(\frac{1}{\sqrt{2}} \sum_{k\in\mathbb{Z}} h_k e^{-i2^{-1}\xi k}\Big) \hat{\varphi}(2^{-1}\xi). \tag{3.9}$$

Let $m_0(\xi) = \frac{1}{\sqrt{2}} \sum_{k\in\mathbb{Z}} h_k e^{-i\xi k}$. Then m_0 is a 2π-periodic function in $L^2(0, 2\pi)$.

To relate this to the filter, take $f \in L^2(\mathbb{R})$ and write $f_{j,k} = \langle f, \varphi_{j,k} \rangle$. The relation (3.8) implies that

$$f_{j,k} = \sum_{n\in\mathbb{Z}} \overline{h}_{n-2k} f_{j+1,n}. \tag{3.10}$$

The right-hand side of (3.10) can be interpreted as follows: Pass $f_{j+1,k}$ through the filter $(\overline{h}_{-n})_{n\in\mathbb{Z}}$ (take the discrete convolution) and then save only the even terms. But this is exactly the original operator T_0, and F_0 is the filter $(\overline{h}_{-n})_{n\in\mathbb{Z}}$. Thus $m_0(\xi)$ and the transfer function $F_0(\xi)$ are related as before by $F_0(\xi) = \sqrt{2}\,\overline{m_0}(\xi)$, and $F_0(\xi) = \sum_{k\in\mathbb{Z}} \overline{h}_{-k} e^{-i\xi k}$.

The hypothesis that F_0 and F_1 are quadrature mirror filters means that

$$|m_0(\xi)|^2 + |m_0(\xi + \pi)|^2 = 1 \tag{3.11}$$

for almost every ξ. But the condition in Theorem 3.2 that the filter F_0 decreases rapidly at infinity implies that m_0 is infinitely differentiable, so (3.11) holds everywhere. The hypothesis $F_0(0) = \sqrt{2}$ implies that $m_0(0) = 1$. Taken together these hypotheses imply that the infinite product $\prod_{k=1}^{\infty} m_0(2^{-k}\xi)$ converges uniformly on compact sets to an infinitely differentiable function.

By iterating (3.9) we have

$$\hat{\varphi}(\xi) = \hat{\varphi}(2^{-N}\xi) \prod_{k=1}^{N} m_0(2^{-k}\xi), \tag{3.12}$$

and from this it follows that

$$\hat{\varphi}(\xi) = \prod_{k=1}^{\infty} m_0(2^{-k}\xi). \tag{3.13}$$

Using (3.13) and the fact that m_0 is infinitely differentiable, it is not difficult to show that φ belongs to $L^2(\mathbb{R})$. On the other hand, showing that $\varphi(x-k)$, $k \in \mathbb{Z}$, is an orthonormal sequence depends on the hypothesis that $F_0(\xi)$ does not vanish on $-\frac{\pi}{2} \leq \xi \leq \frac{\pi}{2}$.

We note that the condition that $m_0(\xi)$ does not vanish on $-\frac{\pi}{2} \leq \xi \leq \frac{\pi}{2}$ is sufficient but not necessary. Necessary and sufficient conditions were discovered by Albert Cohen and are given in [60]. A beautiful application of this result is the construction of the celebrated bases of Ingrid Daubechies.

3.8 The wavelets of Daubechies

These wavelets depend on an integer $N \geq 1$ that is related to the size of the support of the functions φ and ψ, which is $[0, 2N - 1]$. The Hölder regularity of these functions is also determined by N: φ and ψ belong to C^r, where $r = r(N)$ and $\lim_{N \to +\infty} N^{-1}r(N) = \gamma > 0$. The value of γ is about $1/5$. This implies that if a wavelet ψ is to have 10 continuous derivatives, the length of its support must be about 100.

The functions φ and ψ, which ought to be written as φ_N and ψ_N, are the father and mother of the orthonormal wavelet basis. To construct this orthonormal basis, Daubechies applies the method of the last section. One starts with the nonnegative trigonometric sum

$$P_N(t) = 1 - c_N \int_0^t (\sin u)^{2N-1} du = \sum_{|k| \leq 2N-1} \gamma_k e^{ikt}$$

with the constant $c_N > 0$ chosen so that $P_N(\pi) = 0$. There exists (at least one) finite trigonometric sum $m_0(t) = \frac{1}{\sqrt{2}} \sum_0^{2N-1} h_k e^{-ikt}$ with real h_k such that $|m_0(t)|^2 = P_N(t)$ and $m_0(0) = 1$. This classical result is known as the Fejér–Riesz lemma; a proof can be found in [73]. The coefficients h_{-k} are the impulse response of the filter F_0.

Under these conditions, we know from Theorem 3.2 that the functions φ and ψ exist and that they form a multiresolution analysis. We now use these results to construct φ and ψ explicitly. The function φ, which we seek to construct, is given by

$$\hat{\varphi}(\xi) = \prod_{k=1}^{\infty} m_0(2^{-k}\xi). \tag{3.14}$$

One then shows that $\hat{\varphi}$ and all of its derivatives are in $L^2(\mathbb{R})$. By inverting (3.14) as an infinite convolution of distributions, it is almost obvious that the support of φ is in $[0, 2N - 1]$. The fact that $\varphi(x - k)$, $k \in \mathbb{Z}$, is an orthonormal sequence is a direct consequence of Theorem 3.2 and the fact that $m_0(t) \neq 0$ on $[-\frac{\pi}{2}, \frac{\pi}{2}]$.

To determine the Fourier transform $\hat{\psi}$ of the wavelet ψ, we first define m_1 as $m_1(t) = e^{i(1-2N)t} \overline{m_0}(t + \pi)$. Then

$$\hat{\psi}(\xi) = m_1(2^{-1}\xi)\hat{\varphi}(2^{-1}\xi) = m_1(2^{-1}\xi) \prod_{k=2}^{\infty} m_0(2^{-k}\xi), \tag{3.15}$$

and the support of ψ is the interval $[0, 2N - 1]$.

If $N = 1$, φ is the index function of $[0, 1)$, while $\psi(x) = 1$ on $[0, \frac{1}{2})$, $\psi(x) = -1$ on $[\frac{1}{2}, 1)$, and $\psi(x) = 0$ elsewhere. The orthonormal basis $2^{j/2}\psi(2^j x - k)$, $j, k \in \mathbb{Z}$, is then the Haar system.

3.9 Conclusions

The functions $\psi = \psi_N$ used by Daubechies to construct the orthonormal bases named for her are new "special functions." These functions had not appeared in previous work, and their only definition is provided by (3.14) and (3.15). This means that the detour by way of quadrature mirror filters and the corresponding transfer functions was nearly indispensable. In other words, it would hardly have been possible to discover Daubechies's wavelets by trying to solve directly the existence problem: Is there, for each integer $r \geq 0$, a compactly supported function ψ of class C^r such that $2^{j/2}\psi(2^j x - k)$, $j, k \in \mathbb{Z}$, is an orthonormal basis for $L^2(\mathbb{R})$?

On the other hand, the fast convergence of the wavelet decomposition of a function is directly related to the smoothness of the wavelet used [60]. The "good" quadrature mirror filters are those that lead to smooth wavelets, and this leads to a criterion for the selection of filters that would have been difficult to obtain without the detour through wavelets and functional analysis. Quadrature mirror filters will appear again in the algorithms for numerical image processing, which we describe in the next chapter.

Mallat's algorithm for computing the wavelet coefficients of a function in the case of filters of finite length has come to be known as the fast wavelet transform. If the original signal X has N terms, then the cost of computing its fast wavelet transform—measured by the number of additions and multiplications—is of the order N. In contrast, the cost of fast Fourier transform is of the order $N \log N$.

CHAPTER **4**

Pyramid Algorithms for Numerical Image Processing

4.1 Introduction

In his book *Vision* [198, p. 51], David Marr wrote that "although the basic elements in our image are the intensity changes, the physical world imposes on these raw intensity changes a wide variety of spatial organizations, roughly independently at different scales," and later [p. 54] we read that "intensity changes occur at different scales in an image, and so their optimal detection requires the use of operators of different sizes." Adelson, Hingorani, and Simoncelli used the same language in [3]: "Images contain information at all scales."

Cartography illustrates this concept very well. Maps contain different information at different scales. For example, it is impossible to plan a trip to visit the Roman churches in the Poitou-Charentes region using the map of France found on a globe of the earth. Indeed, the villages where these churches are found do not appear on the global representation, whose scale is of the order 1 to 10^7. One can only find these villages on maps whose scale is 1 to 200,000 or smaller.

Cartographers have developed conventions for dealing with geographic information by partitioning it into categories that correspond to the different scales, for example, the scales typically used for a city, a department, a region, a country, a continent, and the whole globe, which may range from 1 to 15,000 to 1 to 10^7. These categories are not entirely independent, and the more important features existing at a given scale are repeated at the next larger scale. Thus, it is sufficient to specify the relations between information given at two adjacent scales to define unambiguously the embedding of the different representations at different scales. Naturally these embedding relations (such as which department belongs to which province, and which province belongs, in turn, to which country, and so on ...) are available to us from our knowledge of geography; however, they could be discovered by merely examining the maps.

We can see from this example the fundamental idea of representing an image by a tree. In the cartographic example, the trunk would be the map of the world. By traveling toward the branches, the twigs, and the leaves, we reach successive maps that cover smaller regions and give more details, details that do not appear at lower levels.

To interpret this cartographic representation using the pyramid algorithm, it will be necessary to reverse the roles of top and bottom, since the pyramid algorithm progresses from "fine to coarse." In cartography, usage and certain conventions determine which details are deleted in going from one scale to another and which significant structures persist across a succession of scales.

In this chapter, we are going to describe the pyramid algorithms of Burt and

Adelson, as well as two important modifications derived from them. The purpose of these algorithms is to provide an automatic process, in the context of digital imagery, to calculate the image at scale 2^{j+1} from the image at scale 2^j. If the original image corresponds to a fine grid with 1024 × 1024 points, the pyramid algorithm first yields a 512 × 512 image, then one 256 × 256, next a 128 × 128 image, and so on until reaching the absurd (in practice) 1 × 1 limit. The interest in pyramid algorithms derives from their iterative structure that uses results from a given scale 2^j to move to the next scale 2^{j+1}.

Returning to our cartographic example, we suppose that we already have maps of the French Departments at a scale of 1 to 200,000. Then it is of no value to refer to the new satellite images to construct a map of France at a scale of 1 to 2,000,000. The information needed to make this new map is already contained in the maps of the departments. The point is that one uses judiciously the work already done without going back to the raw data. We have just outlined the general philosophy of the pyramid algorithms without, however, describing the algorithms that are used to change scale. How, starting with a very precise representation of the Brittany coast at a scale of 1 to 200,000, can we arrive at a more schematic description at a scale of 1 to 2,000,000 without smoothing or softening too much the myriad details and roughness that characterize the Brittany coastline?

The pyramid algorithms of Burt and Adelson [42] and their variants (orthogonal and biorthogonal pyramids) deal with this type of problem. In all cases, this will involve calculating (at each scale) an approximation f_j of a given image by using an iterative algorithm to go from one scale to the next.

4.2 The pyramid algorithms of Burt and Adelson

For the rest of the discussion, $\Gamma_j = 2^{-j}\mathbb{Z}^2$ will denote the sequence of nested grids used for image processing. It often happens that the image is bounded by the unit square, in which case we will speak of a 512 × 512 image to indicate that $j = 9$; similarly, a 1024 × 1024 image will correspond to $j = 10$.

At this point, we are working with images that are already digitized and appear as numerical functions. The raw image that provided these digital images will be a function $f(x, y)$. This function can be very irregular, either because of noise or because of discontinuities in the image itself. For example, discontinuities are often due to the edges of objects in the image.

The sampled images f_j are defined on the corresponding grids $\Gamma_j = 2^{-j}\mathbb{Z}^2$. These sampled images are obtained from the original physical image $f(x, y)$ by the restriction operators $R_j : L^2(\mathbb{R}^2) \to l^2(\Gamma_j)$. These operators R_j will be defined below. They are the same type as those used in numerical analysis to discretize an irregular function or distribution.

The fundamental discovery of Burt and Adelson is the existence of restriction operators R_j with the property that, for all initial images f, the sampled images $f_j = R_j(f)$ are related to each other by extremely simple algorithms. These algorithms, of the type "fine to coarse," allow f_{j-1} to be calculated directly from f_j without having to go back to the original physical image $f(x, y)$.

To define the restriction operators R_j, we first consider the case of a grid given by $x = hk$, $y = hl$, where $h > 0$ is the sampling step and $(k, l) \in \mathbb{Z} \times \mathbb{Z}$.

Very irregular functions should not be sampled directly, and, therefore, the image may have to be smoothed before it is discretized. This leads to the classic scheme illustrated in Figure 4.1.

PYRAMID ALGORITHMS FOR NUMERICAL IMAGE PROCESSING

FIG. 4.1. *F is a low-pass filter prior to the sampling E.*

To determine the characteristics of the filter F, first consider the special case, $f(x, y) = \cos(mx + ny + \varphi)$, $m, n \in \mathbb{N}$. To sample this function correctly on $h\mathbb{Z}^2$, the Nyquist condition must be satisfied. This means that h must be less than $\min\{\frac{\pi}{m}, \frac{\pi}{n}\}$ if we wish to be able to reconstruct f. Another way to interpret the Nyquist condition is that sampling on $h\mathbb{Z}^2$ will lose all information about frequencies higher than $\frac{\pi}{h}$. For the case at hand, the Nyquist condition comes down to suppressing, through the action of the filter F, all the frequencies in f that are greater than $\frac{\pi}{h}$. This is done by smoothing the signal through convolution with $\frac{1}{h^2} g\left(\frac{x}{h}, \frac{y}{h}\right)$, where g is a sufficiently regular function concentrated around zero.

The filtering/sampling scheme maps the physical image f onto a numerical image defined by

$$c(k, l) = \frac{1}{h^2} \iint g\left(k - \frac{x}{h}, l - \frac{y}{h}\right) f(x, y) dx\, dy. \tag{4.1}$$

By writing $\varphi(x, y) = \overline{g}(-x, -y)$ and $\varphi_h(x, y) = \frac{1}{h^2} \varphi\left(\frac{x}{h}, \frac{y}{h}\right)$, we have

$$c(k, l) = \langle f, \varphi_h(\cdot - kh, \cdot - lh)\rangle, \tag{4.2}$$

where $\langle u, v \rangle = \iint u(x, y) \overline{v}(x, y)\, dx\, dy$ and where "$\cdot$" denotes the (dummy) variable of integration. The operator that maps f onto $c(k, l)$ is called the *restriction operator* and is denoted by R_h.

The *extension operator* enables us to extend a sequence $c(k, l)$ defined on $h\mathbb{Z}^2$ to a regular function on $\mathbb{R}^2$. In this sense, it is inverse to the filtering/sampling operation. We define the extension operator to be the *adjoint of the restriction operator*; thus it is given by

$$c(k, l) \mapsto \sum \sum c(k, l) \varphi(h^{-1} x - k, h^{-1} y - l). \tag{4.3}$$

This is an *interpolation operator*, which will be denoted by P_h.

The simplest examples are given by spline functions. We consider the one-dimensional case to simplify the notation. If we let φ be the triangle function $T(x) = \sup(1 - |x|, 0)$, then (4.3) yields the familiar piecewise linear interpolation of a discrete sequence. A second choice is given by $\varphi = T * T$, which is the basic cubic spline.

Returning to the general case, we require that the operator $P_h R_h$, composed of the restriction operator followed by the extension operator, has this property: For all functions $f \in L^2(\mathbb{R}^2)$,

$$P_h R_h(f) \to f \text{ in } L^2(\mathbb{R}^2) \text{ as } h \to 0. \tag{4.4}$$

By assuming, for example, that φ is a continuous function that decreases rapidly at infinity, it is not difficult to show that (4.4) is equivalent to $P_h R_h(\mathbf{1}) = \mathbf{1}$, where $\mathbf{1}$ represents the function identically equal to one. One can also verify that this is equivalent to the Fix and Strang condition [244]:

$$|\hat{\varphi}(0, 0)| = 1, \quad \hat{\varphi}(2k\pi, 2l\pi) = 0 \quad \text{if} \quad (0, 0) \neq (k, l) \in \mathbb{Z}^2. \tag{4.5}$$

In what follows, we assume that $\iint \varphi(x,y)\,dx\,dy = 1$, after possibly multiplying φ by a constant of modulus one.

We return to the fundamental problem posed by Burt and Adelson. Thus we consider the nested sequence $\Gamma_j = 2^{-j}\mathbb{Z}^2$. These grids become finer as $j \to +\infty$ and coarser as $j \to -\infty$.

We begin with a function φ that is continuous on $\mathbb{R}^2$ and decreases rapidly at infinity. We also assume, as above, that $\hat{\varphi}(0,0) = 1$. Denote by R_j and P_j the restriction and extension operators associated with this choice of φ and the grid Γ_j. This means that $h = 2^{-j}$ and that the operators R_h and P_h are denoted by R_j and P_j. (From now on, we will mostly use vector notation for the variables in $\mathbb{R}^2$ and $\mathbb{Z}^2$. Thus, $x \in \mathbb{R}^2$ means $x = (x_1, x_2)$, $k \in \mathbb{Z}^2$ means $k = (k_1, k_2)$, and $k \cdot x = k_1 x_1 + k_2 x_2$, and so on. This will be more compact, and it will better reveal the connection with the one-dimensional case.)

Burt and Adelson's basic idea is that, for certain choices of the function φ, the different sampled images $R_j(f) = f_j$ derived from the same physical image f are necessarily related by extremely simple relations. The dynamic of these relations is from "fine to coarse," which means that a function defined on a fine grid is mapped to one on a coarse grid. To make these relations explicit, we denote by T_j the operators that will eventually be defined by these relations, that is, by $T_j(f_j) = f_{j-1}$, where $f_j = R_j(f)$ and $f_{j-1} = R_{j-1}(f)$. We can summarize this with the two conditions

$$T_j : l^2(\Gamma_j) \to l^2(\Gamma_{j-1}), \tag{4.6}$$

$$R_{j-1} = T_j R_j. \tag{4.7}$$

One might naïvely think that the operator T_j could be defined by inverting the operator R_j in (4.7). But the operator R_j is a smoothing operator, and its inverse is not defined. In terms of images, it is not generally possible to go from a blurred image back to the original image. This means that, in general, we cannot solve (4.7) by elementary algebra. On the other hand, once R_j is restricted to an appropriate closed subspace V_j of $L^2(\mathbb{R}^2)$, $R_j : V_j \to l^2(\Gamma_j)$ becomes, in certain cases, an isomorphism. Then we can solve (4.7) directly.

Burt and Adelson asked how to determine the functions φ such that (4.6) and (4.7) are satisfied. Stated this way, the problem is very difficult, for most of the usual choices of smoothing functions do not have these properties. To resolve this difficulty, Burt and Adelson proceeded the other way around: They sought to construct φ from the operators T_j. For this it is necessary to derive some consequences of (4.7).

The first is that the operator $T_0 : l^2(\mathbb{Z}^2) \to l^2(2\mathbb{Z}^2)$ can be written as $T_0 = DF_0$, where $F_0 : l^2(\mathbb{Z}^2) \to l^2(\mathbb{Z}^2)$ is a filter operator and where $D : l^2(\mathbb{Z}^2) \to l^2(2\mathbb{Z}^2)$ restricts a function defined on $\mathbb{Z}^2$ to $2\mathbb{Z}^2$. D is the decimation operator, which we have already encountered in Chapter 3. That T_0 has this form is a consequence of the fact that T_0 commutes with all even translations (see Theorem A.1). Thus, if $X = (x(k))_{k \in \mathbb{Z}^2}$ is in $l^2(\mathbb{Z}^2)$, we can write

$$T_0(X)(2k) = \sum_{l \in \mathbb{Z}^2} \omega(2k - l) x(l), \quad k \in \mathbb{Z}^2, \tag{4.8}$$

where $\omega(k)$ is the impulse response of the filter F_0. For convenience, we assume (as Burt and Adelson did) that $\omega(k)$ is real.

If we apply T_0 to $x(k) = \int f(t)\overline{\varphi}(t-k)\,dt$, then (4.7) and (4.8) imply that

$$\frac{1}{4}\int f(t)\overline{\varphi}(2^{-1}t - k)\,dt = \sum_{l\in\mathbb{Z}^2} \omega(2k-l) \int f(t)\overline{\varphi}(t-l)\,dt$$

$$= \int f(t)\Big(\sum_{l\in\mathbb{Z}^2} \omega(2k-l)\overline{\varphi}(t-l)\Big)dt.$$

By taking $k = 0$, we conclude that

$$\varphi(t) = 4\sum_{l\in\mathbb{Z}^2} \omega(l)\varphi(2t+l). \tag{4.9}$$

For practical applications, Burt and Adelson were particularly interested in filters with finite length. This means that $\omega(k) = 0$ if $|k| > N$ for some N. By taking Fourier transforms of both sides, (4.9) becomes

$$\hat{\varphi}(\xi) = m_0(2^{-1}\xi)\hat{\varphi}(2^{-1}\xi), \tag{4.10}$$

where

$$m_0(\xi) = \sum_{k\in\mathbb{Z}^2} \omega(k)e^{i(k\cdot\xi)}. \tag{4.11}$$

By iterating (4.10) and passing to the limit (which is possible because $\hat{\varphi}(0) = 1$ and the filter is finite), we see that

$$\hat{\varphi}(\xi) = \prod_{j=1}^{\infty} m_0(2^{-j}\xi). \tag{4.12}$$

The second consequence that we derive from (4.7) is that these conditions for different j are in fact equivalent. This can be seen by making the change of variables $t \mapsto 2^j t$ in (4.9) and integrating both sides against the function f. We then have

$$R_{j-1}(f)(2^{-j+1}k) = \sum_{l\in\mathbb{Z}^2} \omega(2k-l)R_j(f)(2^{-j}l). \tag{4.13}$$

In other words, under our assumptions, the operators T_j are defined by

$$T_j(X)(2^{-j+1}k) = \sum_{l\in\mathbb{Z}^2} \omega(2k-l)x(2^{-j}l) \tag{4.14}$$

when $X = (x(2^{-j}l))$ belongs to $l^2(\Gamma_j)$. The point is that the sequence $\omega(k)$, $k \in \mathbb{Z}^2$, is the same for all the operators T_j. Working backward, Burt and Adelson began with a finite sequence of coefficients $\omega(k)$ with the property that $\sum_{k\in\mathbb{Z}^2} \omega(k) = 1$. They defined m_0 by (4.11) and then $\hat{\varphi}$ by (4.12). The first question to arise is whether the second member of (4.12) defines a square-integrable function. If this is the case, this function is called $\hat{\varphi}$, the restriction operators R_j are defined in terms of the Fourier transform of this function, and the transition operators T_j are defined by (4.14). In this case, $R_{j-1} = T_j R_j$ for all $j \in \mathbb{Z}$.

We have been describing the pyramid algorithms of Burt and Adelson, and much of this description closely resembles what was done in Chapter 3, particularly in sections 3.5, 3.6, and 3.7. However, to avoid confusion between the concepts and notation in the two chapters, we list explicitly some of the similarities and differences:

(a) In both chapters, the mappings $T_j : l^2(\Gamma_j) \to l^2(\Gamma_{j-1})$ are all the same except for a change of scale. This is seen explicitly in equation (4.14).

(b) The mappings T_j were all denoted by T_0 in Chapter 3.

(c) There were two filters, F_0 and F_1, and two corresponding mappings, T_0 and T_1, in Chapter 3. Here, in Chapter 4, only one filter, F_0, has appeared, and while the T_0 of Chapter 3 corresponds to the T_0 of Chapter 4, the same is not true for the two T_1's.

(d) The pyramid algorithm is only a "partial multiresolution analysis." The missing ingredient is orthogonality; so far, we have not encountered the equivalent of equation (3.11). This will appear in section 4.6.

4.3 Examples of pyramid algorithms

Before continuing the presentation of the Burt and Adelson algorithms, we give examples of functions φ that illustrate both the existence and the nonexistence of the transition operators. We also give examples of sequences $\omega(k, l)$ illustrating the existence and nonexistence of the associated function φ.

We begin with two examples where the transition operators do not exist. The first example is the Gaussian $\varphi(x, y) = \frac{1}{\pi} \exp(-x^2 - y^2)$, which plays an important role in Marr's theory of vision (Chapter 8). There are no transition operators in this case because (4.10) implies that $m_0(\xi, \eta) = \exp\left(-\frac{3}{4}(\xi^2 + \eta^2)\right)$, which is clearly not 2π-periodic in ξ and η. In the same way, the transition operators do not exist if $\varphi(x, y) = \frac{1}{4}\exp(-|x| - |y|)$. One senses, justifiably, that the existence of transition operators is exceptional.

Here, however, is an example where the operators do exist. To simplify the discussion, this example (the spline functions) is presented in one dimension. Let $m \geq 0$ be an integer and let χ be the characteristic function of the interval $[0, 1]$. Define φ to be the convolution product $\chi * \cdots * \chi$ where there are m products and $m + 1$ terms. Then

$$\hat{\varphi}(\xi) = \left(\frac{e^{-i\xi} - 1}{-i\xi}\right)^{m+1},$$

and (4.10) is satisfied with

$$m_0(\xi) = \left(\frac{1 + e^{-i\xi}}{2}\right)^{m+1},$$

which is indeed 2π-periodic.

Clearly, there is little chance of finding appropriate φ by guessing; the efficient way is to approach the problem from the other direction. Thus, we begin with a sequence of transition operators (T_j), which is a sequence $\omega(k)$, $k \in \mathbb{Z}^2$, and we propose to reconstruct φ. All the examples that we consider are constructed with separable sequences $\omega(k)$, that is, sequences of the form $\tilde{\omega}(k_1)\tilde{\omega}(k_2)$. The associated function φ will then necessarily be of the form $\tilde{\varphi}(x_1)\tilde{\varphi}(x_2)$. We will be discussing $\tilde{\omega}$ and $\tilde{\varphi}$ in the following examples. We are in the one-dimensional case, and these are functions of the variables $k \in \mathbb{Z}$ and $x \in \mathbb{R}$, respectively.

For the first example take $\tilde{\omega}(k) = 0$ if $k \neq 0$ and $\tilde{\omega}(0) = 1$. In this case the function $\tilde{\varphi}$ defined by (4.12) is the Dirac measure at $x = 0$ and the restriction operators $R_j : L^2(\mathbb{R}^2) \to l^2(\Gamma_j)$ are no longer defined.

In the second example take $\tilde{\omega}(k) = 0$ except for $k = \pm 1$, and $\tilde{\omega}(\pm 1) = \frac{1}{2}$. From this we can deduce that $m_0(\xi) = \cos(\xi)$, $\tilde{\varphi}(x) = \frac{1}{2}$ on the interval $[-1, 1]$, and

$\tilde\varphi(x) = 0$ elsewhere. This choice for $\tilde\varphi$, which is (for the moment) perfectly reasonable, will be excluded when we introduce the concept of multiresolution analysis.

Burt and Adelson proposed a very original function for $\tilde\omega$, and this will be our third example. Take $\tilde\omega(0) = 0.6$; $\tilde\omega(\pm 1) = 0.25$; $\tilde\omega(\pm 2) = -0.05$; and $\tilde\omega(k) = 0$ for $|k| \geq 3$. The corresponding function $\tilde\varphi(x)$ is continuous, its support is $[-2, 2]$, and it resembles $C\exp(-c|x|)$, $C > 0$, $c > 0$, on this interval. The corresponding algorithm is called a Laplacian pyramid. We shall see this example again when we introduce biorthogonal wavelets at the end of the chapter.

The purpose of our last example is to show that the existence of the function $\tilde\varphi$, defined by (4.12), is not a stable property, even in the simplest cases. In fact, we limit our discussion to functions $\tilde\omega(k)$ that are zero except at $k = 0$ and $k = -1$, and here $\tilde\omega(0) = p$, $\tilde\omega(-1) = q$, $0 < p < 1$, $0 < q < 1$, $p + q = 1$. Then the choice $p = q = \frac{1}{2}$ leads to a function $\tilde\varphi$ that is the characteristic function of $[0, 1]$. All other choices imply that the mathematical object on the right side of (4.12) is the Fourier transform of a probability measure μ that is singular with respect to the Lebesgue measure. The support of this probability measure μ is the interval $[0, 1]$. This measure is defined by the following property: If I is a dyadic interval in $[0, 1]$ and if I' is the left half of I and I'' is the right half of I, then $\mu(I') = p\mu(I)$ and $\mu(I'') = q\mu(I)$. This measure is multifractal (see, for example, [7] and Chapters 9 and 10).

We drop for the moment the problem of choosing an optimal filter $\omega(k)$, $k \in \mathbb{Z}^2$. Indeed, such a choice must take into consideration the overall objective. Burt and Adelson's objective was image compression. We are going to present their compression algorithm in the next section. After that we will return to the problem of choosing the sequence $\omega(k)$.

4.4 Pyramid algorithms and image compression

Image compression is one of the uses of the pyramid algorithms. Burt and Adelson's algorithm, which we describe in this section, will later be compared with other algorithms (orthogonal pyramids and biorthogonal wavelets) that perform better. All of the pyramid algorithms act on images that are already sampled and never on the original physical image. In other words, the function φ we have tried to construct using the sequence $\omega(k)$ is never used. Then why have we investigated its properties? The brief answer is that the regularity (or smoothness) of φ influences the efficiency of the compression. More precisely, the regularity is related to the behavior of m_0 at $\xi = 0$, and we will see below how this influences compression. (For a full discussion, see [60].)

The Burt and Adelson pyramid algorithms use only the transition operators $T_j : l^2(\Gamma_j) \to l^2(\Gamma_{j-1})$. All of these operators are the same, except for a change of scale; therefore, we are going to assume that $j = 0$.

The discussion of the algorithm begins with the definition of the trend and the fluctuations around this trend for a sequence f belonging to $l^2(\Gamma_0)$. This trend cannot be $T_0(f)$ because it "lives in a different universe" and cannot be compared with f to obtain a fluctuation. To define the trend, it is necessary to leave the coarse grid $2\Gamma_0$, where $T_0(f)$ is defined, and return to the fine grid Γ_0, where f is defined. This is done by using the adjoint operator $T_0^* : l^2(2\Gamma_0) \to l^2(\Gamma_0)$, and the trend of f is defined by $T_0^* T_0(f)$.

We clearly want the trend of some very regular function such as constants and polynomials of low degree to coincide with these functions. This leads to the requirements that $T_0^* T_0(\mathbf{1}) = \mathbf{1}$ and, more generally, that $T_0^* T_0(x^p y^q) = x^p y^q$ for all p, q with $p + q \leq$ some fixed integer N. This condition is equivalent to the following: The function m_0, defined by (4.11), must vanish, along with all of its derivatives of order less than or equal to N, at points $\varepsilon\pi$, $\varepsilon = (\varepsilon_1, \varepsilon_2)$, $\varepsilon_1, \varepsilon_2 \in \{0, 1\}$, with the exception of the origin. At the origin, one must have

$$|m_0(\xi)|^2 = 1 + \sum_{p+q \geq N+1} c(p,q) \xi_1^p \xi_2^q.$$

The price that must be paid for these regularity conditions is that the length of the filter $\omega(k)$ must be at least proportional to N.

Another observation is that the conditions we have just imposed on m_0 imply, by (4.12), that $\hat{\varphi}(2k\pi) = 0$ if $k \in \mathbb{Z}^2$ and $k \neq 0$. But this last condition is the same as (4.5), which, as we have seen, is necessary and sufficient to have $P_h R_h(f) \to f$ in $L^2(\mathbb{R}^2)$ as $h \to 0$. Since T_0 is the discrete analogue of the restriction operator R_h and since T_0^* corresponds to the extension operator P_h, $T_0^* T_0$ is the "discrete approximation" operator corresponding to the continuous approximation operator $P_h R_h$.

The fluctuation around the trend is $f - T_0^* T_0(f)$ when f belongs to $l^2(\Gamma_0)$. This fluctuation is zero whenever f is a polynomial of degree no greater than N, and one can easily deduce from this that the fluctuation will be very weak in all areas where the image is very regular, since "regular" means being close to a polynomial (recall (2.5)). As we will see, this last property is the key to the success of the Burt and Adelson algorithm.

The trend and the fluctuation around the trend of a sequence f belonging to $l^2(\Gamma_j)$ are defined by a simple change of scale. The trend is $T_j^* T_j(f)$, and the fluctuation is $f - T_j^* T_j(f)$.

If the sequence f is the restriction to the grid Γ_j of a function F that has $N+1$ continuous derivatives in some open region Ω, then

$$|f - T_j^* T_j(f)| \leq C 2^{-(N+1)j} \tag{4.15}$$

at all the points of this region. This means that the Burt and Adelson algorithm becomes more effective as N increases.

To define the coding and compression algorithm of Burt and Adelson, we begin with the fine grid $\Gamma_m = 2^{-m} \mathbb{Z}^2$ and a numerical image f_m sampled on this fine grid. This numerical image is, in fact, the restriction to $\Gamma_m = 2^{-m} \mathbb{Z}^2$ of a physical image $f \in L^2(\mathbb{R}^2)$. This means that f_m is the restriction in the usual sense of the convolution product $f * g_m$, where $g_m(x) = 4^m g(2^m x)$. The properties of the function g were indicated in section 4.2. However, it is not necessary to return to the "physical image" f to use the algorithm.

Burt and Adelson replace f_m by the couple (trend, fluctuation). But the trend, which is given by $T_m^* T_m(f_m)$, is completely defined by $T_m(f_m)$. This means that the trend $T_m^* T_m(f_m)$ can be coded by retaining one pixel in four, and this coding is given by $T_m(f_m)$. In summary, Burt and Adelson code f_m with the couple $[T_m(f_m), f_m - T_m^* T_m(f_m)]$. Then the fluctuation, denoted by r_m, is not processed further. They write $f_{m-1} = T_m(f_m)$ and iterate the procedure. f_{m-1} is coded by (f_{m-2}, r_{m-1}), where $f_{m-2} = T_{m-1}(f_{m-1})$ and $r_{m-1} = f_{m-1} - T_{m-1}^* T_{m-1}(f_{m-1})$. If we suppose that the starting image f_m is bounded on a square of side 1, then the algorithm is stopped on reaching the summit of the pyramid, which is the grid Γ_0.

The image f_m is coded by the sequence $(f_0, S_1, \ldots, r_m)$, where f_0, defined on Γ_0, is a scalar and where the $r_j = (I - T_j^* T_j) f_j$, $1 \leq j \leq m$, are the different fluctuations.

The diagram in Figure 4.2 gives a schematic description of the algorithm.

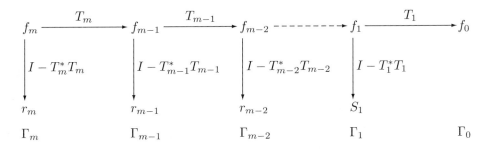

FIG. 4.2. *Burt and Adelson's algorithm.*

Interest in this coding scheme is based on the following two properties:

(1) Going from f_j to f_{j-1} reduces the data one must deal with by a factor of four. Indeed, f_j is defined on Γ_j and f_{j-1} on Γ_{j-1}, which has one-fourth as many points.

(2) In many cases, terms of the fluctuation vector r_m are so small that, after quantization, they are replaced by zero.

Condition (2) is satisfied in regions where the image is sufficiently regular.

Running the algorithm the other way, which is reconstructing f_m from the code, is easy. Begin with f_0 and S_1, and compute $f_1 = T_1^* f_0 + S_1$. In the same way, reconstruct f_2 by $f_2 = T_2^* f_1 + r_2$, and continue until f_m is recovered.

As we will see in section 4.6, this algorithm provides good compression only if most of the fluctuations are small, and hence quantized to zero. Otherwise it is inefficient, since essential information about the image is represented by too much data.

4.5 Pyramid algorithms and multiresolution analysis

Before leaving the first version of Burt and Adelson's algorithms, we are going to describe a continuous version. The interplay between the discrete algorithms and their continuous versions, which is implicit in the work of Burt and Adelson, was made explicit by Stéphane Mallat and Yves Meyer. We consider the general case because dimension two plays no particular role in the following definition.

A multiresolution analysis of $L^2(\mathbb{R}^n)$ is an increasing sequence of closed subspaces $(V_j)_{j \in \mathbb{Z}}$ of $L^2(\mathbb{R}^n)$ having the following three properties:

(1) $\bigcap_{-\infty}^{\infty} V_j = \{0\}$ *and* $\bigcup_{-\infty}^{\infty} V_j$ *is dense in* $L^2(\mathbb{R}^n)$.

(2) *For all functions $f \in L^2(\mathbb{R}^n)$ and all integers $j \in \mathbb{Z}$, $f(x) \in V_0$ is equivalent to $f(2^j x) \in V_j$.*

(3) *There exists a function $\varphi(x) \in V_0$ such that the sequence $\varphi(x - k)$, $k \in \mathbb{Z}^n$, is a Riesz basis for V_0.*

Recall that a *Riesz basis* $(e_j)_{j \in J}$ of a Hilbert space H is, by definition, the image of a Hilbert basis $(f_j)_{j \in J}$ of H under an isomorphism $T : H \to H$. (Note that T is not necessarily an isometry.) Then each vector $x \in H$ is decomposed uniquely in a series

$$x = \sum_{j \in J} \alpha_j e_j, \quad \text{where} \quad \sum_{j \in J} |\alpha_j|^2 < \infty. \tag{4.16}$$

Furthermore, $\alpha_j = \langle x, e_j^* \rangle$, where $e_j^* = (T^*)^{-1}(f_j)$ is the dual basis of e_j, and this dual basis is itself a Riesz basis. The two systems (e_j) and (e_j^*) are said to be biorthogonal. This is the abstract concept that leads to the development of biorthogonal wavelets (section 4.7).

The regularity of a multiresolution analysis is given by the regularity of the functions belonging to V_0. To measure this regularity, we introduce an integer r that can take the values $0, 1, 2, \ldots$, and even $+\infty$.

The multiresolution analysis $(V_j)_{j \in \mathbb{Z}}$ is said to be r-regular if it is possible to choose the function φ in (3) so that for all integers $m \geq 0$ and all $x \in \mathbb{R}^n$,

$$|\partial^\alpha \varphi(x)| \leq C_m (1 + |x|)^{-m}, \tag{4.17}$$

where $\alpha = (\alpha_1, \ldots, \alpha_n)$ is a multi-index satisfying $\alpha_1 + \cdots + \alpha_n \leq r$ and where $\partial^\alpha = (\frac{\partial}{\partial x_1})^{\alpha_1} (\frac{\partial}{\partial x_2})^{\alpha_2} \cdots (\frac{\partial}{\partial x_n})^{\alpha_n}$.

We return to the two-dimensional case. Here, a multiresolution analysis is, in a certain sense, a particular case of a pyramid algorithm. Indeed, suppose that the function φ, which is defined by (4.12), has the following additional property: There exist two constants $C_2 \geq C_1 > 0$ such that for all scalar sequences $(\alpha_k)_{k \in \mathbb{Z}^2}$,

$$C_1 \left(\sum_{k \in \mathbb{Z}^2} |\alpha_k|^2 \right)^{1/2} \leq \left\| \sum_{k \in \mathbb{Z}^2} \alpha_k \varphi(x - k) \right\|_2 \leq C_2 \left(\sum_{k \in \mathbb{Z}^2} |\alpha_k|^2 \right)^{1/2}. \tag{4.18}$$

Let V_0 denote the closed linear subspace of $L^2(\mathbb{R}^2)$ generated by the functions $\varphi(x - k)$, $k \in \mathbb{Z}^2$. Relation (4.18) implies that $\varphi(x - k)$, $k \in \mathbb{Z}^n$, is a Riesz basis for V_0. One can verify that the conditions in (1) hold and that $V_j \subset V_{j+1}$ when the V_j are defined by (2).

The pyramid algorithms associated with multiresolution analyses are the only ones that we will study in the following sections. They have some remarkable properties. For example, the restriction operator R_j is an isomorphism between V_j and $l^2(\Gamma_j)$, and the equation $R_{j-1} = T_j R_j$ can be solved directly. In fact, it is sufficient to restrict the two sides of $R_{j-1} = T_j R_j$ to V_j to invert R_j.

Not all pyramid algorithms are related to a multiresolution analysis. A counterexample is given by one of the pyramid algorithms presented in section 4.3. In this example, $\varphi(x) = \frac{1}{2}$ on $[-1, 1]$ and $\varphi(x) = 0$ elsewhere. Thus,

$$\|\varphi(x) - \varphi(x - 1) + \varphi(x - 2) + \cdots + (-1)^N \varphi(x - N)\|_2 = \frac{1}{\sqrt{2}},$$

whereas, according to (4.18), it should be of the order of $\sqrt{N}$.

4.6 The orthogonal pyramids and wavelets

Shortly after the discovery of quadrature mirror filters by Esteban and Galand, Woods and O'Neil had the idea to apply this technique to image processing [263].

They thus obtained the first example of an orthogonal pyramid. We are going to set aside, for the moment, the specific construction carried out by Woods and O'Neil using separable filters. We will first present the notion of an orthogonal pyramid in complete generality, and then we will return to the particular case where the quadrature mirror filters appear in the construction of an orthogonal pyramid.

The Burt and Adelson algorithm is not particularly efficient because it replaces information coded on N^2 pixels by new information whose description requires $\frac{4}{3}N^2$ pixels. This criticism, which we will analyze in a moment, is not always justified in applications, since in many examples of real images, most of the values of the gray levels on the $\frac{4}{3}N^2$ pixels are in fact small and thus quantized to zero, and the unfavorable pixel count where N^2 becomes $\frac{4}{3}N^2$ rarely occurs.

Let us examine, however, why the information has been wasted or, more precisely, where the inefficient coding occurs. At the start, the image f has been coded on N^2 pixels. Next, we replace this by the couple $[T_0(f), (I-T_0^*T_0)(f)]$, which is composed of the coding for the trend and the complete description of the fluctuations around the trend. The description of $T_0(f)$ requires $\frac{1}{4}N^2$ pixels, whereas the description of $f - T_0^*T_0(f)$ continues to require N^2 pixels. In all, we use $N^2 + \frac{1}{4}N^2$ pixels. At the next step, the pixel count becomes $N^2 + \frac{1}{4}N^2 + \frac{1}{16}N^2$, and so on. At the end, we will have used $N^2 + \frac{1}{4}N^2 + \frac{1}{16}N^2 + \cdots + 1$, or approximately $\frac{4}{3}N^2$ pixels. The "wasted" pixels appear because the fluctuations $f - T_j^*T_j(f)$ have not been coded efficiently.

Orthogonal pyramids are a particular class of pyramid algorithms that code the fluctuations with $\frac{3}{4}N^2$ pixels. With this scheme, there is no waste. When the original image f is replaced by coding the trend and the fluctuations, the required pixels are $\frac{1}{4}N^2$ and $\frac{3}{4}N^2$, respectively, and the volume of data remains constant.

A *pyramid algorithm is said to be orthogonal if the trend $T_0^*T_0(f)$ and the fluctuations $f - T_0^*T_0(f)$ around this trend are orthogonal for each image $f \in l^2(\Gamma_0)$.*

Let $H = l^2(\Gamma_0)$, $H_0 = T_0^*T_0(H)$, and $H_1 = (I - T_0^*T_0)H$. If the pyramid algorithm is orthogonal, then $H = H_0 \oplus H_1$. Since the dimension of H_0 is a quarter that of H, the dimension of H_1 is $\frac{3}{4} \dim H$, as mentioned.

An equivalent definition of orthogonal pyramids requires the adjoint T_0^* of the operator T_0 to be a partial isometry, which means that

$$\langle T_0^*(f), T_0^*(g) \rangle = \langle f, g \rangle$$

for all $f, g \in l^2(\Gamma_{-1})$. (Recall that T_0^* is defined on $l^2(\Gamma_{-1})$ with values in $l^2(\Gamma_0)$.) This takes us back to one of the characteristic properties, in dimension one, of the low-pass filter T_0 in a pair of quadrature mirror filters (T_0, T_1). And this observation prompts us, in dimension two, to look for the corresponding second filter, T_1. We will see in a moment that three filters are necessary in two dimensions. But first, we show how to construct some pyramid algorithms. We return to the transfer function m_0 defined by (4.11). The pyramid algorithm is orthogonal if and only if

$$|m_0(\xi, \eta)|^2 + |m_0(\xi + \pi, \eta)|^2 + |m_0(\xi, \eta + \pi)|^2 + |m_0(\xi + \pi, \eta + \pi)|^2 = 1.$$

This condition is completely analogous to the one on the transfer function m_0 in the case of two quadrature mirror filters (section 3.3).

Continuing this comparison, we consider the function φ in $L^2(\mathbb{R}^2) \cap L^1(\mathbb{R}^2)$ defined by (4.12) and normalized by $\iint \varphi(x)dx = 1$. We might expect that the sequence $\varphi(x - k)$, $k \in \mathbb{Z}^2$, is orthonormal, and this is true in many cases. However, the proof involves a delicate limit process, passing from the discrete to the continuous, and some orthogonal pyramids do not lead to orthogonal sequences of functions $\varphi(x - k)$, $k \in \mathbb{Z}^2$.

This difficulty already appeared in dimension one for the quadrature mirror filters. The condition we assume here about m_0, which is sufficient to allow passage from the discrete to the continuous, is the analogue of the condition we used in dimension one. It is sufficient to assume that m_0 is smooth and that $m_0(\xi, \eta) \neq 0$ if $-\frac{\pi}{2} \leq \xi \leq \frac{\pi}{2}$ and $-\frac{\pi}{2} \leq \eta \leq \frac{\pi}{2}$. Then $\varphi(x - k)$, $k \in \mathbb{Z}^2$, is an orthonormal basis of a closed subspace V_0 of $L^2(\mathbb{R}^2)$. By dilation, we see that $2^j \varphi(2^j x - k)$, $k \in \mathbb{Z}^2$, is an orthonormal basis for the subspace V_j. Furthermore, the extension operator $P_j : l^2(\Gamma_j) \to V_j$ is an isometric isomorphism, and the restriction operator $R_j : L^2(\mathbb{R}^2) \to l^2(\Gamma_j)$ is decomposed into the orthogonal projection operator from $L^2(\mathbb{R}^2)$ onto V_j followed by the inverse isomorphism $P_j^{-1} : V_j \to l^2(\Gamma_j)$. (Recall that the extension operator P_j and the restriction operator R_j were defined on page 51.)

This allows us to define the transition operators $T_j : l^2(\Gamma_j) \to l^2(\Gamma_{j-1})$ explicitly. (They had been defined implicitly by $T_j R_j = R_{j-1}$.) Use the operator P_j to identify $l^2(\Gamma_j)$ with V_j, and similarly use P_{j-1} to identify $l^2(\Gamma_{j-1})$ with V_{j-1}. Having made these identifications, the transition operator $T_j : l^2(\Gamma_j) \to l^2(\Gamma_{j-1})$ corresponds to the orthogonal projection of V_j on V_{j-1}, which is $P_{j-1} T_j P_j^{-1}$ in our notation.

We define W_j to be the orthogonal complement of V_j in V_{j+1}. Thus,

$$V_{j+1} = V_j \oplus W_j.$$

It is easy to verify—by once again using the "isometric interpretations" given by $P_j : l^2(\Gamma_j) \to V_j$ and $P_{j+1} : l^2(\Gamma_{j+1}) \to V_{j+1}$—that this orthogonal decomposition corresponds precisely to the orthogonal decomposition of a function into its trend and fluctuation, and this latter decomposition is the definition of orthogonal pyramids.

We come now to the two-dimensional generalization of the quadrature mirror filters. In dimension two, we consider four operators T_0, S_1, S_2, and S_3. All four are defined on $l^2(\mathbb{Z}^2)$ with values in $l^2(2\mathbb{Z}^2)$. We require that these four operators commute with the even translations $\tau \in 2\mathbb{Z}^2$ and that

$$\|f\|^2 = \|T_0(f)\|^2 + \|S_1(f)\|^2 + \|S_2(f)\|^2 + \|S_3(f)\|^2 \tag{4.19}$$

for all f belonging to $l^2(\mathbb{Z}^2)$. The left term is of course computed in $l^2(\mathbb{Z}^2)$, whereas each term on the right is computed in $l^2(2\mathbb{Z}^2)$.

One of the important results in the theory of orthogonal pyramids is the existence of the operators S_1, S_2, and S_3 and the ability to construct them. Furthermore, if the impulse response $\omega(k)$ of T_0 decreases rapidly at infinity, the operators S_1, S_2, and S_3 can be constructed to have this same property.

Once S_1, S_2, and S_3 are constructed, we can construct the corresponding wavelets ψ_1, ψ_2, and ψ_3. Assuming that $m_0(\xi, \eta) \neq 0$ if $|\xi| \leq \frac{\pi}{2}$ and $|\eta| \leq \frac{\pi}{2}$, we define these three wavelets by

$$\psi_j(x) = 4 \sum_{k \in \mathbb{Z}^2} \omega_j(k) \varphi(2x + k), \qquad j = 1, 2, \text{ or } 3, \tag{4.20}$$

where $\omega_j(k)$ denotes the impulse response of S_j.

Thus, under quite general conditions, the orthogonal pyramids lead to orthonormal wavelet bases, and this development proceeds by way of the two-dimensional generalization of quadrature mirror filters.

We move on to the two-dimensional generalization of Mallat's algorithm. The exact reconstruction identity,

$$I = T_0^* T_0 + S_1^* S_1 + S_2^* S_2 + S_3^* S_3, \tag{4.21}$$

is deduced from (4.19). Identity (4.21) provides a particularly elegant solution to the problem of coding the fluctuation $f - T_0^* T_0(f)$. This fluctuation is exactly

$$S_1^* S_1(f) + S_2^* S_2(f) + S_3^* S_3(f).$$

The three operators S_1^*, S_2^*, and S_3^* are partial isometries, and this allows us to code the fluctuation $f - T_0^* T_0(f)$ with the three sequences $S_1(f)$, $S_2(f)$, $S_3(f)$. These three sequences belong to $l^2(2\Gamma_0)$ when $f \in l^2(\Gamma_0)$, and thus the coding of each of them uses only one pixel out of four. Hence, three-fourths of the pixels are used to code the fluctuation, whereas one-fourth of them are used to code the trend. Consequently there are no longer any wasted pixels.

We can now return to the algorithm and give it a much more precise formulation. This is illustrated in Figure 4.3, where $T_j(f_j) = f_{j-1}$ and $S_{j,i}(f_j) = s_{j-1,i}$, and where $i = 1$, 2, or 3.

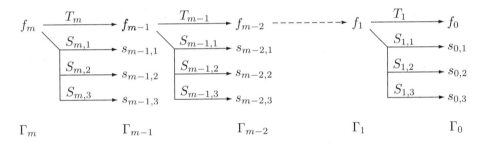

FIG. 4.3. *Two-dimensional generalization of Mallat's algorithm.*

The wavelets appear in the asymptotic limit of this scheme, which is the two-dimensional analogue of Figure 3.4. This limit is taken on the number of steps m, which must tend to infinity. We start with a fixed function f belonging to $L^2(\mathbb{R}^2)$. We restrict f to the fine grid Γ_m using the classic scheme. This means we have a fixed regular function g, which decreases rapidly at infinity and whose Fourier transform $\hat{g}$ satisfies $\hat{g}(0) = 1$ and $\hat{g}(2k\pi) = 0$ if $k \neq 0$. We write $g_m(x) = 4^m g(2^m x)$. Finally, f_m is the restriction to the (fine) grid Γ_m of the (filtered) function $f * g_m$. We emphasize that what follows does not depend on the particular g that is used. (Recall that for the development of the pyramid algorithms of Burt and Adelson in section 4.2, we had $g(x,y) = \overline{\varphi}(-x,-y)$. This is not the case here; g and φ are completely independent, as were g and the filters in Theorem 3.2.)

If we still assume that m_0 does not vanish on $\left[-\frac{\pi}{2}, \frac{\pi}{2}\right] \times \left[-\frac{\pi}{2}, \frac{\pi}{2}\right]$ and that the pyramid is orthogonal as defined by (4.25), then Mallat's algorithm converges as the number of steps tends to infinity. The limit of this process is another algorithm, namely, the decomposition of the original function f in the orthonormal basis composed of the following four sequences:

$$\varphi(x-k), \quad 2^j \psi_1(2^j x - k), \quad 2^j \psi_2(2^j x - k), \quad 2^j \psi_3(2^j x - k),$$

where $x \in \mathbb{R}^2$, $k \in \mathbb{Z}^2$, $j \in \mathbb{N}$.

This means that if we fix the index j of the grid Γ_j, and if we examine the "outputs" of Mallat's algorithm that are defined on this grid, then the limits of their coefficients are, respectively,

$$2^j \int f(x)\varphi(2^j x - k)\, dx, \qquad 2^j \int f(x)\psi_1(2^j x - k)\, dx,$$

$$2^j \int f(x)\psi_2(2^j x - k)\, dx, \qquad 2^j \int f(x)\psi_3(2^j x - k)\, dx.$$

Albert Cohen established this result under very general hypotheses [60]; of these, the most convenient is that $m_0(\xi, \eta) \neq 0$ if $|\xi| \leq \frac{\pi}{2}$ and $|\eta| \leq \frac{\pi}{2}$. The beauty of this theory leads one to think that it provides the correct response to the image-processing problem. Indeed, the image is decomposed by wavelet analysis into information that is independent (orthogonal) from one scale to another, and this agrees with the general philosophy expressed in the introduction. These independent packets of information are represented by the trend in V_0 and the fluctuations $f_j \in W_j$, whose orthogonal sum is equal to f. The characteristic scale of W_j is 2^j, and each $f_j \in W_j$ is itself decomposed into orthogonal components according to the basis $2^j \psi(2^j x - k)$, $k \in \mathbb{Z}^2$, $\psi = \psi_1, \psi_2$, or ψ_3.

The Haar system provides the simplest example of two-dimensional orthogonal wavelets. This version is constructed as follows: Let φ and ψ be the one-dimensional Haar functions; then $\varphi(x,y) = \varphi(x)\varphi(y)$, $\psi_1(x,y) = \varphi(x)\psi(y)$, $\psi_2(x,y) = \psi(x)\varphi(y)$, and $\psi_3(x,y) = \psi(x)\psi(y)$. This system has been used for image processing for a long time, and it is still used in astronomy (Chapter 12). However, the Haar system has the disadvantage that, following quantization, it introduces rather harsh edge effects, thus producing unpleasant images (see Figure 2.1).

This prompts us to say a few words about the quantization problem. If we stay in the L^2 setting, all orthonormal bases allow the signal to be reconstructed exactly. This is not the point of view of the numerical analyst or image specialist. In practice, the coefficients from the decomposition must be quantized, whether we like it or not. These approximations arise from the machine accuracy or are imposed by a desire to compress the data. If it is true that

$$f(x,y) = \sum_{j \in J} \alpha_j e_j(x,y),$$

what happens to f if the α_j are replaced by coefficients $\tilde{\alpha}_j$ satisfying $|\tilde{\alpha}_j - \alpha_j| \leq \varepsilon$, where $\varepsilon > 0$ is related to the machine precision? If we use a discontinuous wavelet, one bad thing that happens is that spurious edges will appear, and even though the L^2 error is small, the visual effect can be very disturbing (see Figure 2.1). The use of smooth wavelets produces a much better result.

In spite of this, orthogonal wavelets (and the corresponding pyramid algorithms) have not completely satisfied the experts in image processing. One criticism is the lack of symmetry. The function φ ought to be even, while the function ψ ought to be symmetric in the sense that $\psi(1-x) = \psi(x)$. These properties are satisfied by certain orthogonal wavelets, but they do not hold for wavelets with compact support. The Haar system, which is antisymmetric about $\frac{1}{2}$, is the only exception. This lack of symmetry leads to visible defects, again following quantization. These defects do not appear when one uses symmetric, biorthogonal wavelets having compact support. We introduce these wavelets in the next section.

4.7 Biorthogonal wavelets

Following the pioneering work of Philippe Tchamitchian [246], Albert Cohen, Ingrid Daubechies, and Jean-Christophe Feauveau [57] studied a remarkable generalization of the notion of orthonormal wavelet bases, namely, biorthogonal systems of wavelets (see also [107]). We begin with the one-dimensional case.

In place of an orthonormal basis of the form $2^{j/2}\psi(2^j x - k)$, $j, k \in \mathbb{Z}$, we use two Riesz bases, each the dual of the other, denoted by $\psi_{j,k}$ and $\tilde{\psi}_{j,k}$. The first is used for synthesis, and the second is used for analysis. This means that for all f belonging to $L^2(\mathbb{R})$,

$$f(x) = \sum_{j=-\infty}^{\infty} \sum_{k=-\infty}^{\infty} \alpha_{j,k} \psi_{j,k}(x), \qquad (4.22)$$

where $\|f\|_2$ and $(\sum_{j=-\infty}^{\infty} \sum_{k=-\infty}^{\infty} |\alpha_{j,k}|^2)^{1/2}$ are equivalent norms on $L^2(\mathbb{R})$ and where the coefficients are defined by

$$\alpha_{j,k} = \int_{-\infty}^{\infty} f(x) \overline{\tilde{\psi}_{j,k}(x)} \, dx. \qquad (4.23)$$

As before, we define

$$\psi_{j,k}(x) = 2^{j/2}\psi(2^j x - k) \quad \text{and} \quad \tilde{\psi}_{j,k}(x) = 2^{j/2}\tilde{\psi}(2^j x - k).$$

Up to this point we have only weakened the definition of the orthonormal wavelet bases. But what we gain in flexibility by not requiring that $\psi = \tilde{\psi}$ allows us to make considerably stronger demands on ψ. For example, we can require that ψ be the function in Figure 4.4.

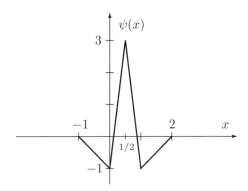

FIG. 4.4. *An example of ψ with $\psi(1 - x) = \psi(x)$.*

The general theory of Cohen, Daubechies, and Feauveau tells us that, for this choice of ψ, $\psi_{j,k}$ is a Riesz basis for $L^2(\mathbb{R})$ and the dual basis has the same structure, which is given by $2^{j/2}\tilde{\psi}(2^j x - k)$, $j, k \in \mathbb{Z}$. In this special case, the dual wavelet is not a continuous function. This is not necessarily a problem, but if we want more regularity, we need to take a more general approach.

We will select ψ from a set of functions that are continuous, have compact support, are linear on each interval $[\frac{k}{2}, \frac{k+1}{2}]$, $k \in \mathbb{Z}$, and are symmetric with respect to

$x = \frac{1}{2}$; that is, $\psi(1-x) = \psi(x)$. We can do this so that the dual wavelet $\tilde{\psi}$ is a function in the class C^r and has compact support.

We are going to outline how ψ and $\tilde{\psi}$ are constructed. Start with the triangle function $\varphi(x) = \sup(1 - |x|, 0)$, which was mentioned in section 4.2. Then define $m_0(\xi) = (\cos 2^{-1}\xi)^2$; by construction, $\hat{\varphi}(\xi) = m_0(2^{-1}\xi)\hat{\varphi}(2^{-1}\xi)$. Next, consider $g_N(\xi) = c_N \int_\xi^\pi (\sin t)^{2N+1} dt$, where $c_N > 0$ is chosen so that $g_N(0) = 1$.

If $\tilde{m}_0$ is defined by $m_0(\xi)\tilde{m}_0(\xi) = g_N(\xi)$, then

$$m_0(\xi)\tilde{m}_0(\xi) + m_0(\xi + \pi)\tilde{m}_0(\xi + \pi) = 1. \tag{4.24}$$

(In the construction of the Daubechies wavelets, one imposed the condition $|m_0(\xi)|^2 = g_N(\xi)$.) Define $\tilde{\varphi} \in L^2(\mathbb{R})$ by its Fourier transform

$$\hat{\tilde{\varphi}}(\xi) = \prod_{j=1}^\infty \tilde{m}_0(2^{-j}\xi). \tag{4.25}$$

Then the identity (4.24) is equivalent to

$$\int \tilde{\varphi}(x)\varphi(x-k)dx = \begin{cases} 0 & \text{if } k \neq 0, \\ 1 & \text{if } k = 0. \end{cases} \tag{4.26}$$

The function $\tilde{\varphi}$ is even, its support is the interval $[-2N, 2N]$, and $\tilde{\varphi}$ is in the Hölder space C^r for all sufficiently large N.

It is clear that

$$\left\| \sum_{k=-\infty}^\infty \alpha_k \tilde{\varphi}(x-k) \right\|_2 \leq C \left(\sum_{k=-\infty}^\infty |\alpha_k|^2 \right)^{1/2},$$

but (4.26) implies that the inverse inequality also holds. Thus one can consider the closed subspace $\tilde{V}_0 \subset L^2(\mathbb{R})$ for which $\tilde{\varphi}(x-k)$, $k \in \mathbb{Z}$, is a Riesz basis. If the subspaces $\tilde{V}_j$ are defined by

$$f(x) \in \tilde{V}_0 \iff f(2^j x) \in \tilde{V}_j,$$

then this sequence forms a multiresolution analysis of $L^2(\mathbb{R})$. In the same way, let V_0 be the closed subspace of $L^2(\mathbb{R})$ for which $\varphi(x-k)$, $k \in \mathbb{Z}$, is a Riesz basis and construct the V_j similarly.

The two multiresolutions (V_j) and $(\tilde{V}_j)$ are the duals of each other. This duality is used to define the subspaces W_j and $\tilde{W}_j$: f belongs to W_j if f belongs to V_{j+1} and if $\int_{-\infty}^\infty f(x)\overline{u}(x)\, dx = 0$ for all $u \in \tilde{V}_j$. The wavelets ψ and $\tilde{\psi}$ will be constructed so that $\psi(x-k)$, $k \in \mathbb{Z}$, is a Riesz basis for W_0 and, similarly, $\tilde{\psi}(x-k)$, $k \in \mathbb{Z}$, is a Riesz basis for $\tilde{W}_0$. For this, we define

$$m_1(\xi) = e^{-i\xi}\overline{\tilde{m}_0}(\xi + \pi) \quad \text{and} \quad \tilde{m}_1(\xi) = e^{-i\xi}\overline{m_0}(\xi + \pi),$$

and we define the Fourier transforms of ψ and $\tilde{\psi}$ by

$$\hat{\psi}(\xi) = m_1(2^{-1}\xi)\hat{\varphi}(2^{-1}\xi) \quad \text{and} \quad \hat{\tilde{\psi}}(\xi) = \tilde{m}_1(2^{-1}\xi)\hat{\tilde{\varphi}}(2^{-1}\xi).$$

Write $\psi_{j,k}(x) = 2^{j/2}\psi(2^j x - k)$, and define $\tilde{\psi}_{j,k}(x)$ similarly.

The only properties that are difficult to prove are that the family $\psi_{j,k}$, $j,k \in \mathbb{Z}$, is a Riesz basis for $L^2(\mathbb{R})$ and that the same is true for the $\tilde{\psi}_{j,k}$. These Riesz bases are the duals of each other. This means that $\langle \psi_{j,k}, \tilde{\psi}_{j',k'} \rangle = \delta_{j,j'} \delta_{k,k'}$ and that $f \in L^2(\mathbb{R})$ can be represented as

$$f(x) = \sum_{j,k \in \mathbb{Z}} \langle f, \tilde{\psi}_{j,k} \rangle \psi_{j,k}(x)$$

and as

$$f(x) = \sum_{j,k \in \mathbb{Z}} \langle f, \psi_{j,k} \rangle \tilde{\psi}_{j,k}(x).$$

Furthermore, the function ψ is as simple as it is explicit: It is continuous; it has compact support; it is linear on each interval $[\frac{k}{2}, \frac{k+1}{2}]$; and the values $\psi(2^{-1}k)$ are explicit rational numbers. Finally, we have $\psi(1-x) = \psi(x)$, and the symmetry, which Daubechies's wavelets lack, is reestablished. In dimension two, we use the wavelets $\varphi(x)\psi(y)$, $\varphi(y)\psi(x)$, and $\psi(x)\psi(y)$, as in the orthogonal case. Then the dual wavelets are $\tilde{\varphi}(x)\tilde{\psi}(y)$, $\tilde{\varphi}(y)\tilde{\varphi}(x)$, and $\tilde{\psi}(x)\tilde{\psi}(y)$.

While the JPEG committee is still working on developing the upcoming JPEG-2000 standard for still image compression, at the time of writing, it is very likely that the JPEG-2000 standard will be based on biorthogonal filters and bitplane coding [197].

The flexibility offered by biorthogonal wavelet expansions is not limited to filter applications but extends to other areas. For instance, if D^s denotes the fractional powers of $D = -i\frac{d}{dx}$, we can require the two Riesz bases $\psi_{j,k}$ and $\tilde{\psi}_{j,k}$, $j,k \in \mathbb{Z}$, to be orthogonal with respect to the scalar product

$$\langle f, g \rangle = \int (D^s f)(x) \overline{(D^s g)}(x) \, dx.$$

Fabrice Sellan has shown that such wavelets "decorrelate" fractional Brownian motion (fBm) of order $H = s - \frac{1}{2}$. This means that this process can be written as

$$\sum 2^{-sj} g_{j,k} \psi_{j,k}(x),$$

where the $g_{j,k}$ are independent, identically distributed Gaussian random variables with mean 0 and variance 1. This decomposition involves small scales ($j \to +\infty$) and large scales ($j \to -\infty$), and it is easily seen that this second half of the series is divergent whenever $H = s - \frac{1}{2} > 0$. This divergence must be fixed. One option is to replace $\psi_{j,k}(t)$ by $\psi_{j,k}(t) - \psi_{j,k}(0)$. A second option consists in introducing a scaling function φ such that

$$\sum_{j<0} 2^{-sj} g_{j,k} \psi_{j,k}(t) = \sum_k c(k, H) \varphi(t - k),$$

where $c(k, H)$ is a FARIMA process. This is currently the best way to simulate accurately the long-range correlations in fBm (see [2]). P. G. Lemarié-Rieusset used the same idea to construct divergence-free biorthogonal wavelets in $\mathbb{R}^3$, which may prove to be useful in the study of turbulence [171]. (There will be much more about wavelets and turbulence in Chapter 9.)

CHAPTER 5

Time-Frequency Analysis for Signal Processing

5.1 Introduction

Time-frequency analysis for signal processing is an active field of research. Here, as in many domains, heuristic concepts structure and guide the work. The heuristic notions that will serve us in this and the following three chapters are (1) time-frequency atoms, (2) the optimal decomposition of a signal into time-frequency atoms, (3) instantaneous frequency, (4) the time-frequency plane, (5) the optimal representation of a signal in the time-frequency plane, and (6) optimal partitioning of the time-frequency plane. In this and the following chapters, we will try to give precise scientific meaning to these heuristic ideas. We add, however, that this is a large field of research and that our exposition is by no means exhaustive.

Dennis Gabor [124] and Jean Ville [254] both addressed the problem of developing a mixed representation of a signal in terms of a double sequence of elementary signals, each of which occupies a certain domain in the time-frequency plane. In the following sections we will define what is meant by time-frequency plane and mixed representation, and we will suggest several choices for the elementary signals, or atoms.

Roger Balian tackled the same problem and expressed the motivation for his work in these terms [17, p. 1357]:

> One is interested, in communication theory, in representing an oscillating signal as a superposition of *elementary wavelets*, each of which has a rather well defined *frequency* and *position in time*. Indeed, useful information is often conveyed by both the emitted frequencies and the signal's temporal structure (music is a typical example). The representation of a signal as a function of time provides a poor indication of the spectrum of frequencies in play, while, on the other hand, its Fourier analysis masks the point of emission and the duration of each of the signal's elements. An appropriate representation ought to combine the advantages of these two complementary descriptions; at the same time, it should be discrete so that it is better adapted to communication theory.[4]

Similar criticism of the usual Fourier analysis, as applied to acoustic signals, is found in the celebrated work of Ville [254, p. 63]:

> If we consider a passage [of music] containing several measures (which is the least that is needed) and if a note, *la* for example, appears once in

---
[4] Here and elsewhere, the translations from French are ours.

the passage, harmonic analysis will give us the corresponding frequency with a certain amplitude and a certain phase, without localizing the *la* in time. But it is obvious that there are moments during the passage when one does not hear the *la*. The [Fourier] representation is nevertheless mathematically correct because the phases of the notes near the *la* are arranged so as to destroy this note through interference when it is not heard and to reinforce it, also through interference, when it is heard; but if there is in this idea a cleverness that speaks well for mathematical analysis, one must not ignore the fact that it is also a distortion of reality: indeed when the *la* is not heard, the true reason is that the *la is not* emitted.

Thus it is desirable to look for a mixed definition of a signal of the sort advocated by Gabor: at each instance, a certain number of frequencies are present, giving volume and timbre to the sound as it is heard; each frequency is associated with a certain partition of time that defines the intervals during which the corresponding note is emitted. One is thus led to define an *instantaneous spectrum as a function of time*, which describes the structure of the signal at a given instant; the spectrum of the signal, in the usual sense of the term, which gives the frequency structure of the signal based on its total duration, is then obtained by putting together all of the instantaneous spectrums in a precise way by integrating them with respect to time. In a similar way, one is led to a distribution of frequencies with respect to time; by integrating these distributions, one reconstructs the signal.

Ville thus proposed to unfold the signal in the time-frequency plane in such a way that this development would lead to a mixed representation in time-frequency atoms. The choice of these time-frequency atoms would be guided by an energy distribution of the signal in the time-frequency plane.

The time-frequency atoms proposed by Gabor are constructed from the Gaussian $g(t) = \pi^{-1/4} e^{-t^2/2}$ and are defined by

$$w(t) = h^{-1/2} e^{i\omega t} g\left(\frac{t - t_0}{h}\right). \tag{5.1}$$

The parameters ω and t_0 are arbitrary real numbers, whereas h is positive. The meaning of these three parameters is the following: ω is the average frequency of w, $h > 0$ is the duration of w, and $t_0 - h$ and $t_0 + h$ are the start and finish of the "note" w. Naturally, this depends on the convention used to define the width of g.

The essential problem is to describe an algorithm that allows a given signal to be decomposed, in an optimal way, as a linear combination of judiciously chosen time-frequency atoms.

The set of all time-frequency atoms (with ω and t_0 varying arbitrarily in the time-frequency plane and $h > 0$ covering the whole scale axis) is a collection of elementary signals that is much too large to provide a unique representation of a signal as a linear combination of time-frequency atoms. Each signal admits an infinite number of representations, and this leads us to choose the best among them according to some criterion.

A similar program (the definition of time-frequency atoms, analysis, and synthesis) was proposed by Jean-Sylvain Liénard in [173, pp. 948, 949], where he wrote:

> We consider the speech signal to be composed of elementary waveforms, wf, (windowed sinusoids), each one defined by a small number of parameters.

A waveform model (wfm) is a sinusoidal signal multiplied by a windowing function. It is not to be confused with the signal segment, wf, that it is supposed to approximate. Its total duration can be decomposed into attack (before the maximum of the envelope), and decay. In order to minimize spectral ripples, the envelope should present no 1st or 2nd order discontinuity.

The initial discontinuity is removed through the use of an attack function (raised sinusoid) such that the total envelope is null at the origin, and maximum after a short time.

Although exponential damping is natural in the physical world, we choose to model the decaying part of the wfs with another raised sinusoid. Actually we see the wf as a perceptual unit, and not necessarily as the response of a format filter to a voicing impulse.

Liénard's time-frequency atoms (Figure 5.1) are different from those used by Gabor. They are, however, based on analogous principles. The Liénard atoms are of the form $w(t) = A(t)\cos(\omega t + \varphi)$, where ω represents the average frequency of the emitted "note" and where the envelope A incorporates the attack and decay. The principal difference is that, in the atoms of Liénard, the duration of the attack and that of the decay are independent. Thus Liénard's atoms depend on four independent parameters, and the optimal representation of a speech signal as a linear combination of time-frequency atoms is more difficult to obtain. Some empirical methods exist, and they lead to wonderful results for synthesizing the singing voice. For example, the Queen of the Night's grand aria from Mozart's *Magic Flute* has been interpreted by time-frequency atoms. This was not a copy of the human voice; it involved the creation of a purely numerical (superhuman) voice. This was commissioned by Pierre Boulez, the director of the Institut de Recherches Coordonnées Acoustique-Musique, and achieved by X. Rodet of that institute (see [231]).

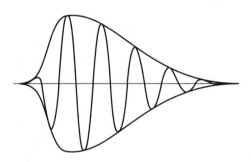

FIG. 5.1. *A Liénard time-frequency atom.*

5.2 The collections Ω of time-frequency atoms

The time-frequency atoms of Gabor defined by (5.1) (which are also called Gabor wavelets and Gaborlets) and the waveforms of Liénard are two examples of what we call *a collection Ω of time-frequency atoms*. This concept will play an essential role in this and the next two chapters.

Mathematically, a collection of time-frequency atoms Ω is a subset of $L^2(\mathbb{R})$ that is complete. This means that the finite linear combinations $\sum \alpha_j w_j$, $w_j \in \Omega$, are dense in $L^2(\mathbb{R})$. We also assume that if $w \in \Omega$, then $\|w\|_2 = 1$. But this definition is much too general to serve in practice for signal processing. Thus, in addition, we require that the elements w of Ω have a simple algorithmic structure and that the elements $w \in \Omega$ are optimally localized in the time-frequency plane. Obviously these last two requirements have not yet been made mathematically precise; instead, they will be illustrated by the various examples we discuss. We will use the Wigner–Ville transform (section 5.5) to study localization in the time-frequency plane.

Here are some of the collections Ω that are available to us today:

(1) The Gabor wavelets where ω and t_0 are arbitrary but $h = 1$.
(2) The complete collection of Gabor time-frequency atoms where ω, t_0, $h > 0$ are arbitrary.
(3) The waveforms of J.-S. Liénard and X. Rodet.
(4) Malvar–Wilson wavelets (Chapter 6).
(5) Chirplets (Chapter 6).
(6) Wavelet packets (Chapter 7).

Given that we have these collections at our disposal, two problems arise:

(a) What collection Ω should one choose to study a given signal or a given class of signals?
(b) Having chosen Ω, how is one to decompose a signal f optimally in a series $\alpha_0 w_0 + \alpha_1 w_1 + \cdots + \alpha_j w_j + \cdots$, where $w_j \in \Omega$ and the α_j are scalars?

There is no general answer to the first question. A current point of view in signal analysis could be called a resemblance criterion. It holds that the time-frequency atom should "look like" the signal (or pieces of the signal) that is being analyzed. This is the point of view taken by Liénard, but intuition can be misleading. As an example, we mention the problem of storing fingerprints. The first compression algorithm depended on the optimal use of wavelet packets. This choice seemed natural, since the structure of fingerprints exhibits certain textures that one feels ought to be analyzed by a time-frequency algorithm rather than by a time-scale algorithm. However, to general surprise, it appears that the biorthogonal wavelets of Cohen, Daubechies, and Feauveau provide the best results. This conclusion was not obtained by theoretical considerations; it resulted from experimentation [41].

The fingerprint example brings us back to statistical modeling, which was briefly mentioned in section 1.2. If one is studying a large collection of signals or images that exhibit common features and if one wishes to choose a collection Ω, then we believe one should first develop a statistical model of the collection. Such a model should include random variables that model the intrinsic variability within the data set. The goal is to find an algorithm that produces signals or images that have the same "look and feel" as the ones in the data set. This in turn should point the way to choosing an appropriate collection Ω.

We move on to question (b), for which we have several pieces of an answer. Having chosen a collection of time-frequency atoms Ω, we must find a decomposition

$$f = \alpha_0 w_0 + \alpha_1 w_1 + \cdots + \alpha_j w_j + \cdots, \qquad w_j \in \Omega, \tag{5.2}$$

that is in some sense optimal. In signal processing, the notion of optimality should be defined in terms of some goal, and as discussed in Chapter 1, the most important ones currently are analysis, compression, transmission, storage, restoration, denoising, and some specific diagnostic. We conclude by emphasizing once again that when dealing with a large data set a reliable diagnosis depends on the efficacy of the statistical model of the data.

5.3 Mallat's matching pursuit algorithm

One of the most elegant of the algorithms that lead to an optimal decomposition of the form (5.2) is Mallat's matching pursuit algorithm. Its goal is analysis or diagnosis.

Mallat's algorithm can be applied to any collection of time-frequency atoms Ω that satisfies a certain compactness property. In all of the cases we have in mind, the time-frequency atoms w are functions of a parameter $\lambda \in \Lambda$, and the forms $\langle f, w_\lambda \rangle$ are continuous functions of λ. For example, in the case of the Gabor wavelets with arbitrary duration, $\Lambda = \mathbb{R} \times \mathbb{R} \times (0, \infty)$. Furthermore, in this case, the functions $\langle f, w_\lambda \rangle$ tend to zero when λ is in the complement of the compact set K_N defined by $[-N, N] \times [-N, N] \times [1/N, N]$ and $N \to +\infty$. Another way to say this is that the functions $\langle f, w_\lambda \rangle$ are continuous on the one-point compactification $\overline{\Lambda} = \Lambda \cup \{\infty\}$ of Λ and vanish at the ideal point ∞. It follows that for each $f \in L^2(\mathbb{R})$, the function $|\langle f, w_\lambda \rangle|$ attains its maximum value at some point in Λ. The general property we require is that the functions $\langle f, w_\lambda \rangle$ "vanish at infinity" or, more precisely, that they are continuous on the one-point compactification of Λ and vanish at the ideal point. This property can be verified for all of the examples discussed.

Mallat's algorithm consists in solving the optimization problem

$$\beta_0 = \sup_{w \in \Omega} |\langle f, w \rangle|. \tag{5.3}$$

This problem has at least one solution $w_0 \in \Omega$ by virtue of the assumption. We write $\alpha_0 = \langle f, w_0 \rangle$, and assuming that $\|w\|_2 = 1$ for all $w \in \Omega$, we define

$$f_1(t) = f(t) - \alpha_0 w_0(t).$$

Note that there is no reason for w_0 to be unique, and thus there is no reason for f_1 to be unique. By iterating the process we obtain

$$f_{j+1}(t) = f_j(t) - \alpha_j w_j(t), \qquad j \in \mathbb{N}.$$

It is proved in [183] that this algorithm converges and gives a representation (5.2).

For an elementary discussion of this algorithm and pursuit algorithms in general, we suggest Mallat's book [181]. One learns there that these algorithms were used in statistics in the early 1980s [117], and that convergence in case Ω is infinite was first proved by L. K. Jones in 1987 [162] (see also [75] and [163]). DeVore's paper [79] contains a review of pursuit algorithms in the context of nonlinear approximation, which is to say, with an emphasis on the rate at which these algorithms converge.

The optimization problem (5.3) is inherently unstable, and a solution can be costly. These shortcomings have inspired the development of faster, more robust algorithms, whose names typically include "pursuit." This is an active field of research. As an example of the kind of work being done, we recommend the recent thesis by Rémi Gribonval where pursuit algorithms are used to analyze acoustic signals [132].

5.4 Best-basis search

Those who use Malvar–Wilson wavelets or wavelet packets have adopted a different point of view toward optimality. In both of these cases, the time-frequency atoms in Ω can be regrouped to form a set $\mathcal{A}$ of orthonormal bases α. In other words, Ω is replaced by a "library" $\mathcal{A}$ whose "books" α are orthonormal bases.

In this case, Mallat's matching pursuit algorithm is replaced by an algorithm that looks for the best basis. Said differently, in place of looking for "the best atom" $w_0 \in \Omega$, one looks for "the best basis" $\alpha_0 \in \mathcal{A}$. This optimal basis is defined in terms of an entropy criterion that leads to "the most compact representation" of the given signal. We note that this algorithm is not iterative. However, there is a "Mallat" version of this algorithm that is used for denoising. One looks for the best basis $\alpha_0 \in \mathcal{A}$, but one retains in the corresponding decomposition of the signal f only those terms whose energy exceeds a certain threshold (which is related to the assumed level of the noise). If the sum of these terms is called f_0, one considers the function $f - f_0$ and repeats the process. As an example, this denoising technique has been applied to the 1889 recording of Johannes Brahms performing his Hungarian Dance no. 1 in G Minor [34].

The connections between time-frequency atoms, the time-frequency plane, and the optimal representation of the analyzed signal in the time-frequency plane will be developed in the following sections. But at this point we are going to pause and discuss the special case of the Gabor wavelets.

For the moment, the time-frequency plane will be the usual $\mathbb{R}^2$ plane. The idea behind calling this the time-frequency plane is based on the following heuristic: One looks for an algorithm that allows one to "write," in the time-frequency plane, a "partition" of a given signal f. The "notes" used to write this partition should be the time-frequency atoms found in one of the decompositions (5.2). We hope that these "notes" are simple and convenient; this means that they are accessible via a simple algorithm working in real time and that they are optimally localized in the time-frequency plane. This localization will be defined in the following sections using the Wigner–Ville transform.

In the case of Gabor wavelets, the disc

$$\{(t,\xi) \mid (\xi - \omega)^2 + (t - t_0)^2 \leq 1\}$$

is associated with the wavelet $e^{i\omega t} g(t - t_0)$, and, more generally, the elliptical domain

$$E = \left\{(t,\xi) \mid h^2(\xi - \omega)^2 + \frac{(t - t_0)^2}{h^2} \leq 1\right\}$$

is associated with the wavelet $e^{i\omega t} g_h(t - t_0)$, where $g_h(t) = \frac{1}{\sqrt{h}} g(\frac{t}{h})$. Later we will replace these elliptical domains with the corresponding rectangles

$$\left\{(t,\xi) \mid |t - t_0| \leq h, \; |\xi - \omega| \leq \frac{\pi}{2h}\right\},$$

which are called *Heisenberg boxes*. We note that these Heisenberg boxes can be horizontal, vertical, or square depending on the value of h. The localization of the time-frequency atoms depends on the Wigner–Ville transform, which we discuss in the next few sections.

5.5 The Wigner–Ville transform

In work that is still stimulating to read [254], Ville set himself the task of studying three topics and of relating them to each other: (1) the distribution of energy of a signal in the time-frequency plane, (2) the definition of instantaneous frequency, and (3) the optimal decomposition of a signal in a series of Gabor wavelets. In this and the following sections we will study the Wigner–Ville transform. Later we will indicate its use for studying the problems posed by Ville.

We begin by presenting the point of view of Ville. We will then indicate how to interpret the results in terms of the theory of pseudodifferential operators as expressed in Hermann Weyl's formalism. This will bring us back to work done by the physicist Eugene Wigner in the 1930s.

Ville, searching for an "instantaneous spectrum," wanted to display the energy of a signal in the time-frequency plane and to obtain an energy density $W(t, \xi)$ having (at least) the following properties:

$$\int_{-\infty}^{\infty} W(t, \xi) \frac{d\xi}{2\pi} = |f(t)|^2, \tag{5.4}$$

$$\int_{-\infty}^{\infty} W(t, \xi) \, dt = |\hat{f}(\xi)|^2, \tag{5.5}$$

where $\hat{f}$ denotes the Fourier transform of f. The heuristic behind this research is the following: $W(t, \xi)$ should represent "the square of the modulus of the instantaneous Fourier transform of f at the instant t," so that if a theory of the instantaneous Fourier transform existed, (5.4) would look like the Plancherel identity. Similarly, (5.5) would mean that the various instantaneous spectral contributions are summed to form the square of the modulus of the Fourier transform. An exhaustive description of densities $W(t, \xi)$ that satisfy (5.4) and (5.5) can be found in [112].

Ville made the following choice, which is now called the Wigner–Ville transform:

$$W(t, \xi) = \int_{-\infty}^{\infty} f\left(t + \frac{\tau}{2}\right) \overline{f}\left(t - \frac{\tau}{2}\right) e^{-i\xi\tau} d\tau. \tag{5.6}$$

We now look at how well $W(t, \xi)$ fulfills the notion of an energy density. If the signal f has finite energy ($\int_{-\infty}^{\infty} |f(t)|^2 dt < \infty$), then $W(t, \xi)$ exists as a real-valued, continuous function in the time-frequency plane. As we will see in a moment, the converse is completely false. Even if rather restrictive conditions are imposed, such as belonging to the Schwartz class, a real function $W(t, \xi)$ is not in general the Wigner–Ville transform of a signal with finite energy.

If $\|f\|_2 = 1$, it is clear from either (5.4) or (5.5) that

$$\iint W(t, \xi) \, dt \, \frac{d\xi}{2\pi} = 1,$$

but it is not true that $W(t, \xi)$ is always nonnegative.

Another property of $W(t, \xi)$ concerns localization in the time-frequency plane. If f vanishes outside an interval $[t_0, t_1]$, then the same is true for $W(t, \xi)$. Similarly, if the Fourier transform of f is zero outside $[\omega_0, \omega_1]$, then $W(t, \xi) = 0$ if $\xi \notin [\omega_0, \omega_1]$. If—and this is of course impossible—f were zero outside $[t_0, t_1]$ while its Fourier transform $\hat{f}$ were zero outside $[\omega_0, \omega_1]$, then the Wigner–Ville transform of f would be zero outside the rectangle $[t_0, t_1] \times [\omega_0, \omega_1]$. It is this "property" that speaks in favor of Ville's interpretation of $W(t, \xi)$.

Unfortunately, we immediately encounter a trap. If the Fourier transform of f is supported on $\omega_0 \leq |\omega| \leq \omega_1$, where $0 < \omega_0 < \omega_1$, the same is not true for

$W(t,\xi)$, which, in general, is supported on $|\omega| \leq \omega_1$. The formerly empty interval $|\omega| \leq \omega_0$ can be filled with $W(t,\xi)$. This phenomenon causes $W(t,\xi)$ to be difficult to interpret: The Wigner–Ville transform can take nonzero values in regions of the time-frequency plane having nothing to do with the spectral properties of the signal. In spite of these artifacts and the fact that $W(t,\xi)$ can take negative values—and thus is an imperfect energy density—the Wigner–Ville transform has an important role in signal processing.

It is interesting to note that the Wigner–Ville transform did not originate in signal processing but rather in quantum mechanics, and the technology transfer to signal processing was begun by Ville in the 1950s. At the time Ville did his work, there were no heuristics originating from signal processing that would lead to this specific quadratic transformation. Today, however, the Wigner–Ville transform appears naturally in signal processing because it is related to the *ambiguity function* of a signal f, which is defined by

$$A(\tau,\omega) = \int_{-\infty}^{\infty} f\left(t+\frac{\tau}{2}\right) \overline{f}\left(t-\frac{\tau}{2}\right) e^{-i\omega t}\, dt.$$

The ambiguity function is a two-dimensional Fourier transform of the Wigner–Ville transform of f, and it is widely used in signal processing for radar (see, for example, [37]). On the other hand, if we abandon signal processing and instead move to the theory of pseudodifferential operators (section 5.7), then the Wigner–Ville transform appears naturally in quantum mechanics. (We find it remarkable that so much progress in signal processing has been realized by experts in quantum mechanics.)

5.6 Properties of the Wigner–Ville transform

We begin with the case of signals with finite energy. If $W(f;t,\xi)$ denotes the Wigner–Ville transform of f, we need to compute $W(Tf;t,\xi)$ when T is a linear operator. Since the mapping $f \mapsto W(Tf;t,\xi)$ is quadratic, it is not clear that, given T, there will exist a linear operator $\tilde{T}$ such that

$$W(Tf;t,\xi) = \tilde{T}[W(f;t,\xi)].$$

This is the case, however, in a number of important examples. We consider the problem for the following operators:

Fourier transform $\hat{f}$, $\qquad \hat{f}(\xi) = \int e^{-it\xi} f(t)\, dt;$

unitary dilation D_a, $a > 0$, $\qquad (D_a f)(t) = \dfrac{1}{\sqrt{a}} f\!\left(\dfrac{t}{a}\right);$

symmetry S, $\qquad (Sf)(t) = f(-t);$

translation R_b, $b \in \mathbb{R}$, $\qquad (R_b f)(t) = f(t-b);$

modulation M_ω, $\omega \in \mathbb{R}$, $\qquad (M_\omega f)(t) = e^{i\omega t} f(t);$

multiplication operator P_ω, $\omega \in \mathbb{R}$, $\qquad (P_\omega f)(t) = e^{i\omega t^2} f(t).$

With this notation we have the following relations:

$$W(\hat{f}; t, \xi)/2\pi = W(f; -\xi, t), \tag{5.7}$$
$$W(D_a f; t, \xi) = W(f; t/a, a\xi), \tag{5.8}$$
$$W(Sf; t, \xi) = W(f; -t, -\xi), \tag{5.9}$$
$$W(R_b f; t, \xi) = W(f; t - b, \xi), \tag{5.10}$$
$$W(M_\omega f; t, \xi) = W(f; t, \xi - \omega), \tag{5.11}$$
$$W(P_\omega f; t, \xi) = W(f; t, \xi - 2\omega t). \tag{5.12}$$

Moreover, $W(\bar{f}; t, \xi) = W(f; t, -\xi)$, and

$$W(e^{i\omega H} f; t, \xi) = W(f; t \cos 2\omega + \xi \sin 2\omega, \xi \cos 2\omega - t \sin 2\omega),$$

where $H = -\frac{d^2}{dt^2} + t^2 - 1$ (the harmonic oscillator) and ω is any real number.

A consequence of these relations is that the collection of all Wigner–Ville transforms of signals with finite energy is invariant under the Euclidean group of the time-frequency plane. This observation has some crucial consequences. If one truly trusts signal processing based on the Wigner–Ville transform, it implies that one should pave the time-frequency plane with Heisenberg boxes with arbitrary directions and eccentricities. We will return to this point later.

We list two more properties of the Wigner–Ville transform. If

$$f(t) = \int g(t-s) h(s) \, ds,$$

where g and h are signals (or functions) for which the integrals make sense, then

$$W(f; t, \xi) = \int_{-\infty}^{\infty} W(g; t - s, \xi) W(h; s, \xi) \, ds.$$

This is easily checked, but it is not intuitive since the Wigner–Ville transform is quadratic.

The Moyal identity for functions f and g having finite energy is

$$\iint W(f; t, \xi) W(g; t, \xi) \, dt \, d\xi = 2\pi \left| \int f(t) \overline{g}(t) \, dt \right|^2. \tag{5.13}$$

We now indicate how some of these properties can be used. Suppose that

$$Q(t, \xi) = p\xi^2 + 2r\xi t + qt^2, \quad p > 0, \quad q > 0, \quad pq > r^2.$$

Then $W(t, \xi) = 2\exp(-Q(t, \xi))$ is the Wigner–Ville transform of a signal f such that $\int |f(t)|^2 \, dt = 1$ if and only if $\iint W(t, \xi) \, dt \, d\xi = 2\pi$. This last condition is obviously necessary since, by (5.4),

$$\iint W(t, \xi) \, dt \, d\xi = 2\pi \int |f(t)|^2 \, dt.$$

To prove the result in the other direction, we first write Q as

$$Q(t, \xi) = p\left(\xi + \frac{r}{p} t\right)^2 + \left(\frac{pq - r^2}{p}\right) t^2.$$

Using this, it is easy to compute $\iint W(t,\xi)\,dt\,d\xi$ and thus to see that the condition $\iint W(t,\xi)\,dt\,d\xi = 2\pi$ implies that $pq - r^2 = 1$. Thus, $Q(t,\xi) = p(\xi + \frac{r}{p}t)^2 + \frac{1}{p}t^2$.

If there is a function f such that $W(f;t,\xi) = 2\exp(-Q(t,\xi))$, then by (5.12) $W(f_1;t,\xi) = 2\exp(-p\xi^2 - \frac{1}{p}t^2)$ where $f_1(t) = f(t)\exp(i\frac{r}{2p}t^2)$. Similarly, using (5.8) we see that $W(f_2;t,\xi) = 2\exp(-\xi^2 - t^2)$ with $f_2(t) = p^{-1/4}f_1(\frac{1}{\sqrt{p}}t)$.

Another computation shows that the Wigner–Ville transform of the Gaussian $g(t) = \pi^{-1/4}\exp(-\frac{1}{2}t^2)$ is $W(g;t,\xi) = 2\exp(-\xi^2 - t^2)$. By taking the transformations (5.8) and (5.12) in the other direction, we see that the transform of the function

$$f(t) = p^{-1/4} g\left(\frac{t}{\sqrt{p}}\right)\exp\left(-i\frac{r}{2p}t^2\right)$$

is exactly $W(f;t,\xi) = 2\exp(-Q(t,\xi))$.

We have already stressed that the Wigner–Ville transform $W(f;t,\xi)$ is not always a nonnegative function. In fact, the only cases where $W(f;t,\xi)$ is nonnegative are $W(t-t_0, \xi-\xi_0)$, where $W(t,\xi) = 2\exp(-Q(t,\xi))$ is defined as above with $pq-r^2 = 1$ [112].

Finally, there are several averaging procedures that allow one to eliminate the negative values of Wigner–Ville transforms. It suffices to consider

$$\frac{1}{\pi}\iint W(t-\tau, \xi-\eta)\exp(-Q(\tau,\eta))\,d\tau\,d\eta, \tag{5.14}$$

where $pq - r^2 = 1$. Roughly speaking (5.14) amounts to averaging $W(t,\xi)$ over generalized Heisenberg boxes with arbitrary directions and eccentricities.

5.7 The Wigner–Ville transform and pseudodifferential calculus

The following considerations allow us to relate the Wigner–Ville transform to quantum mechanics and the work of Wigner. We are going to forget signal processing for the moment and go directly to dimension n. The analogue of the time-frequency plane is the phase space $\mathbb{R}^n \times \mathbb{R}^n$ whose elements are pairs (x,ξ), where x is a position and ξ is a frequency.

We start with a *symbol* $\sigma(x,\xi)$ defined on phase space. Certain technical hypotheses have to be made about this symbol to ensure convergence of the following integral when f belongs to a reasonable class of test functions. We will deal with this point in a moment.

Following the formalism of Weyl, we associate with the symbol $\sigma(x,\xi)$ the pseudodifferential operator $\sigma(x,D)$ defined by

$$(2\pi)^n \sigma(x,D)[f](x) = \iint \sigma\left(\frac{x+y}{2},\xi\right)e^{i(x-y)\cdot\xi}f(y)\,dy\,d\xi, \tag{5.15}$$

where the integral is over $\mathbb{R}^n \times \mathbb{R}^n$. Define the kernel $K(x,y)$ associated with the symbol $\sigma(x,\xi)$ by

$$\begin{aligned}(2\pi)^n K(x,y) &= \int \sigma\left(\frac{x+y}{2},\xi\right)e^{i(x-y)\cdot\xi}d\xi \\ &= (2\pi)^n L\left(\frac{x+y}{2}, x-y\right).\end{aligned} \tag{5.16}$$

The symbol $\sigma(x,\xi)$ is thus the partial Fourier transform, in the variable u, of the function $L(x,u)$, and the kernel $K(x,y)$ that interests us is $L\bigl(\frac{x+y}{2}, x-y\bigr)$. We can also write $L(x,y) = K\bigl(x+\frac{y}{2}, x-\frac{y}{2}\bigr)$, and this allows us to recover the symbol $\sigma(x,\xi)$ by writing

$$\sigma(x,\xi) = \int K\left(x+\frac{y}{2}, x-\frac{y}{2}\right) e^{-iy\cdot\xi} dy. \tag{5.17}$$

Thus we are led to hypotheses about the symbols that are the reflections, through the partial Fourier transform, of hypotheses that we may wish to make about the kernels. If we admit all the distribution kernels $K(x,y)$ belonging to the space of tempered distributions on $\mathbb{R}^n \times \mathbb{R}^n$ (which we denote by $\mathcal{S}'(\mathbb{R}^n \times \mathbb{R}^n)$), then there will be no restrictions on $\sigma(x,\xi)$ other than the condition that

$$\sigma(x,\xi) \in \mathcal{S}'(\mathbb{R}^n \times \mathbb{R}^n).$$

An immediate consequence of (5.17) is this: If $\sigma(x,\xi)$ is the symbol for the operator T, then $\overline{\sigma}(x,\xi)$ is the symbol for the adjoint operator T^*.

Finally, we consider a function f belonging to $L^2(\mathbb{R}^n)$ and satisfying $\|f\|_2 = 1$. Let P_f denote the orthogonal projection operator that maps $L^2(\mathbb{R}^n)$ onto the linear span of f. Then the kernel $K(x,y)$ of P_f is $f(x)\overline{f}(y)$ and the corresponding Weyl symbol is

$$\sigma(x,\xi) = \int f\left(x+\frac{y}{2}\right) \overline{f}\left(x-\frac{y}{2}\right) e^{-iy\cdot\xi} dy. \tag{5.18}$$

Returning to dimension one, we have the following result: *The Wigner–Ville transform of the function f is the Weyl symbol of the orthogonal projection operator onto the linear span of f.* From this it is clear that the Wigner–Ville transform of f characterizes f, up to multiplication by a constant of modulus one. The following result is an important consequence of the preceding remarks.

THEOREM 5.1. *Let f_j, $j \in \mathbb{N}$, be a sequence of functions in $L^2(\mathbb{R})$ and let $W_j(t,\xi)$ be the Wigner–Ville transforms of the f_j. Then the following two properties are equivalent:*

$$f_j,\ j \in \mathbb{N},\ \text{is an orthonormal basis for } L^2. \tag{5.19}$$

$$\begin{cases} \sum_{j=0}^{\infty} W_j(t,\xi) = 1, \\ \iint W_j(t,\xi) W_{j'}(t,\xi)\, dt\, d\xi = 2\pi \delta_{j,j'}. \end{cases} \tag{5.20}$$

If P_j denotes the projection associated with f_j, then (5.19) amounts to writing $\langle f_j, f_{j'} \rangle = \delta_{j,j'}$ and $\sum_{j=0}^{\infty} P_j = I$. Since $W_j(t,\xi)$ is the Weyl symbol of P_j, $\sum_{j=0}^{\infty} P_j = I$ is equivalent to the first equation of (5.20). More precisely, $\sum_{j=0}^{\infty} W_j(t,\xi)$ should converge to one in the sense of distributions. On the other hand, Moyal's identity yields the second equation of (5.20).

This simple and elegant theorem led to the following heuristic: Orthonormal bases for $L^2(\mathbb{R})$ consisting of time-frequency atoms a_j are in one-to-one correspondence with partitionings of the time-frequency plane with horizontal or vertical Heisenberg boxes. For example, orthonormal wavelet bases correspond to the now-familiar paving of the time-frequency plane (Figure 5.2) that James Glimm and

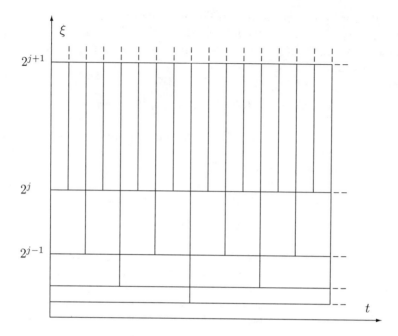

FIG. 5.2. *Dyadic paving of the time-frequency plane.*

Arthur Jaffe were using in quantum physics before wavelets existed [130]. Such pictures should not be taken too literally, however. Consider, for example, the Lemarié–Meyer wavelets where the mother wavelet ψ belongs to the Schwartz class and satisfies

$$\hat{\psi}(\xi) = 0 \quad \text{if} \quad |\xi| \leq \frac{2\pi}{3} \quad \text{or} \quad |\xi| \geq \frac{8\pi}{3}. \tag{5.21}$$

The corresponding orthonormal basis is $\psi_{j,k}(t) = 2^{j/2}\psi(2^j t - k)$, $j, k \in \mathbb{Z}$. We then observe that the Fourier transform of $\psi_{j,k}$ is $2^{-j/2}e^{-ik2^{-j}\xi}\hat{\psi}(2^{-j}\xi)$, so it is natural to label $\psi_{j,k}$ by the Heisenberg boxes $R_{j,k}^{\varepsilon}$ defined as

$$R_{j,k}^{\varepsilon} = \{(t, \xi) \mid k2^{-j} \leq t < (k+1)2^{-j},\ \pi 2^j \leq \varepsilon \xi \leq 2\pi 2^j\},$$

where $\varepsilon = \pm 1$.

However, this labeling is not consistent with the definition of the Wigner–Ville transform of $\psi_{j,k}$. Indeed, the Wigner–Ville transform of $\psi_{j,k}$ is not supported by $R_{j,k}^+ \cup R_{j,k}^-$. This is clear in the time domain, but of course the rapid decay of $\psi_{j,k}$ is a substitute for the lack of compact support. The situation in the frequency domain is more serious. Here, the transform $W_{j,k}(t, \xi)$ of $\psi_{j,k}$ is supported on $|\xi| \leq \frac{8\pi}{3}$, but it takes large values on $|\xi| \leq \frac{2\pi}{3}$, where it should vanish. These large values are related to the artifacts that prevent the interpretation of $W_{j,k}$ as an energy density. Also, the sums in the fundamental identity

$$\sum_j \sum_k W_{j,k}(t, \xi) = 1 \tag{5.22}$$

converge only in the sense of distributions: The series $\sum_j \sum_k |W_{j,k}(t, \xi)|$ does not converge, but the large oscillations of $W_{j,k}(t, \xi)$ cancel each other in the left-hand side of (5.22). This bad news means that the supports of the $W_{j,k}(t,\xi)$ in the time-frequency plane are not "almost disjoint Heisenberg boxes" as was expected from the classical representations.

5.8 Return to the definition of time-frequency atoms

Consider the following problem: For functions $f \in L^2(\mathbb{R})$ with $\|f\|_2 = 1$, we would like to have a "measure" of how $W(f;t,\xi)$ is distributed in the time-frequency plane. Is $W(f;t,\xi)$ concentrated or is it spread out? Since $|W(f;t,\xi)| \leq 1$ whenever $\|f\|_2 = 1$, the maximum of $W(f;t,\xi)$ is not a good measure. Although f may have most of its energy concentrated in frequencies around ξ_0 near time t_0, $|W(f;t_0,\xi_0)|$ is always bounded by one. Similarly, the Moyal identity (5.13) implies that

$$\iint W^2(f;t,\xi)\,dt\,d\xi = 2\pi. \qquad (5.23)$$

This means that the concentration of W in the time-frequency plane cannot be measured in the L^2 norm. The L^1 norm of W is not always finite, and since W is bounded, this is due to the behavior of W at infinity. For a fixed t, $W(f;t,\xi)$ is the Fourier transform of the L^1 function $f(t+\frac{\tau}{2})\overline{f}(t-\frac{\tau}{2})$, and, as such, it is sensitive to the smoothness of f. Thus, if f contains many high frequencies, then we may expect $\iint |W(w;t,\xi)|\,dt\,d\xi$ to be large. For example, if $f = \chi_{[0,1]}$, then $\iint |W(w;t,\xi)|\,dt\,d\xi = +\infty$. In view of these considerations, it is reasonable to ask that the time-frequency atoms $w \in \Omega$ are such that

$$\iint |W(w;t,\xi)|\,dt\,d\xi \leq C \qquad (5.24)$$

for some constant C, uniformly in w. This relation will hold when the time-frequency atoms are derived from a smooth "mother function" ψ by the operations (5.7) through (5.11) in section 5.6. This holds for Daubechies's wavelets, but it is not true for the Haar wavelets because they are not smooth.

With this motivation, we are ready to give a more precise definition of time-frequency atoms: *A collection Ω of functions w is a collection of time-frequency atoms if the finite linear combinations of the functions $w \in \Omega$ ($\|w\|_2 = 1$) are dense in $L^2(\mathbb{R})$, and there exists a constant C such that (5.24) holds uniformly for $w \in \Omega$.*

5.9 The Wigner–Ville transform and instantaneous frequency

In Ville's fundamental work, which has essentially been the source for this chapter, he makes a careful distinction between the instantaneous frequency of a signal (assumed to be real) and the instantaneous spectrum of frequencies given by the Wigner–Ville transform. More precisely, let f be a real-valued signal with finite energy. Then Ville writes $f(t) = \operatorname{Re} F(t)$, where F is the corresponding analytic signal: $F(t)$ is the restriction to the real axis of a function $F(z)$ that is holomorphic in the upper half-plane $\operatorname{Im} z > 0$ and belongs to the Hardy space $H^2(\mathbb{R})$. Ville writes

$$F(t) = A(t)e^{i\varphi(t)}, \qquad (5.25)$$

where $A(t)$ is the modulus of $F(t)$ and $\varphi(t)$ is its argument. (If F vanishes at some [isolated] points, one needs to add conditions to preserve smoothness of the functions A and φ, but we will set aside this issue for the moment.) Ville then defines the instantaneous frequency of f at t to be

$$\omega = \frac{1}{2\pi} \frac{d}{dt} \varphi(t), \tag{5.26}$$

and the local pseudoperiod is its inverse $2\pi/\varphi'(t)$. The idea we wish to capture is that of a slowly varying envelope $A(t)$ inside which the "true" oscillations are modeled by $e^{i\varphi(t)}$. For this model to make sense, we need to introduce conditions to ensure that the variations in $A(t)$ do not interfere with the determination of the instantaneous frequency. Specifically, we require that $A(t)$ change very little on a scale given by the local pseudoperiod $2\pi/\varphi'(t)$. We express this semiquantitatively by the relation

$$|(\log A(t))'| \ll |\varphi'(t)|, \quad \text{or} \quad \left| \frac{A'(t)}{A(t)\varphi'(t)} \right| \ll 1. \tag{5.27}$$

Furthermore, the pseudoperiod should vary slowly:

$$\left| \left(\frac{1}{\varphi'(t)} \right)' \right| \ll 1, \quad \text{or} \quad \left| \frac{\varphi''(t)}{(\varphi'(t))^2} \right| \ll 1. \tag{5.28}$$

These two conditions are precisely those given for the definition of a *chirp* in signal processing (see [49]).

In the same spirit, the instantaneous spectrum of f is defined as the Wigner–Ville transform of the analytic signal F. Ville discovered a beautiful relation between these two concepts. In fact, the instantaneous frequency is the weighted average of the frequency when the weight is the instantaneous spectrum:

$$\frac{1}{2\pi} \int_{-\infty}^{\infty} \xi W(t,\xi) \, d\xi = \varphi'(t)|F(t)|^2 = \varphi'(t) \left(\frac{1}{2\pi} \int_{-\infty}^{\infty} W(t,\xi) \, d\xi \right). \tag{5.29}$$

If $W(t,\xi)$ were nonnegative, then $\frac{1}{2\pi} W(t,\xi)|F(t)|^2$ would be a probability density, and the instantaneous frequency would be the expectation of the frequency ξ with respect to this probability density. As many authors have stressed (see, for example, [112]), the definition of instantaneous frequency proposed by Ville works well only for very special signals. These are the asymptotic signals whose precise analytic definition is given by $f_\lambda(t) = a(t)\cos(\lambda\varphi(t))$, where a and φ are regular, real-valued functions of time and where λ is a large parameter, which, mathematically speaking, tends to infinity. We will write $\lambda \gg 1$ to express this notion. The application of Ville's program to asymptotic signals will be studied in the following sections. We will show, in contrast to what Ville thought, that the time-frequency atoms adapted to this analysis are not the Gabor wavelets. In fact, they are not the Malvar–Wilson wavelets discussed in Chapter 6; they are the *chirplets* introduced for this purpose by Richard Baraniuk and Steve Mann. We will return to this in Chapter 6.

We stress here, however, that the definition of instantaneous frequency given by Ville loses all sense when the signal f_λ consists of two spectral lines, that is, when

$$f_\lambda(t) = a_1(t)\cos(\lambda\varphi_1(t)) + a_2(t)\cos(\lambda\varphi_2(t)), \tag{5.30}$$

where $\lambda \gg 1$. In this case, if we follow Ville, we are led to assign a single instantaneous frequency to f_λ, which is absurd. Later, in section 5.13, we will look at what the instantaneous spectrum $\xi \mapsto W(f_\lambda; t, \xi)$ provides in this case.

5.10 The Wigner–Ville transform of asymptotic signals

The purpose of this section is to show exactly how well one can determine the instantaneous frequency of a signal f (as defined by (5.26)) from the Wigner–Ville transform of the signal. We wish to answer this question: To what extent does the Wigner–Ville transform reveal the instantaneous frequency of a signal f? As indicated in the last section, Ville's definition of instantaneous frequency works well only for asymptotic signals. Thus our analysis is limited to signals of the form

$$f_\lambda(t) = A(t)\cos(\lambda\varphi(t)), \tag{5.31}$$

where A and φ are real-valued and belong to C^∞, and where $\varphi'(t) \geq 1$. This last assumption, which has not been mentioned, will play an essential role. To simplify the computations, we will also assume that A is in the Schwartz class $\mathcal{S}(\mathbb{R})$. We will study the asymptotic behavior as $\lambda \to +\infty$. (For readers unfamiliar with this kind of analysis, it is perhaps useful to visualize the function as oscillating within the envelope A. By letting λ become large, the influence of the envelope on the "frequency" becomes negligible.)

With these assumptions, it is possible to show that the analytic signal associated with f_λ, namely, $F_\lambda = (I + iH)f_\lambda$, has the form

$$F_\lambda(t) = A(t)e^{i\lambda\varphi(t)} + R_\lambda(t), \tag{5.32}$$

where $R_\lambda(t) = O(\lambda^{-N})$ for all $N \geq 1$. It goes without saying that if we change φ to $-\varphi$ the original signal is unchanged. Thus, if the assumption about φ was $\varphi'(t) \leq -1$, the corresponding analytic signal would be

$$F_\lambda(t) = A(t)e^{-i\lambda\varphi(t)} + R_\lambda(t).$$

Ville's definition of the instantaneous frequency gives

$$\omega = \lambda\varphi'(t) + O(\lambda^{-N})$$

if $A(t) \neq 0$, which agrees with our intuition. As a consequence of Rouché's theorem, the frequencies appearing in the analytic signal must have a positive average, and hence we come back to the constraint $\varphi'(t) \geq 1$. If $A(t_0) = 0$, where t_0 is an isolated zero of A, computing the instantaneous frequency according to Ville no longer makes sense. However, it is natural to compute the instantaneous frequencies at neighboring points and to pass to the limit, defining the instantaneous frequency at t_0 as $\omega = \lim_{t \to t_0} \lambda\varphi'(t) = \lambda\varphi'(t_0)$. On the other hand, if $A(t) = 0$ in a neighborhood of t_0, then this computation no longer makes sense, since the signal does not exist; it has vanished.

We are now going to compare this direct approach with the analysis of the analytic signal F_λ using the Wigner–Ville transform. We wish to determine if this Wigner–Ville transform is essentially concentrated on the curve Γ that is defined by $\xi = \lambda\varphi'(t)$ in the time-frequency plane.

The Wigner–Ville transform of F_λ is given by the oscillatory integral

$$W(t,\xi) = \int_{-\infty}^{\infty} \alpha(t,\tau)e^{i(\lambda\beta(t,\tau) - \xi\tau)}d\tau + O(\lambda^{-N}), \tag{5.33}$$

where

$$\alpha(t,\tau) = A\left(t + \frac{\tau}{2}\right)A\left(t - \frac{\tau}{2}\right),$$

and where
$$\beta(t,\tau) = \varphi\left(t+\frac{\tau}{2}\right) - \varphi\left(t-\frac{\tau}{2}\right).$$

We used the asymptotic expansion (5.32), and we wish to find the asymptotic behavior of $W(t,\xi)$ when λ is large. For this, we will use the stationary phase method.

To simplify the discussion, we assume that $\varphi'(t)$ is strictly convex on the whole real line and that $\lim_{|t|\to\infty} \varphi'(t) = +\infty$. The stationary phase method proceeds by supposing that $\xi = \lambda p$ where p is a constant and by solving the equation

$$p = \frac{1}{2}\left(\varphi'\left(t+\frac{\tau}{2}\right) + \varphi'\left(t-\frac{\tau}{2}\right)\right) \qquad (5.34)$$

for τ when t and p are fixed. It is necessary to distinguish three separate cases:

(a) If $p > \varphi'(t)$, then (5.34) has two solutions τ and $-\tau$, and this leads to an asymptotic expansion whose dominant term is $O(\lambda^{-1/2})$, which will be explained in a moment.

(b) If $p = \varphi'(t)$, the unique solution of (5.34) is $\tau = 0$, and the dominant term is $O(\lambda^{-1/3})$.

(c) If $p < \varphi'(t)$, the dominant term is $O(\lambda^{-N})$ for all $N \geq 1$.

In the first case, the dominant term of the asymptotic expansion is

$$4\lambda^{-1/2} B(t,\tau) \cos\left\{\lambda\left(\varphi\left(t+\frac{\tau}{2}\right) - \varphi\left(t-\frac{\tau}{2}\right)\right) + \frac{\pi}{4}\right\}, \qquad (5.35)$$

where τ is defined by (5.34) and where

$$B(t,\tau) = A\left(t+\frac{\tau}{2}\right) A\left(t-\frac{\tau}{2}\right) \left|\varphi''\left(t+\frac{\tau}{2}\right) - \varphi''\left(t-\frac{\tau}{2}\right)\right|^{-1/2}.$$

As often happens in applied mathematics, we have solved an academic problem: ξ and λ tend to infinity simultaneously while $\xi\lambda^{-1} = p$ is constant. The real problem is different: $\lambda \gg 1$ is fixed and (t,ξ) ranges over the time-frequency plane. In this case, the situation is quite different, and it is not discussed in the classic texts. In fact, the three cases (a), (b), and (c) must be modified. The new regimes (a'), (b'), and (c') are defined by

(a') $\xi > \lambda\varphi'(t) + \lambda^{1/3}$; (b') $|\xi - \lambda\varphi'(t)| \leq \lambda^{1/3}$; (c') $\xi < \lambda\varphi'(t) - \lambda^{1/3}$.

In the first case, the asymptotic term (5.35) can be used. One observes that $|\tau| \geq c\lambda^{-1/3}$, where $c > 0$ is a constant. This implies that $|B(t,\tau)| \leq C\lambda^{1/6}$, and we have $|W(t,\xi)| \leq C'\lambda^{-1/3}$. In the second case, $|W(t,\xi)|$ is of the order $\lambda^{-1/3}$. Finally, in the third case, we have

$$W(t,\xi) = \lambda^{-1/3} \omega[\lambda^{-1/3}(\xi - \lambda\varphi'(t))], \qquad (5.36)$$

where $|\omega(x)| \leq C_N(1+|x|)^{-N}$ for all integers $N \geq 1$. These three behaviors agree at the boundaries of the three regions.

Here is a simple example where the three regimes arise. If $f(t) = \exp(i\lambda t^3)$, $\lambda \geq 1$, then the Wigner–Ville transform of f is

$$W(t,\xi) = 2\pi \left(\frac{4}{3\lambda}\right)^{1/3} A\left[\left(\frac{4}{3\lambda}\right)^{1/3}(\xi - 3\lambda t^2)\right], \qquad (5.37)$$

where A is the Airy function defined by

$$A(\xi) = \frac{1}{2\pi} \int_{-\infty}^{\infty} e^{i(\xi s - s^3/3)} ds. \tag{5.38}$$

The Airy function decreases exponentially as $\xi \to -\infty$ and oscillates within an envelope of order $O(|\xi|^{-1/4})$ as $\xi \to +\infty$. In this case, one finds all three regimes $\xi > \lambda \varphi'(t) + \lambda^{1/3}$, $|\xi - \lambda \varphi'(t)| \leq \lambda^{1/3}$, and $\xi < \lambda \varphi'(t) - \lambda^{1/3}$ as indicated in the general discussion.

The conclusion is this: The investigation of the large values taken by the modulus of the Wigner–Ville transform of an asymptotic signal $f(t) = A(t)e^{i\lambda\varphi(t)}$ does not allow one to isolate the instantaneous frequency $\xi = \lambda \varphi'(t)$ with a precision better than $\lambda^{1/3}$. The best one can hope to obtain is $|\xi - \lambda \varphi'(t)| \leq \lambda^{1/3}$.

5.11 Instantaneous frequency and the matching pursuit algorithm

Mallat's matching pursuit algorithm provides a third approach to the instantaneous frequency. This is the reason: If ω_0 is the instantaneous frequency of a signal f when $t = t_0$, this means that there is a "confidence interval" $[t_0 - h, t_0 + h] = I$ in which the analytic signal F associated with f behaves like $A_0(t) \exp(i\omega_0(t_0 - t))$ where $A_0(t)$ is a regular function of the auxiliary variable $s = \frac{t-t_0}{h}$. This assumes that $\omega_0 h \gg 1$. A function behaving this way is strongly correlated with a Gabor wavelet

$$w_{h,\omega}(t) = h^{-1/2} g\left(\frac{t-t_0}{h}\right) e^{i\omega(t-t_0)}$$

whose average frequency ω is near the instantaneous frequency ω_0 that one is trying to evaluate.

The Mallat algorithm amounts to optimizing $|\langle F, w_{h,\omega}\rangle|$ over $h > 0$ and $\omega \in \mathbb{R}$. The hope is that the pair (h_0, ω_0) where the maximum is attained would provide the length of the confidence interval and the instantaneous frequency. However, by going back to the case of asymptotic signals, it is easy to show that this approach yields a value h_0 of the order $\lambda^{-1/2}$. Thus the precision with which ω is determined is no better than $\lambda^{1/2}$, which is less precise than that given by the algorithm based on the Wigner–Ville transform.

To bring the performance of the matching pursuit algorithm up to that of the Wigner–Ville transform, it is necessary to enlarge the collection of time-frequency atoms. We define Ω as the set of *linear chirps*. These are the functions of the form

$$w(t) = h^{-1/2} g\left(\frac{t-t_0}{h}\right) \exp(i[\alpha(t-t_0) + \beta(t-t_0)^2]), \tag{5.39}$$

where α, β, and t_0 are three arbitrary real numbers and where $h > 0$ is also arbitrary. The function g is still the Gaussian $g(t) = \pi^{-1/4} e^{-t^2/2}$.

Applying the Mallat algorithm using this extended collection of time-frequency atoms, one finds, for t_0 fixed, an optimal value $h = \lambda^{-1/3}$. This value is much larger than the value $h = \lambda^{-1/2}$ that is obtained when the atoms are limited to the Gabor wavelets. At the same time, with this extended set of atoms, the frequency resolution in this case is $O(\lambda^{1/3})$ rather than $O(\lambda^{1/2})$.

5.12 Matching pursuit and the Wigner–Ville transform

We are going to interpret the matching pursuit algorithm in terms of the Wigner–Ville transform. More precisely, we start with the Moyal identity,

$$|\langle f, w \rangle|^2 = \frac{1}{2\pi} \iint W(f; t, \xi) W(w; t, \xi) \, dt \, d\xi. \tag{5.40}$$

If w is defined by (5.39), then $W(w; t, \xi)$ is essentially the characteristic function of the oblique Heisenberg box B defined by

$$|t - t_0| \leq h, \qquad |\xi - [\alpha + 2\beta(t - t_0)]| \leq \frac{\pi}{2h}. \tag{5.41}$$

One can then write w_B in place of w because the definition of B provides all of the parameters used to define w.

Thus, to optimize $|\langle f, w \rangle|^2$, one must skew B so that $W(f; t, \xi)$ is, on average, as large as possible on B. But $W(f; t, \xi)$ attains its maximum when $|\xi - \lambda\varphi'(t)| \leq \lambda^{1/3}$. This leads to an oblique Heisenberg box that is aligned with the instantaneous frequency of the signal; it is defined by

$$\alpha = \lambda\varphi'(t_0), \quad \beta = \frac{\lambda}{2}\varphi''(t_0), \quad \text{and} \quad h = \lambda^{-1/3}.$$

These choices have a couple of explanations. A purely geometric explanation is furnished by the following problem: Fix t_0, and let $\alpha, \beta \in \mathbb{R}$ and $h > 0$ vary arbitrarily. Find the largest value of h, and the corresponding pair (α, β), such that the Heisenberg box B defined by these parameters contains the arc defined by

$$|t - t_0| \leq h, \qquad \xi = \lambda\varphi'(t).$$

Figure 5.3 illustrates this problem. It is not difficult to see that the solution to this problem is again given by the orders of magnitude found before, namely, $h = \lambda^{-1/3}$ and the slope of the Heisenberg box B is $\lambda\varphi''(t_0)$.

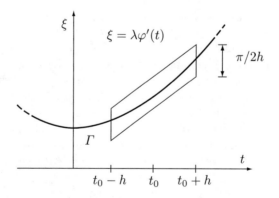

FIG. 5.3. *Γ and an oblique Heisenberg box.*

We will show in Chapter 6 that the search for the optimal decomposition of the signal f in an adapted modulated Malvar–Wilson basis comes down to finding the optimal covering of the graph Γ of $\xi = \lambda\varphi'(t)$ by oblique Heisenberg boxes B. In a practical, nonacademic situation, one studies the signal f on a given interval $[-T, T]$. In this case, the optimal covering of Γ by oblique Heisenberg boxes B also can be described as the covering that minimizes the number of boxes used.

5.13 Several spectral lines

All of the preceding discussion is based on the fundamental assumption that

$$f(t) = A(t)\cos(\lambda\varphi(t)),$$

where A and φ are real-valued, regular functions and where $\lambda \gg 1$ is a large parameter. The case

$$f(t) = A(t)\cos(\lambda\varphi(t)) + B(t)\sin(\lambda\varphi(t))$$

reduces to the former one. If we define $\alpha(t)$ by

$$\begin{cases} A(t) = \sqrt{A^2+B^2}\cos\alpha(t), \\ B(t) = \sqrt{A^2+B^2}\sin\alpha(t), \end{cases}$$

then $f(t) = \sqrt{A^2+B^2}\cos(\lambda\theta(t))$, where $\theta(t) = \varphi(t) - \lambda^{-1}\alpha(t)$ is as regular as φ.

On the other hand, consideration of the finite sum

$$f(t) = A_1\cos(\lambda\varphi_1(t)) + \cdots + A_n\cos(\lambda\varphi_n(t)), \tag{5.42}$$

where $\varphi_1, \ldots, \varphi_n, A_1, \ldots, A_n$ are smooth, real-valued functions, is a step in another direction. Here we speak of "several spectral lines," and the task of the algorithm we wish to describe is to extract each of these spectral lines from the noisy signal $f + \sigma z$ where z is a white noise and $\sigma > 0$ is a small parameter.

As Patrick Flandrin has explained in [112], looking for *the* instantaneous frequency of a signal having several spectral lines does not make sense, and the search must be abandoned.

If we again assume that $\varphi_1'(t) \geq 1, \ldots, \varphi_n'(t) \geq 1$, the analytic signal F associated with f is

$$F(t) = A_1(t)e^{i\lambda\varphi_1(t)} + \cdots + A_n(t)e^{i\lambda\varphi_n(t)} + O(\lambda^{-N}), \tag{5.43}$$

where $N \geq 1$ is arbitrary. Then the Wigner–Ville transform of F can be written

$$\sum_{j=1}^{n} W_j(t,\xi) + \sum\sum_{1\leq j<k\leq n} W_{j,k}(t,\xi), \tag{5.44}$$

where $W_j(t,\xi)$ is the Wigner–Ville transform of $A_j(t)e^{i\lambda\varphi_j(t)}$ and the $W_{j,k}(t,\xi)$ represent the "cross terms."

From what we have seen in section 5.10, we know that $W_j(t,\xi)$ is "essentially" concentrated on the curvilinear band $|\xi - \lambda\varphi_j'(t)| \leq \lambda^{1/3}$. This first part of the discussion leads us to assume that these bands are disjoint.

By doing computations based on the stationary phase method, one can show that the "cross terms" $W_{j,k}(t,\xi)$ are "concentrated" in the bands

$$\left|\xi - \frac{\lambda}{2}(\varphi_j'(t) + \varphi_k'(t))\right| \leq \lambda^{1/3}.$$

These cross terms play the role of artifacts and should be eliminated. They act like noise in the image of the signal represented in the time-frequency plane. To eliminate this noise, one takes advantage of the fact that these cross terms oscillate as a function of time. The farther the curves Γ_j (defined by $\xi = \lambda \varphi_j'(t)$) are separated from each other, the greater will be these oscillations. These parasitic terms are eliminated by appropriately averaging the Wigner–Ville transform $W(t, \xi)$ of F. One can prove that this averaging algorithm is equivalent to using Mallat's algorithm, as in the last section. Of course, this averaging entails a loss of localization, and in practice, the Wigner–Ville transform is used mainly to detect a single spectral line in the presence of noise. This is precisely the setting for the detection of gravitational waves. We will say more about gravitational waves in section 6.11.

5.14 Conclusions

We currently have several tools for doing time-frequency analysis. These tools are of three kinds: (a) the Wigner–Ville transform, (b) Mallat's matching pursuit algorithm, and (c) the best-basis algorithms of Coifman and Wickerhauser, which we will discuss in Chapter 6.

The scientific problem that has motivated the development of these algorithms is the search for an "instantaneous Fourier transform" or an "instantaneous frequency," or for an optimal decomposition of the signal in time-frequency atoms, or finally for an optimal representation of the signal in the time-frequency plane.

Today we are faced with a paradox. The three algorithms (a), (b), and (c) provide three responses to the scientific problem. However, we cannot decide if these responses are pertinent to the problem since, in fact, the scientific problem has no precise meaning.

Even if we stay within the context of the Wigner–Ville transform, there are an infinite number of choices because for certain applications it is necessary to "smooth" this transform either in time, in frequency, or in both variables simultaneously. These smoothings "erase" the undesirable artifacts (for example, the cross terms that appear in (5.44)). The choices for the windows used for smoothing and the sizes of these windows will clearly depend on the signal being studied, and we do not now have an algorithm that leads to objective choices.

The situation is even worse. The Wigner–Ville transform is one among a vast collection of quadratic transforms known as the Cohen class. If the Wigner–Ville transform ideally compresses the linear chirps of the form $f(t) = \exp(i(\alpha t + \beta t^2))$, $\alpha, \beta \in \mathbb{R}$, one can hope to have a quadratic transform that ideally compresses *hyperbolic chirps*, which are the signals of the form $f(t) = \exp(i\lambda \log(t))$, λ real. This problem is treated in [112], and one finds there the classical vicious circle: The analytic tool depends on the a priori information one has. If one wishes to analyze the sounds that bats emit, which are essentially hyperbolic chirps, then the Wigner–Ville transform is probably not the optimal tool.

The other two algorithms are just as nonobjective. It goes without saying that Mallat's matching pursuit algorithm depends critically on the choice of time-frequency atoms in Ω, and similarly the Coifman–Wickerhauser algorithm depends on the bases one chooses for the library of bases.

Thus it appears that things are in a state of disorder and confusion. The asymptotic signals have served as a test case to clarify the relations between the different algorithms.

5.15 Historical remarks

Eugene Wigner was motivated by problems in quantum mechanics when he introduced what is called the Wigner transform of a function ψ [260]. J. E. Moyal elaborated on Wigner's work, proved the identity (5.12) that bears his name, and obtained Theorem 5.1 [214]. The connection between signal processing and quantum mechanics was discovered by Jean Ville [254, p. 65]. He tells us that "la fréquence est, à proprement parler, un opérateur," and of course he meant the operator $\frac{1}{2\pi i}\frac{d}{dt}$. N. G. de Bruijn [76, p. 59] also noted the connection between signal processing and quantum mechanics:

> Both in music and in quantum mechanics we have the situation of a function of a single variable, which appears to be a function of two variables as long as the observation is not too precise. The parallel between quantum mechanics and music can be carried a little further by comparing the composer to the classical physicist. The way the composer writes an isolated note as a dot, and thinks of it as being completely determined in time and frequency, is similar to the classical physicist's conception of a particle with a well-determined position and momentum.

These quotations are to support our observations concerning time-frequency analysis versus time-scale analysis: The former is a result of cross-fertilization between signal processing and quantum mechanics, and mathematicians took little interest in these ideas until recently. The latter was pioneered by mathematicians long before it was adopted in physics and signal processing.

The situation is different today, and time-frequency analysis is widely used in mathematics under the name of microlocal analysis and its variants. We mention, as examples, the theory of wave packets by Cordoba and Fefferman [67] and the Fourier–Bros–Iagolnitzer transform [77]. The Cordoba–Fefferman wave packets generalize Gabor wavelets. In the n-dimensional case they are defined by

$$g_{\xi_0, x_0}(x) = e^{i\xi_0 \cdot x} h^{-n/2} g\left(\frac{x - x_0}{h}\right),$$

where $g(x) = e^{-|x|^2/2}$ and $h = |\xi_0|^{1/2}$. One is interested in the asymptotics as h tends to zero. Cordoba and Fefferman then study the action of Fourier integral operators on such wave packets. This action mimics the well-known action on Gabor wavelets of the unitary group generated by the harmonic oscillator. The Cordoba–Fefferman wave packets motivated the construction of wavelet packets by Coifman, Meyer, and Wickerhauser. Anticipating the discussion in Chapter 7, we note that the orthonormal basis for $L^2(\mathbb{R})$ consisting of the functions $w_0(x - k)$, $k \in \mathbb{Z}$, together with $2^{j/2} w_n(2^j x - k)$, $2^j \leq n < 2^{j+2}$, $j \geq 0$, $n \geq 1$, $k \in \mathbb{Z}$, mimics the Cordoba–Fefferman wave packets. If we accept that $w_n(x - k)$, $k \in \mathbb{Z}$, is centered around the frequency n, then the "main frequency" of $w_n(2^j - k)$ will be $4^j = h^{-2}$, where h is the "length" of $w_n(2^j - k)$.

These remarks illustrate the interactions between quantum mechanics, signal processing, and, eventually, mathematics.

CHAPTER 6

Time-Frequency Algorithms Using Malvar–Wilson Wavelets

6.1 Introduction

This chapter continues the time-frequency analysis of Chapter 5. We will introduce algorithms that allow us to decompose a given signal s into a linear combination of time-frequency atoms. The time-frequency atoms that we use are denoted by f_R and are "coded" by Heisenberg rectangles R with sides parallel to the axes and with area 1 or 2π, depending on the normalization. If $R = [a, b] \times [\alpha, \beta]$, we require that the function f_R be essentially supported on the interval $[a, b]$ and that its Fourier transform $\hat{f}_R$ be essentially supported on $[\alpha, \beta]$ and the opposite frequencies $[-\beta, -\alpha]$. We also want the algorithmic structure of f_R to be simple and explicit to facilitate numerical processing in real time. The decomposition

$$s(t) = \sum_{j=0}^{\infty} \alpha_j f_{R_j}(t) \tag{6.1}$$

cannot be unique, and we take advantage of this flexibility by looking for optimal decompositions, which for our purposes means that they contain the fewest possible terms.

The point of view of Ville (and of numerous other signal-processing experts) is that it is first necessary to understand the physics of the process and that "the algorithms will follow." A careful reading of Ville's fundamental paper [254] suggests the following algorithm for finding the optimal decomposition (6.1): (1) Compute the Wigner–Ville transform $W(t, \xi)$ of f; (2) define the domains Ω_j of the time-frequency plane by $2^{-j-1} \leq |W(t, \xi)| < 2^{-j}$, $j \geq 0$; and (3) optimally cover Ω_j with Heisenberg boxes $R_{j,k}$. One should then use these boxes to write the optimal decomposition (6.1). This program appears unrealistic, and one of the main objections is this: The domains Ω_j may have complicated structures, and thus the Heisenberg boxes may provide poor coverings for the Ω_j. This situation can be improved if the set of horizontal and vertical Heisenberg boxes is enlarged to include oblique boxes, but this means that we will need other time-frequency atoms. These new atoms will be introduced in section 6.11.

For the time being we will be less ambitious and stay with horizontal and vertical Heisenberg boxes. The time-frequency atoms that we use are completely explicit. They are either Malvar–Wilson wavelets or wavelet packets, and we will immediately write down the atomic decompositions of the type (6.1). This means that the synthesis will be direct, whereas the analysis will consist of choosing—with the use of an entropy criterion—the most effective synthesis, which is the one that leads to optimal compression. Thus the analysis proceeds according to algorithmic criteria and not according to physics, and it is not at all clear that this approach leads to a signal analysis that reveals physical properties having real significance. For example, Marie Farge had the idea to apply the algorithm to simulated two-dimensional turbulence. The algorithm extracted various coherent structures, beginning with the larger ones and continuing down the scale to a cutoff point. This shows that coherent structures can be well represented using a few wavelets. However, this remarkable work calls for an interpretation: What does this tell us about coherent structures?

After these general remarks, it is time to specify the algorithms. There are two options: Malvar–Wilson wavelets and wavelet packets. With the first option, the signal is segmented adaptively and optimally, and then the segments are analyzed using classical Fourier analysis. The second option, wavelet packets, reverses the order of these operations: The signal is first filtered adaptively; then the analysis in the time variable is imposed by the algorithm.

Ville proposed two types of analysis [254, p. 64]: "We can either first cut the signal into slices (in time) with a switch and then pass these different slices through a system of filters to analyze them, or we can first filter different frequency bands and then cut these bands into slices (in time) to study their energy variations." The first approach leads to Malvar–Wilson wavelets and the second to wavelet packets. As mentioned above, a third option will be proposed in section 6.11. This option provides a better fit between the Heisenberg boxes R_j, which appear in (6.1), and the level sets Ω_j associated with the Wigner–Ville transform of f.

6.2 Malvar–Wilson wavelets: A historical perspective

The scientific program that led to adaptive Malvar–Wilson wavelets was initiated by the physicist Kenneth Wilson (Nobel laureate in physics, 1982) [261]. These time-frequency wavelets were later discovered independently by the signal processing expert Henrique Malvar [185] (see also [186], [187], [188], and [189]). Malvar–Wilson wavelets fall within the general framework of windowed Fourier analysis. The window is denoted by w, and it allows the signal s to be cut into "slices" that are regularly spaced in time $w(t - bl)s(t)$, $l \in \mathbb{Z}$. The parameter $b > 0$ is the nominal length of these slices. Next, following Ville, one does a Fourier analysis on these slices, which reduces to calculating the coefficients $\int e^{-iakt} w(t - bl) s(t) dt$, where $a > 0$ must be related to b and where $k \in \mathbb{Z}$. This is thus the same as taking the scalar products of the signal s with the "wavelets"

$$w_{k,l}(t) = e^{iakt} w(t - bl).$$

This analysis technique was proposed by Gabor [124], in which case the w was the Gaussian. The Gabor wavelets lead to serious algorithmic difficulties. More generally, Low and Balian showed in the early 1980s that if w is sufficiently regular and well localized, then the functions $w_{k,l}$, $k, l \in \mathbb{Z}$, can never be an orthonormal basis for $L^2(\mathbb{R})$ [17]. More precisely, if the two integrals $\int_{\mathbb{R}} (1 + |t|)^2 |w(t)|^2 dt$ and $\int_{\mathbb{R}} (1 + |\xi|)^2 |\hat{w}(\xi)|^2 d\xi$ are both finite, the functions $w_{k,l}$, $k, l \in \mathbb{Z}$, cannot be an orthonormal basis of $L^2(\mathbb{R})$.

The crude window defined by $w(t) = 1$ on the interval $[0, 2\pi)$ and $w(t) = 0$ elsewhere escapes this criterion. By choosing $a = 1$ and $b = 2\pi$, the windowed analysis consists of restricting the signal to each interval $[2l\pi, 2(l+1)\pi)$ and using the Fourier transform (in this case, Fourier series) to analyze each of the corresponding functions. But even if one starts with a smooth signal, the functions obtained by this crude segmentation are not the restrictions of smooth 2π-periodic functions, and the Fourier analysis will highlight this lack of periodicity and interpret it as a discontinuity in the signal.

One way to attenuate these numerical artifacts, which does not eliminate them completely, is to use the discrete cosine transform (DCT). We will describe the continuous version of this transform. On each interval $[2l\pi, 2(l+1)\pi)$, the signal $s(t)$ is analyzed using the orthonormal basis composed of the functions $\frac{1}{\sqrt{2\pi}}$ and $\frac{1}{\sqrt{\pi}}\cos\frac{k}{2}t$, $k \in \mathbb{N}^*$. If s is a very regular function, this segmentation introduces discontinuities only in the derivative of the signal, and the numerical artifacts produced by the segmentation are reduced from the order of magnitude $\frac{1}{k}$ to $\frac{1}{k^2}$.

Wilson was the first to have the idea that one could get around the problem presented in the Balian–Low theorem by imitating the DCT and using a segmentation created with very regular windows. Wilson proposed to alternate the DCT with the discrete sine transform (DST) according to whether l is even or odd, where l denotes the position of the interval. The DST uses the orthonormal basis consisting of the functions $\frac{1}{\sqrt{\pi}}\sin\frac{k}{2}t$, $k \in \mathbb{N}^*$.

Wilson's ideas have been the point of departure for numerous efforts, the most notable of which is due to Ingrid Daubechies, Stéphane Jaffard, and Jean-Lin Journé [74]. They used a window w having the property that both it and its Fourier transform decay exponentially, and they constructed w so that the functions $u_{k,l}$, $k \in \mathbb{N}^*$, $l \in \mathbb{Z}$, and $u_{0,l}$, $l \in 2\mathbb{Z}$, defined by

$$u_{k,l}(t) = \sqrt{2}\, w(t - 2l\pi)\cos\frac{k}{2}t, \quad l \in 2\mathbb{Z}, \quad k = 1, 2, \ldots, \tag{6.2}$$

$$u_{0,l}(t) = w(t - 2l\pi), \quad l \in 2\mathbb{Z}, \quad k = 0, \tag{6.3}$$

$$u_{k,l}(t) = \sqrt{2}\, w(t - 2l\pi)\sin\frac{k}{2}t, \quad l \in 2\mathbb{Z}+1, \quad k = 1, 2, \ldots, \tag{6.4}$$

constitute an orthonormal basis for $L^2(\mathbb{R})$. Exponential decay for both w and $\hat{w}$ is an essential requirement for the applications that Wilson had in mind in renormalization theory.

Malvar did not know about Wilson's work. He discovered a family of orthonormal bases $u_{k,l}$ whose algorithmic structure is the same as that described by (6.2), (6.3), and (6.4), but where the choice of the window w is simpler and more explicit. In fact, Malvar had only these hypotheses:

$$w(t) = 0 \quad \text{if} \quad t \leq -\pi \quad \text{or} \quad t \geq 3\pi; \tag{6.5}$$

$$0 \leq w(t) \leq 1 \quad \text{and} \quad w(2\pi - t) = w(t); \tag{6.6}$$

$$w^2(t) + w^2(-t) = 1 \quad \text{if} \quad -\pi \leq t \leq \pi. \tag{6.7}$$

Then the construction is the same, and the sequence $u_{k,l}$ defined by (6.2), (6.3), and (6.4) is an orthonormal basis for $L^2(\mathbb{R})$. In Malvar's construction, the window w can be very regular (infinitely differentiable, for example), but the Fourier transform of w cannot have exponential decay. Condition (6.5) prevents it, and this condition plays an essential role in the proofs.

The Malvar basis can be incorporated into a general framework developed by Daubechies, Jaffard, and Journé. It appears there as a simple example in a systematic construction. It can, however, be developed directly, and in this way Malvar's construction happens to be more flexible than the Daubechies–Jaffard–Journé approach. This remark will become clear in the next section.

6.3 Windows with variable lengths

Coifman and Meyer modified the preceding constructions to create windows with arbitrary, variable lengths [64]. The construction by Daubechies, Jaffard, and Journé does not extend to this context, while that of Malvar generalizes to the case of arbitrary windows without the slightest difficulty.

We begin with an arbitrary partition of the real line into adjacent intervals $[a_j, a_{j+1}]$, where $\ldots < a_{-1} < a_0 < a_1 < a_2 < \ldots$, $\lim_{j \to +\infty} a_j = +\infty$, and $\lim_{j \to -\infty} a_j = -\infty$. Write $l_j = a_{j+1} - a_j$ and let $\alpha_j > 0$ be positive numbers that are small enough so $l_j \geq \alpha_j + \alpha_{j+1}$ for all $j \in \mathbb{Z}$.

The windows w_j that we use will be essentially the characteristic functions of the intervals $[a_j, a_{j+1}]$; the role played by the disjoint intervals $(a_j - \alpha_j, a_j + \alpha_j)$ is to allow the windows to overlap, which is necessary if we want the windows to be regular (Figure 6.1). More precisely, we impose the following conditions:

$$0 \leq w_j(t) \leq 1 \quad \text{for all} \quad t \in \mathbb{R}, \tag{6.8}$$

$$w_j(t) = 1 \quad \text{if} \quad a_j + \alpha_j \leq t \leq a_{j+1} - \alpha_{j+1}, \tag{6.9}$$

$$w_j(t) = 0 \quad \text{if} \quad t \leq a_j - \alpha_j \quad \text{or} \quad t \geq a_{j+1} + \alpha_{j+1}, \tag{6.10}$$

$$w_j^2(a_j + \tau) + w_j^2(a_j - \tau) = 1 \quad \text{if} \quad |\tau| \leq \alpha_j, \tag{6.11}$$

$$w_{j-1}(a_j + \tau) = w_j(a_j - \tau) \quad \text{if} \quad |\tau| \leq \alpha_j. \tag{6.12}$$

Note that these conditions allow the windows w_j to be infinitely differentiable. It is clear that $\sum_{-\infty}^{\infty} w_j^2(t) = 1$, identically on the whole real line.

Finally, we come to the Malvar–Wilson wavelets. They appear in two distinct forms. The first is given by

$$u_{j,k}(t) = \sqrt{\frac{2}{l_j}} w_j(t) \cos\left[\frac{\pi}{l_j}\left(k + \frac{1}{2}\right)(t - a_j)\right], \quad j \in \mathbb{Z}, \quad k \in \mathbb{N}. \tag{6.13}$$

The second form consists of alternating the cosines and sines according to whether j is even or odd. Thus we have three distinct expressions for the second form:

$$u_{j,k}(t) = \sqrt{\frac{2}{l_j}} w_j(t) \cos \frac{k\pi}{l_j}(t - a_j), \quad j \in 2\mathbb{Z}, \quad k = 1, 2, \ldots, \tag{6.14}$$

$$u_{j,k}(t) = \sqrt{\frac{1}{l_j}} w_j(t), \quad j \in 2\mathbb{Z}, \quad k = 0, \tag{6.15}$$

$$u_{j,k}(t) = \sqrt{\frac{2}{l_j}} w_j(t) \sin \frac{k\pi}{l_j}(t - a_j), \quad j \in 2\mathbb{Z} + 1, \quad k = 1, 2, \ldots. \tag{6.16}$$

The functions $u_{j,k}$, $j \in \mathbb{Z}$, $k \in \mathbb{N}$, given by (6.13) are an orthonormal basis for $L^2(\mathbb{R})$, and so are the functions defined by (6.14), (6.15), and (6.16).

Two Malvar–Wilson wavelets of the form (6.13) with $k = 8$ are shown in Figure 6.2. Note the similarity between these wavelets and the time-frequency atoms proposed by Liénard: The Malvar–Wilson wavelets are constructed with an attack (whose duration is $2\alpha_j$), a stationary period (which lasts $l_j - \alpha_j - \alpha_{j+1}$), and then a decay (which lasts $2\alpha_{j+1}$). The ability to choose, arbitrarily and independently, the

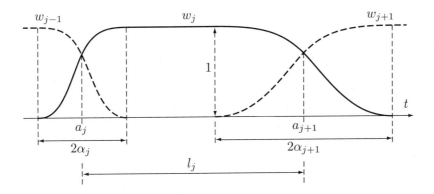

FIG. 6.1. *A typical Malvar window.*

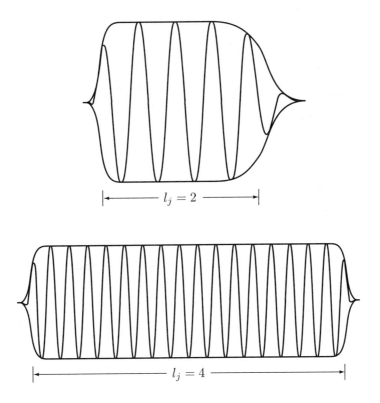

FIG. 6.2. *Two Malvar–Wilson wavelets.*

duration of the attack, then that of the stationary section, and finally the duration of the relaxation is precisely what differentiates the Malvar–Wilson wavelets from the preceding constructions (Gabor or Daubechies–Jaffard–Journé). It is, of course, important to make good use of the choices at our disposal, and we will see how to do this in the following sections.

6.4 Malvar–Wilson wavelets and time-scale wavelets

In 1985, Yves Meyer constructed a function ψ belonging to the Schwartz class $\mathcal{S}(\mathbb{R})$ such that $2^{j/2}\psi(2^j t - k)$, $j, k \in \mathbb{Z}$, is an orthonormal basis for $L^2(\mathbb{R})$. In addition, the Fourier transform of ψ is zero outside the intervals $\left[-\frac{8\pi}{3}, -\frac{2\pi}{3}\right]$ and $\left[\frac{2\pi}{3}, \frac{8\pi}{3}\right]$. We will see that these wavelets $2^{j/2}\psi(2^j t-k)$, $j, k \in \mathbb{Z}$, constitute a particular case of the general Malvar construction. This is quite surprising because the Lemarié–Meyer wavelets constitute a time-scale algorithm, whereas the Malvar–Wilson wavelets are a time-frequency algorithm. There is thus an apparent incompatibility. In fact, it is by analyzing the Fourier transform $\hat{f}$ of an arbitrary function f in an appropriate Malvar–Wilson basis that we arrive at the analysis by Lemarié–Meyer wavelets.

We begin with the following observation: The Malvar–Wilson wavelets let us analyze functions defined on a half-line. The segmentation of $(0, \infty)$ we use is the "natural" division into dyadic intervals $[2^j, 2^{j+1}]$, $j \in \mathbb{Z}$. Then it is natural to choose the windows w_j, associated with these intervals, to be of the form $w_j(x) = w(2^{-j}x)$. Thus the whole construction rests on the precise choice for the function w. For this, we make the following choices in accordance with conditions (6.8)–(6.12): $w(x) = 0$ outside the interval $\left[\frac{2}{3}, \frac{8}{3}\right]$, $w(2x) = w(2-x)$ for $\frac{2}{3} \leq x \leq \frac{4}{3}$, and $w^2(x) + w^2(2-x) = 1$ on the same interval. Then $a_j = 2^j$, $\alpha_j = \frac{1}{3}2^j$, and $l_j = 2^j = \alpha_j + \alpha_{j+1}$. This is illustrated in Figure 6.3.

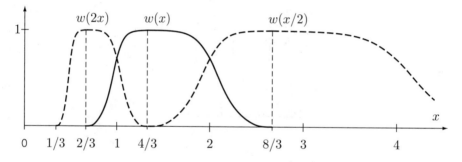

FIG. 6.3. *Malvar windows of the form $w_j(x) = w(2^{-j}x)$.*

Using these parameters, the Malvar–Wilson wavelets of type (6.13) are, up to an irrelevant power of -1,

$$u_{j,k}(x) = \sqrt{2}\, 2^{-j/2} w(2^{-j}x) \sin\left[\pi\left(k + \frac{1}{2}\right) 2^{-j} x\right]. \tag{6.17}$$

If we replace the cosines in (6.13) by sines, we obtain a second orthonormal basis for $L^2[0, \infty]$ of the form

$$v_{j,k}(x) = \sqrt{2}\, 2^{-j/2} w(2^{-j}x) \cos\left[\pi\left(k + \frac{1}{2}\right) 2^{-j} x\right]. \tag{6.18}$$

We next extend w to the whole line by making it an even function: $w(-x) = w(x)$. This gives a natural odd extension for the functions $u_{j,k}$ and an even extension for the $v_{j,k}$. Finally, the complete collection of extended functions

$$\left\{\frac{1}{\sqrt{2}} u_{j,k}(x), \frac{1}{\sqrt{2}} v_{j,k}(x);\ j \in \mathbb{Z},\ k = 0, 1, \ldots \right\} \tag{6.19}$$

is an orthonormal basis for $L^2(\mathbb{R})$. It follows that the set of functions $\frac{1}{2}(v_{j,k} - iu_{j,k})$, $\frac{1}{2}(v_{j,k} + iu_{j,k})$ is also an orthonormal basis for $L^2(\mathbb{R})$. Next, we observe that

$$\frac{1}{2}(v_{j,k} - iu_{j,k})(x) = 2^{-(j+1)/2} w(2^{-j}x) e^{i\pi(k+1/2)2^{-j}x} \tag{6.20}$$

and that by letting $k^* = -1 - k$, we have

$$\frac{1}{2}(v_{j,k} + iu_{j,k})(x) = 2^{-(j+1)/2} w(2^{-j}x) e^{i\pi(k^*+1/2)2^{-j}x}. \tag{6.21}$$

The conclusion is that the sequence

$$2^{-(j+1)/2} w(2^{-j}x) \exp\left[i\pi\left(k + \frac{1}{2}\right) 2^{-j}x\right], \qquad j, k \in \mathbb{Z}, \tag{6.22}$$

is an orthonormal basis for $L^2(\mathbb{R})$.

Denote the Fourier transform of the function $\frac{1}{\sqrt{2}} w(x) e^{i\pi x/2}$ by θ. This function is real-valued and satisfies $\theta(\pi - t) = \theta(t)$. Then the sequence

$$\frac{1}{\sqrt{2\pi}} 2^{j/2} \theta(2^j t - k\pi), \qquad j, k \in \mathbb{Z}, \tag{6.23}$$

is an orthonormal basis for $L^2(\mathbb{R})$. By defining $\psi(t) = \frac{1}{\sqrt{2}} \theta(\pi t)$ we regain the usual form, and $2^{j/2} \psi(2^j t - k)$, $j, k \in \mathbb{Z}$, is also an orthonormal basis for $L^2(\mathbb{R})$.

It is clearly possible to require that w be an infinitely differentiable function, in which case ψ will be a function in the Schwartz class $\mathcal{S}(\mathbb{R})$.

Recall the program of Ville. There were two possible approaches: Either segment the signal appropriately and follow this by Fourier analysis, or pass the signal through a bank of filters and then study the individual outputs of the filter banks. Here we have taken the second approach. The filter bank was defined by the transfer functions $w(2^{-j}x)$, where w is the even window used above.

6.5 Adaptive segmentation and the split-and-merge algorithm

From now on, we will give up trying to find an optimal segmentation. Instead, we will only consider a quite specific collection of segmentations and find the optimal segmentation within this collection. This collection will be fixed, and we note that there is no reason to believe that the solution within this collection will be related to the "physics" of the problem. For example, there is no reason to believe that this segmentation of a speech signal will have any relation to objects intrinsic to speech such as phonemes.

We are not going to create the best segmentation all at once. We will modify an existing segmentation to produce a new one, and by iterating this procedure we approach an optimal segmentation. The modification operation is described in this section. A segmentation is modified by adjusting the partition (a_j) that defines the segmentation, and this is done by iterating the following elementary modifications: An elementary modification consists of suppressing a point a_j of the partition; this means that the two intervals $[a_{j-1}, a_j]$ and $[a_j, a_{j+1}]$ are combined into a single interval, namely, $[a_{j-1}, a_{j+1}]$. The other intervals remain unchanged. This operation is called *merging*. The inverse operation consists of adding an extra point α between the points a_j and a_{j+1}, which results in replacing the interval

$[a_j, a_{j+1}]$ by the two intervals $[a_j, \alpha]$ and $[\alpha, a_{j+1}]$. This inverse operation is called *splitting*, but, in fact, we will be using only the merging half of the algorithm. A split-and-merge algorithm provides a criterion to decide when and where to use one or the other of these elementary operations.

We are going to examine the effect of these operations on a Malvar–Wilson basis. We will show that an elementary operation induces an elementary modification of the basis that is easy to calculate. The following observation is the point of departure for this discussion.

For each fixed j, let W_j denote the closed subspace of $L^2(\mathbb{R})$ generated by the functions $u_{j,k}$, $k \in \mathbb{N}^*$ described by (6.13). Then f belongs to W_j if and only if $f(t) = w_j(t)q(t)$, where q belongs to $L^2[a_j - \alpha_j, a_{j+1} + \alpha_{j+1}]$ and satisfies the following two conditions:

$$q(a_j + \tau) = q(a_j - \tau) \quad \text{if} \quad |\tau| \leq \alpha_j,$$
$$q(a_{j+1} + \tau) = -q(a_{j+1} - \tau) \quad \text{if} \quad |\tau| \leq \alpha_{j+1}.$$

There are no conditions that need to be satisfied on the interval $[a_j + \alpha_j, a_{j+1} - \alpha_{j+1}]$.

From here, the merging algorithm is quite simple. Removing the point a_j of the partition amounts to replacing the two subspaces W_{j-1} and W_j by their direct orthogonal sum $W_{j-1} \oplus W_j$ without disturbing any of the other spaces $W_{j'}$, $j' \neq j-1$ and $j' \neq j$. But this, in turn, comes down to replacing the two windows w_{j-1} and w_j by the new window $\tilde{w}_j$ defined by $\tilde{w}_j(t) = (w_{j-1}^2(t) + w_j^2(t))^{1/2}$. The two lengths l_{j-1} and l_j are replaced by $\tilde{l}_j = l_{j-1} + l_j$, which changes the fundamental frequency in (6.13).

We consider a simple example to fix our ideas. Start with a segmentation with intervals of length 1, $a_j = j$, and choose $w_j(t) = w(t - j)$ with $\alpha_j = \frac{1}{3}$. We wish to examine the windows that can appear as a result of the merging algorithm. These windows and the corresponding wavelets will look like centipedes (see the second Malvar–Wilson wavelet in Figure 6.2). The localization of these centipedes in the time-frequency plane is not optimal. This is because, in using the merging algorithm, we never change the values of the numbers α_j. In our example, we always keep $\alpha_j = \frac{1}{3}$.

We must now provide the criterion that allows us to decide when to use the dynamic split-and-merge algorithm. This means that we need to establish a numerical value to measure what is gained or lost by adding or deleting a point in the subdivision. This is the purpose of the next section.

6.6 The entropy of a vector with respect to an orthonormal basis

Let H denote a Hilbert space and let $(e_j)_{j \in J}$ be an orthonormal basis for H. Let x be a vector of H of norm 1 and write $x = \sum_{j \in J} \alpha_j e_j$. The entropy of x relative to the basis e_j is defined by $\exp(-\sum_{j \in J} |\alpha_j|^2 \log |\alpha_j|^2)$. Roughly, this entropy measures the number of significant terms in this decomposition. In information theory it measures the quantity of information needed to store these coefficients. Note that it is minimal in the simplest case where x is one of the e_j, and it becomes large when many of the α_j are of the same order of magnitude.

If we have a collection $(e_j^\omega)_{j \in J}$ of orthonormal bases where ω ranges over a set Ω, we will choose for the analysis of x the particular basis (indexed by ω_0) that yields the minimum entropy. This point of view poses three problems:

(1) Does an optimal basis exist?

(2) It is not clear that a compression algorithm whose only objective is efficiency can also be used for diagnostic tasks.

(3) The underlying energy criterion (the square of the norm in the Hilbert space H) can cause certain information in the signal to be given low priority, and this information can subsequently disappear in the compression even though it may be crucial for the diagnostic.

Until recently, the algorithms used in image analysis were based on an energy function that is defined as the quadratic mean value of the gray levels. The algorithm used to search for an optimal basis for compression does not escape this difficulty. The search for a norm that is better adapted to the structure of images is still an open problem, but some progress has been made using Besov spaces. This will be discussed in Chapter 11.

6.7 The algorithm for finding the optimal Malvar–Wilson basis

We will examine in detail the particular case where the Hilbert space H is the space of signals f with finite energy, which is defined by $\int_{-\infty}^{\infty} |f(t)|^2 dt$. The quality of the compression will be measured only by this L^2 criterion. The algorithm looks for "the best basis"; this is the one that optimizes compression based on the reduction of transmitted data. The search is done by comparing the scores of a whole family of orthonormal bases of $L^2(\mathbb{R})$. These are Malvar–Wilson bases, and they are obtained from segmentations of the real line into dyadic intervals.

The decision to use only dyadic intervals is a poor man's limitation to save search time. Indeed, it would be impossible to scan all Malvar–Wilson bases. Note, however, that the decision to limit the search to dyadic intervals may introduce artifacts. For example, in speech processing, one goal of optimal segmentation is to extract the phonemes. It is clear that phonemes are not subject to the condition that they begin and end on dyadic intervals. It is rather surprising that this limited search for a best basis has proved to be interesting for speech processing [258]. (While on this subject, we mention that X. Fang has developed a segmentation algorithm that can be used to partition a speech signal so that the signal in each segment is "almost" a phoneme. This is not a wavelet algorithm, but once Fang's algorithm is used for preprocessing the signal, a wavelet algorithm can be used to analyze the individual phonemes. This is discussed in [257]. We note that the best-basis algorithm also played a role in the development of the standard for fingerprint compression [41].)

The dyadic intervals are systematically constructed in a scheme that moves from "fine" to "coarse." One begins with a segmentation having intervals of length 2^{-q}, where $q \geq 0$ is large enough to capture the finest details appearing in the signal. By a change of scale, we may assume that $q = 0$. The process consists of removing, if necessary, certain points in the segmentation and in replacing, at the same time, two contiguous dyadic intervals I' and I'' (appearing in the former segmentation) with the dyadic interval $I = I' \cup I''$. The desire to have a fast algorithm dictates that the merge algorithm be limited to situations where $[a_{j-1}, a_j]$ and $[a_j, a_{j+1}]$ are the left (I') and right (I'') halves of a dyadic interval I. For example, $[2, 3]$ and $[3, 4]$ can become $[2, 4]$ with the disappearance of 3, but $[3, 4]$ and $[4, 5]$ can never become $[3, 5]$. For the point 4 to disappear, it would be necessary to wait for the possible merging of the intervals $[0, 4]$ and $[4, 8]$.

Having set $q = 0$, we start with the segmentation where the "fine grid" is $\mathbb{Z}$. The intervals $[a_j, a_{j+1}]$ of section 6.5 are now $[j, j+1]$, and the first orthonormal basis to participate in the competition will be

$$u_{j,k}(t) = \sqrt{2} w(t-j) \cos\left[\pi\left(k+\frac{1}{2}\right)(t-j)\right], \tag{6.24}$$

where $j \in \mathbb{Z}$, $k \in \mathbb{N}$. The other orthonormal bases that participate in the competition will all be obtained from this first one by merging. The algorithm that merges two orthonormal bases into one was described in section 6.5.

Each partition of the real line into dyadic intervals of length greater than or equal to one defines one of the orthonormal bases that are allowed to participate in the competition. One reaches all of these partitions by iterating those elementary operations that combine the left and right halves of a dyadic interval and by traversing this tree structure, starting from the "fine grid" $\mathbb{Z}$.

We will show how the competition proceeds in a moment, but first we establish a handy notation and make some simplifying assumptions.

The collection of all the dyadic intervals I of length $|I| \geq 1$ will be denoted by $\mathcal{I}$, and if $I = [a_j, a_{j+1}]$ is one of these dyadic intervals, w_I denotes the window that was denoted by w_j in section 6.5. In the same way, W_I denotes the closed subspace of $L^2(\mathbb{R})$ that was denoted by W_j; $w_I^{(k)}$ denotes the orthonormal sequence defined by (6.13), which is now an orthonormal basis for W_I. If I' and I'' are, respectively, the left and right halves of the dyadic interval I, then $W_I = W_{I'} \oplus W_{I''}$, and this direct sum is orthogonal.

The signal f that we wish to analyze optimally is normalized by $\|f\|_2 = 1$. To simplify the following discussion, we assume in addition that $f(t)$ is zero outside the interval $[1, T]$ for some sufficiently large T. Then f belongs to W_L if $L = [0, 2^l]$ and l is large enough.

It can be shown that if m tends to infinity, the entropy of f in the orthonormal basis $w_I^{(k)}$ of W_I, $I = [0, 2^m]$, also tends to infinity. Thus there exists some value of m after which $w_I^{(k)}$ no longer enters into the competition. In other words, the dyadic partitions that come into play will, in fact, be the partitions of $L = [0, 2^m]$ (for sufficiently large m) into dyadic intervals I of length $|I| \geq 1$. The number of partitions is thus finite, but it can be incredibly large, the order of magnitude being $2^{(2^m)}$. It remains to find a fast algorithm to search for the "best basis." This is the algorithm that we are now going to describe.

If I belongs to $\mathcal{I}$, then we will write $c_I^{(k)} = \int f(t) w_I^{(k)}(t) dt$,

$$\varepsilon(I) = -\sum_{k=0}^{\infty} |c_I^{(k)}|^2 \log |c_I^{(k)}|^2, \tag{6.25}$$

and

$$\varepsilon^*(I) = \inf \sum_p \varepsilon(J_p), \tag{6.26}$$

where the lower bound is taken over all the partitions (J_p) of the interval I into dyadic intervals J_p belonging to $\mathcal{I}$. If $I = [j, j+1]$, then clearly $\varepsilon^*(I) = \varepsilon(I)$.

The problem that we must solve is thus reduced to finding the optimal partition (J_p) when $I = L = [0, 2^m]$, the largest of the dyadic intervals involved in the competition. The calculation of $\varepsilon^*(L)$ and the determination of the optimal partition cannot be done directly because the number of cases to be considered is too large. We will calculate $\varepsilon^*(I)$ for $|I| = 2^n$ by induction on n. For $n = 0$, we must calculate $\varepsilon^*(I) = \varepsilon(I)$ for all intervals $I = [j, j+1]$ in $[0, 2^m]$. Next we proceed by induction on n, assuming that we have calculated $\varepsilon^*(I)$ for $|I| = 2^n$ and that we have determined the corresponding covering (J_p).

Suppose that $|I| = 2^{n+1}$ and let I' and I'' be the left and right halves of I. There are two cases:

> If $\varepsilon(I) \leq \varepsilon^*(I') + \varepsilon^*(I'')$, keep I and forget all the preceding information about I' and I''; define $\varepsilon^*(I) = \varepsilon(I)$ and the partition of I is the trivial partition consisting of only I.

> If $\varepsilon(I) > \varepsilon^*(I') + \varepsilon^*(I'')$, set $\varepsilon^*(I) = \varepsilon^*(I') + \varepsilon^*(I'')$ and the partition of I is obtained by combining the partitions of I' and I'' that were used to calculate $\varepsilon^*(I')$ and $\varepsilon^*(I'')$.

Arriving at the "summit of the pyramid," that is to say, at L, we expect to have found the minimal entropy and the optimal partition of L, which leads to the optimal basis.

We have just described the dyadic version of the optimal basis search, but as we indicated above, the restriction to certain dyadic intervals is unrealistic. A translation-invariant algorithm that avoids this restriction has been developed by Coifman and Donoho [62].

6.8 An example where this algorithm works

Consider a signal $f(t) = g(t) + \frac{1}{\sqrt{h}} e^{i\omega t} g\left(\frac{t-t_0}{h}\right)$, where $g(t) = e^{-t^2/2}$, the real number ω can be arbitrarily large, and $0 < h < 1$. We will be concerned with the limiting situation where h is very small.

If f is analyzed using the Malvar–Wilson basis associated with a regularly segmented grid $(a_j = ja)$, then the entropy of the decomposition is necessarily greater than $C \log \frac{1}{h}$. Indeed, if the grid mesh is of order 1, the term $\frac{1}{\sqrt{h}} e^{i\omega t} g\left(\frac{t-t_0}{h}\right)$ is very poorly represented, whereas if the mesh is of order h, the term $g(t)$ is very poorly represented.

The entropy of a decomposition of f can decrease to C (a constant) by using the adaptive segmentation of the last section. Assume that $h = 2^{-q}$ and that the initial grid is $2^{-q}\mathbb{Z}$. The optimal partition in dyadic intervals is then formed from the sequence of nested dyadic intervals $J_q \subset J_{q-1} \subset \cdots \subset J_0$ containing t_0 and having lengths $2^{-q}, 2 \cdot 2^{-q}, \ldots, 1$. To each J_n we associate the two contiguous intervals of the same length to the left and right of J_n. The extremities of the dyadic intervals thus defined constitute the optimal segmentation for f.

It is not difficult to show that the entropy of f in the Malvar–Wilson basis corresponding to this segmentation does not exceed a certain constant C. The adaptive segmentation has allowed us to "zoom in" on the singularity of f, which is located at $t = t_0$. Thus, in this example, the optimal segmentation algorithm has provided an interesting analysis of the signal f.

6.9 The discrete case

We replace the real line $\mathbb{R}$ by the grid $h\mathbb{Z}$, where $h > 0$ is the sampling step. Thus the signal f is given by a sampling denoted $f(hk)$, $k \in \mathbb{Z}$, but we will not discuss here the technique used to arrive at this sampling. We will forget h in all that follows and assume that f is sampled on $\mathbb{Z}$.

A partition of $\mathbb{Z}$ is defined by the intervals $[a_j, a_{j+1}]$, where $a_j - \frac{1}{2}$ is an integer and where a_j is not an integer. (This construction often has been adopted for the DCT.) Denote the number of points belonging to $[a_j, a_{j+1}] \cap \mathbb{Z}$ by $l_j = a_{j+1} - a_j$, and let the numbers $\alpha_j > 0$ be small enough so that $\alpha_j + \alpha_{j+1} \leq l_j$.

The windows w_j will be subject to exactly the same conditions as in the continuous case. This means that

$$w_j(t) = 0 \text{ outside the interval } [a_j - \alpha_j, a_{j+1} + \alpha_{j+1}]; \tag{6.27}$$

$$w_j(t) = 1 \text{ on the interval } [a_j + \alpha_j, a_{j+1} - \alpha_{j+1}]; \tag{6.28}$$

$$0 \leq w_j(t) \leq 1 \text{ and } w_{j-1}(a_j + \tau) = w_j(a_j - \tau) \text{ if } |\tau| \leq \alpha_j; \tag{6.29}$$

$$w_j^2(a_j + \tau) + w_j^2(a_j - \tau) = 1 \text{ if } |\tau| \leq \alpha_j. \tag{6.30}$$

Then the double sequence

$$\sqrt{\frac{2}{l_j}} w_j(t) \cos\left[\frac{\pi}{l_j}\left(k + \frac{1}{2}\right)(t - a_j)\right], \quad 0 \leq k \leq l_j - 1, \ j \in \mathbb{Z}, \tag{6.31}$$

is an orthonormal basis for $l^2(\mathbb{Z})$.

Nothing prevents us from considering a finite interval of integers and replacing $l^2(\mathbb{Z})$ by $l^2\{1, \ldots, N\}$. Start with $a_0 = \frac{1}{2}$ and end with $a_{j_0+1} = N + \frac{1}{2}$. We require that $w_0(t)$ be equal to 1 on $[\frac{1}{2}, a_1 - \alpha_1]$, and there is no other constraint on this interval. Similarly, $w_{j_0}(t) = 1$ on $[a_{j_0+1} - \alpha_{j_0+1}, a_{j_0+1}]$ with no other constraint on the interval.

This shows that the Malvar–Wilson bases exist in very different algorithmic settings, and it is this that makes them more flexible than other analytic techniques such as Gabor wavelets and Grossmann–Morlet wavelets, for example.

6.10 Modulated Malvar–Wilson bases

As indicated in the introduction, the use of Malvar–Wilson bases comes down to covering the time-frequency plane with Heisenberg boxes whose sides are parallel to the coordinate axes. More precisely, the boxes are defined by an adaptive segmentation of the time axis; once this is done, the partition of the frequency axis follows automatically from the uncertainty principle (the area of each box being 2π). The use of wavelet packets is based on a similar, but inverse, approach; in this case the adaptive filtering precedes the segmentation.

Unfortunately, these two options are incompatible with the use of more elaborate time-frequency algorithms such as the Wigner–Ville transform. This incompatibility was stressed in the first edition of this book. Since then the situation has changed considerably, and today we have orthonormal bases that are adapted to frequency modulated signals. These new results were reported in [63], and the rest of this section is based on that article.

Our story begins with work by Richard Baraniuk, Simon Haykin, Douglas Jones, and Steve Mann (see [18], [19], [20], [21], [194], [195], and [196]). The time-frequency atoms that are adapted to frequency modulated signals are called *chirplets* by these authors. Their chirplets are Gabor-type wavelets with an extra frequency modulation that is given by a linear chirp. The weakness of the original approach is the lack of specific orthonormal chirplet bases that are flexible enough to be used in the context of the Coifman–Wickerhauser best-basis algorithm. To achieve this goal, the original chirplets will be reshaped in a way that mimics the construction of the Malvar–Wilson bases.

We will now show how to construct orthonormal chirplet bases. The signals for which such a construction might be useful are quasi-stationary signals that can be partitioned into a sequence of pieces with specific frequency modulation laws. This segmentation is provided by an arbitrary increasing sequence t_j, $j \in \mathbb{Z}$, of real numbers. The best-basis algorithm will be looking for the optimal partition.

Since we want to avoid abrupt discontinuities, the segmentation of the signal is given by a sequence of bell-shaped functions w_j that mimic the characteristic functions of the intervals $[t_j, t_{j+1}]$. More precisely, we assume that $\lim_{j \to \pm\infty} t_j = \pm\infty$, and we choose $\alpha_j > 0$ such that

$$\alpha_j + \alpha_{j+1} \leq l_j = t_{j+1} - t_j, \qquad j \in \mathbb{Z}. \tag{6.32}$$

We require that the bell-shaped functions w_j have the following properties:

$$0 \leq w_j(t) \leq 1, \qquad w_j \in C_0^\infty(\mathbb{R}), \tag{6.33}$$

$$w_j(t) = 0 \text{ if } t \leq t_j - \alpha_j \text{ or } t \geq t_{j+1} + \alpha_{j+1}, \tag{6.34}$$

$$w_{j-1}(t_j + s) = w_j(t_j - s), \quad |s| \leq \alpha_j, \text{ and} \tag{6.35}$$

$$\sum_{j=-\infty}^{\infty} w_j^2(t) = 1 \quad \text{for all } t. \tag{6.36}$$

These are exactly the conditions we used for constructing the Malvar–Wilson bases.

We can now introduce the modulation "law" that did not exist in the standard Malvar–Wilson bases. The functions φ that provide the frequency modulation are real-valued quadratic spline functions whose knots are exactly the segmentation points t_j, $j \in \mathbb{Z}$. In other words,

$$\varphi(t) = \frac{a_j}{2} t^2 + b_j t + c_j \quad \text{if } t_j \leq t \leq t_{j+1},$$

and φ is continuously differentiable on the real line.

The orthonormal bases we will construct are adapted to frequency modulated signals of the type $f(t) = A(t) \exp(i\varphi(t))$ where A is smooth. Let Γ be the graph of $\xi = \varphi'(t)$ in the time-frequency plane. The class of signals we want to treat is illustrated by Γ in the time-frequency plane (see Figure 5.3).

THEOREM 6.1. *The collection of functions*

$$\tilde{w}_{j,k}(t) = \sqrt{\frac{2}{l_j}} e^{i(b_j t + a_j t^2/2)} \sin\left[\pi\left(k + \frac{1}{2}\right) \frac{t - t_j}{l_j}\right] w_j(t), \tag{6.37}$$

where $j \in \mathbb{Z}$ and $k \in \mathbb{N}$, is an orthonormal basis for $L^2(\mathbb{R})$.

This is proved in [62]. Note that we have not defined $\tilde{w}_{j,k}$ as $e^{i\varphi(t)} w_{j,k}(t)$, where $w_{j,k}$ is the standard Malvar–Wilson basis. We have chosen (6.37) instead because we wish to mimic the standard linear chirps, which are the functions $e^{i(ut+vt^2/2)} g_h(t - t_0)$, where $g_h(t) = h^{-1/2} g\left(\frac{t}{h}\right)$ and where g is the Gaussian $g(t) = \pi^{-1/4} e^{-t^2/2}$. Recall that these are the only functions for which the Wigner–Ville transform is nonnegative. Note, however, that the functions $e^{i\varphi(t)} w_{j,k}(t)$ also form an orthonormal basis, since they are obtained from the Malvar–Wilson basis by a unitary mapping.

6.11 Examples

Frequency modulated signals play an important role in signal processing. One of the more scientifically interesting examples is given by the gravitational waves that are predicted by Einstein's general relativity. Although these waves have not yet been observed, two international programs were launched to obtain evidence of their existence. One process that is predicted to produce these waves is the collapse of binary stars. In this case, the analytic description is given explicitly by

$$f(t) = (t - t_0)^{-1/4} \cos[\omega(t_0 - t)^{5/8} + \theta], \tag{6.38}$$

where t_0 is the time when the collapse occurred, θ is a parameter, and ω is a large constant that depends on the masses of the two stars. Since there is currently so much scientific interest in detecting gravitational waves, these signals are ideal for testing and comparing various time-frequency algorithms. Recalling the definition of a chirp that was given in section 5.9, we see that the two conditions $\left|\frac{A'(t)}{A(t)\varphi'(t)}\right| \ll 1$ and $\left|\frac{\varphi''(t)}{(\varphi'(t))^2}\right| \ll 1$ become $|t - t_0| \gg (\frac{1}{\omega})^{1/\beta}$.

The signals that experiments seek to measure are considerably corrupted by noise. If one follows Donoho's paradigm (discussed in Chapter 11), one is led—in the ideal case of Gaussian white noise—to build orthonormal bases in which these gravitational waves have a minimal description length [89]. This is what we are going to do now.

We begin with a textbook example. Define $f(t) = w(t) \exp(i\lambda\varphi(t))$ where φ is a smooth, real-valued function with $\varphi''(t) \geq 1$, λ is a large parameter, and the window is a smooth function with compact support. Then we use the best-basis algorithm. However, we will be looking for a suboptimal basis since the optimal one is out of reach. A suboptimal basis is a basis for which the entropy is of the same order of magnitude as the absolute minimum that would be reached as λ tends to infinity.

The search for a suboptimal basis inside the unmodified Malvar–Wilson library leads to a segmentation of $w(t) \exp(i\lambda\varphi(t))$ with a uniform step size $h = c\lambda^{-1/2}$, $c > 0$. On the other hand, if the chirplet library is used, the segmentation is still uniform but with a larger step size $h = c\lambda^{-1/3}$, $c > 0$, and this means better compression. The constant c is the order of magnitude of the inverse of the cube root of the third derivative of φ. This implies that c is infinite if the signal happens to be a linear chirp.

This discussion leads to the following conclusion: For the class of frequency modulated signals we are studying, a Wigner–Ville transform performs no better than a best-basis search inside the chirplet library. In both cases, the frequency resolution is $O(\lambda^{1/3})$.

The second example is perhaps more interesting, since the optimal segmentation is no longer uniform. We consider a signal of the form $f(t) = w(t) \cos(\omega t^{1/2})$, where w is again a smooth window with compact support and ω is a large parameter. To find the suboptimal segmentation in the chirplet library, we use a new variable $x = \omega^2 t$. This leads to the segmentation of the function $\cos(x^{1/2})$ over the large interval $[0, \omega^2]$. Then a suboptimal segmentation is given by $x_k = ck^6$ where c is a positive constant. The values of the integers k are $0, 1, \ldots, k_0$ where $k_0 = c^{-1/6}\omega^{1/3}$. Returning to the t variable, we see that this nonuniform segmentation becomes $t_k = c\omega^{-2}k^6$, $1 \leq k \leq k_0$.

Finally, we consider gravitational waves. We assume that the parameter ω is large, that we are using the chirplet library, and that we are looking for a suboptimal basis. In this situation, the suboptimal segmentation is no longer uniform, and in fact it becomes finer and finer as one approaches the blowup of the instantaneous frequency, which is the time when the binary star collapses. The segmentation of the signal $f(t)$ on $[t_0 - 1, t_0]$ is highly nonuniform. Without loss of generality, we are assuming that $t_0 = 0$. Then (up to an obvious sign change), the segmentation is given by $t_k = c\omega^{-8/5} k^{24/5}$ where $1 \leq k \leq k_0$ and $k_0 = c^{-1/6} \omega^{1/3}$. This means that the size of the segmentation step ranges from $\omega^{-1/3}$ to $\omega^{-8/5}$ when one reaches 0, which is when the star collapses.

If we are looking for gravitational waves, the Wigner–Ville transform does not lead to very sharp time-frequency localization. However, in this case, it is possible to take advantage of the knowledge of the exact form of the chirp being sought to construct a quadratic transform, different from the Wigner–Ville transform, that is fashioned to detect optimally this particular kind of chirp. This transform is chosen from a large collection of quadratic transforms called Cohen's class. (For more information see [112] and [49].)

To complete this discussion, we will outline a wavelet technique proposed by J. M. Innocent and B. Torrésani [148] for detecting the chirps described by (6.38). Their technique is based on a "ridge" detection. Consider the half-plane $a > 0$, $b \in \mathbb{R}$ where the continuous wavelet transform is defined. The "ridge" is the region near $b = t_0$ where the wavelet transform of a chirp will be large. This is explained informally as follows: Consider the chirp $f(t) = A(t) e^{i\varphi(t)}$. Its wavelet transform

$$W(f; a, b) = \frac{1}{a} \int f(t) \psi\left(\frac{t-b}{a}\right) dt \qquad (6.39)$$

will be small due to cancellations if the chirp and the wavelet do not oscillate at the same frequency. By the same reasoning, the wavelet transform will be large if the pseudoperiod of the chirp, $\frac{2\pi}{\varphi'(b)}$, coincides with the pseudoperiod of the wavelet, which is a. Thus the wavelet transform will be large near the curve defined by

$$a = \frac{2\pi}{\varphi'(b)}.$$

There is no cancellation on this curve, and the computation of the integral in (6.39) looks like this:

$$W(f; a, b) = \frac{1}{a} \int f(t) \psi\left(\frac{t-b}{a}\right) dt$$

$$\approx \frac{1}{a} \int |f(t)| \left|\psi\left(\frac{t-b}{a}\right)\right| dt$$

$$= \frac{1}{a} \int A(t) \left|\psi\left(\frac{t-b}{a}\right)\right| dt.$$

In view of condition (5.27), we expect that $A(t)$ does not vary much on the support of the wavelet, so that

$$\frac{1}{a} \int A(t) \left|\psi\left(\frac{t-b}{a}\right)\right| dt \approx \|\psi\|_1 A(b).$$

This argument leads to the following heuristic: The continuous wavelet transform of a chirp is large in a neighborhood of the curve $a = \frac{2\pi}{\varphi'(b)}$, where

$$W(f; a, b) \approx \|\psi\|_1 A(b).$$

This idea was first developed by Tchamitchian and Torrésani in [247] and independently by Hunt, Kevlahan, Vassilicos, and Farge in [147].

In the case of chirps generated by the collapse of binary stars, $\varphi(t) = (t - t_0)^{5/8}$ and $A(t) = (t - t_0)^{-1/4}$, and the ridge is located near the curve

$$a = \frac{16\pi}{5\omega}(t_0 - b)^{3/8}.$$

If we take $\|\psi\|_1 = 1$, $|W(a,b)| \approx (t_0 - b)^{-1/4}$ near this curve. This shows that the ridge depends on only two parameters. Innocent and Torrésani propose a parametric statistical test to identify these two parameters, and thus to locate the ridge: Find t_0 and determine the characteristic mass parameter ω.

6.12 Conclusions

The examples mentioned above suggest the following heuristic: If the Wigner–Ville transform $W(t, \xi)$ of a given signal f is sharply concentrated in the time-frequency plane, then f has a compact decomposition in a suitable modulated Malvar–Wilson basis. This is too ambitious as stated, but it is an idea that opens the way to further study. At the present stage of research, a best-basis decomposition in either a Malvar–Wilson library or a wavelet packet library (Chapter 7) is a quick and efficient processing for a given signal or image. If one considers this analysis as a *preprocessing* of an image, one is led to the concept of a multilayered analysis. The idea is that the initial processing with, say, a Malvar–Wilson basis reveals aspects of the image, such as textures, that can be further analyzed with more refined tools. This idea of multilayered analysis is illustrated by the denoising of the Brahms live recording. (See section 5.4 and [34].) We will return to this subject in the next chapter, where once again the available analytic tools will be expanded, this time to include wavelet packet bases.

CHAPTER 7

Time-Frequency Analysis and Wavelet Packets

7.1 Heuristic considerations

A time-frequency analysis of a signal is a representation of the signal as a linear combination of time-frequency atoms. These time-frequency atoms are essentially characterized by an arbitrary duration $t_2 - t_1$ and an arbitrary frequency ω. The instant t_1 is the moment when the signal is first heard (if it is a speech signal, for example), and t_2 is the instant when it ceases to be heard. The frequency ω is an average frequency; this is the frequency of the emitted tone in the case of a musical signal, while the frequency spectrum given by Fourier analysis takes into consideration the parasitic frequencies created by the note's attack and decay.

We also think of a time-frequency atom as occupying a symbolic region in the time-frequency plane (Figure 7.1). This symbolic region is a rectangle R with area 2π, which expresses the Heisenberg uncertainty principle.

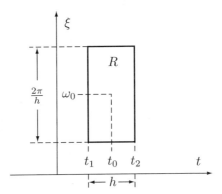

FIG. 7.1. *A Heisenberg box in the time-frequency plane.*

The most famous example of time-frequency atoms is that of the Gabor wavelets. For these we have $f_R(t) = e^{i\omega_0 t} g_h(t - t_0)$, where $t_0 = \frac{1}{2}(t_1 + t_2)$ is the center of the time-frequency atom and $g_h(t) = h^{-1/2} g\big(\frac{t}{h}\big)$, $g(t) = \pi^{-1/4} e^{-t^2/2}$. In this case, the "size" of the time-frequency atom is approximately h, and h is approximately equal to the duration $t_2 - t_1$.

To say that the time-frequency atom f_R occupies the symbolic region R of the time-frequency plane means that f_R is essentially supported by the interval $[t_1, t_2]$ and that the Fourier transform $\hat{f}_R$ of f_R is essentially supported by the interval $[\omega_0 - \frac{\pi}{h}, \omega_0 + \frac{\pi}{h}]$. It is well known that there does not exist a function with compact support whose Fourier transform also has compact support. This leads one to consider the following, less stringent conditions:

$$\int_{-\infty}^{\infty} (t-t_0)^2 |f_R(t)|^2 dt \leq C^2 h^2, \tag{7.1}$$

$$\int_{-\infty}^{\infty} (\xi - \omega_o)^2 |\hat{f}_R(\xi)|^2 d\xi \leq 2\pi C^2 h^{-2}. \tag{7.2}$$

The time-frequency atom that optimizes this criterion (that is, for which the constant C is the smallest possible) is precisely the Gabor wavelet, and the Gabor wavelet owes it success to this optimal localization in the time-frequency plane. On the other hand, we will see below that the Gabor wavelets have a disagreeable property that makes them unsuitable for the time-frequency signal analysis.

If the time-frequency atoms f_R were actually concentrated on rectangles R in the time-frequency plane, they would enjoy the following property: If R_1 and R_2 are disjoint rectangles in the time-frequency plane, then

$$\int_{-\infty}^{\infty} f_{R_1}(t)\overline{f_{R_2}(t)} dt = 0. \tag{7.3}$$

We indicate the "proof" of this property. If R_1 and R_2 are disjoint, then either the horizontal sides of the rectangles are disjoint or the vertical sides are disjoint. In the first case, the supports of f_{R_1} and f_{R_2} (in t) are disjoint, and the integral (7.3) is zero. In the second case, the supports of the Fourier transforms $\hat{f}_{R_1}$ and $\hat{f}_{R_2}$ (in ξ) are disjoint, and the integral (7.3) is still zero, as we see by applying Parseval's identity. We know, in fact, that this cannot happen, and if f_{R_1} and f_{R_2} are Gabor wavelets, the integral (7.3) is never zero. But this integral is small if R_1 and R_2 are "remote," that is, if the rectangles mR_1 and mR_2 are disjoint. Here $m \geq 1$ is an integer, and mR is the rectangle that has the same center as R and whose sides are m times the length of the sides of R. If m is large, remoteness becomes a very strong condition.

Eric Séré has shown that remoteness of the rectangles $R_0, R_1, R_2, \ldots$ does not imply that the corresponding Gabor wavelets $f_{R_0}, f_{R_1}, f_{R_2}, \ldots$ are well separated from each other [235]. More precisely, for every m (no matter how large), there exist rectangles $R_0, R_1, \ldots$ in the time-frequency plane such that the rectangles mR_j are pairwise disjoint and coefficients $\alpha_0, \alpha_1, \ldots$, such that

$$\sum_{j=0}^{\infty} |\alpha_j|^2 = 1, \quad \text{and} \quad \int_{-\infty}^{\infty} \left| \sum_{j=0}^{\infty} \alpha_j f_{R_j}(t) \right|^2 dt = +\infty. \tag{7.4}$$

Thus remoteness of the rectangles in the time-frequency plane does not even imply that the corresponding Gabor wavelets are almost orthogonal, and consequently the apparent heuristic simplicity of the time-frequency plane is completely misleading.

This phenomenon results from the arbitrariness of the $h > 0$ that are used in the definition of the time-frequency atoms. The rectangles $R_0, R_1, \ldots$ in Séré's result have arbitrarily large eccentricity. When $h = 1$ all is well, and the corresponding situation has been studied extensively. This is then a form of windowed Fourier analysis where the sliding window is a Gaussian [72].

Once we abandon the Gabor wavelets, we have two options: Malvar–Wilson wavelets and wavelet packets. We will briefly indicate the advantages and disadvantages of these two options.

If we use Malvar–Wilson wavelets, then, by their nature, the duration of the attack or of the decay is not necessarily related to the duration of the stationary part. We can, for example, have a Malvar–Wilson wavelet for which the durations of the attack and of the decay are of order 1, while the stationary part lasts $T \gg 1$. If ω_0 is the frequency corresponding to this stationary part, then the Fourier transform of the wavelet will be, at best, of the form

$$\frac{\sin T(\xi - \omega_0)}{\sqrt{T}(\xi - \omega_0)} \hat{\varphi}(\xi - \omega_0),$$

and it cannot satisfy (7.2) because h is of the order of magnitude of T. Furthermore, this is true even if we allow in (7.2) the concentration around two frequencies of the same amplitude but opposite signs, that is, if we replace (7.2) with

$$\int_{-\infty}^{\infty} (|\xi| - \omega_o)^2 |\hat{f}_R(\xi)|^2 d\xi \leq 2\pi C^2 h^{-2}.$$

This wavelet looks like the second wavelet in Figure 6.2, the centipede referred to in the text on page 96. On the other hand, the Malvar–Wilson wavelets are constructed to be exactly orthogonal. The implication of these observations is that the orthogonality of the Malvar–Wilson wavelets has been won at the price of their frequency localization, a localization that no longer guarantees the "minimal conditions" (7.1) and (7.2). One last remark about the Malvar–Wilson wavelets is obvious but significant: Although they are given by a simple formula, they are not obtained by translation, change of scale, and modulation (or modulation by $e^{i\omega t}$) of a fixed function g.

The option we propose in this chapter is that of wavelet packets. Here are a few advantages of wavelet packets:

(a) Daubechies's orthogonal wavelets (Chapter 3) are a particular case of wavelet packets.

(b) Wavelet packets are organized naturally into collections, and each collection is an orthonormal basis for $L^2(\mathbb{R})$.

(c) One can compare the advantages and disadvantages of the various possible decompositions of a given signal in these orthonormal bases and select the optimal collection of wavelet packets for representing the given signal.

(d) Wavelet packets are described by a simple algorithm $2^{j/2} w_n(2^j x - k)$, where $j, k \in \mathbb{Z}$, $n \in \mathbb{N}$, and where the supports of the w_n are in the same fixed interval $[0,L]$.

The integer n plays the role of a frequency, and it can be compared with the integer k that occurs in the definition of the Malvar wavelets.

The price paid for these advantages is the same as that associated with the Malvar–Wilson wavelets. Indeed, if to facilitate intuition we associate the rectangle R defined by $k2^{-j} \leq t < (k+1)2^{-j}$ and $n2^j \leq \xi < (n+1)2^j$ in the time-frequency plane with the wavelet packet $2^{j/2} w_n(2^j t - k)$, then this choice does not meet conditions (7.1) and (7.2). Furthermore, we cannot do better by assigning a frequency different from n to w_n, for although $\|w_n\|_2 = 1$,

$$\varlimsup_{n \to +\infty} \left\{ \inf_{\omega \in \mathbb{R}} \int_{-\infty}^{\infty} (\xi - \omega)^2 |\hat{w}_n(\xi)|^2 d\xi \right\} = +\infty. \tag{7.5}$$

The frequency localization of wavelet packets is relatively poor, except for certain values of n, and hence the "lim sup" in (7.5) (see [102]).

7.2 The definition of basic wavelet packets

We begin by defining a special sequence of functions w_n, $n \in \mathbb{N}$, supported by the interval $[0, 2N-1]$, where $N \geq 1$ is fixed at the outset. If $N = 1$, these functions w_n constitute the Walsh system, which is a well-known orthonormal basis for $L^2[0,1]$. (The Walsh system is discussed below.) If $N \geq 2$, the functions w_n are no longer supported by $[0,1]$; however, the double sequence

$$w_n(x-k), \quad n \in \mathbb{N}, \quad k \in \mathbb{Z}, \tag{7.6}$$

will be an orthonormal basis for $L^2(\mathbb{R})$. This orthonormal basis will allow us to do an orthogonal windowed Fourier analysis. Thus, for the moment, this construction is similar to the Malvar–Wilson wavelets. The difference occurs when the dilations enter, the changes of variable of the form $x \mapsto 2^j x$.

We start with an integer $N \geq 1$ and consider two finite trigonometric sums,

$$m_0(\xi) = \frac{1}{\sqrt{2}} \sum_{k=0}^{2N-1} h_k e^{-ik\xi} \quad \text{and} \quad m_1(\xi) = \frac{1}{\sqrt{2}} \sum_{k=0}^{2N-1} g_k e^{-ik\xi}, \tag{7.7}$$

that satisfy the following familiar conditions:

$$g_k = (-1)^{k+1} \overline{h}_{2N-1-k} \quad \text{or} \quad m_1(\xi) = e^{-i(2N-1)\xi} \overline{m_0}(\xi + \pi), \tag{7.8}$$

$$m_0(0) = 1 \quad \text{and} \quad m_0(\xi) \neq 0 \text{ for } \xi \in \left[-\frac{\pi}{3}, \frac{\pi}{3}\right], \tag{7.9}$$

and

$$|m_0(\xi)|^2 + |m_0(\xi + \pi)|^2 = 1. \tag{7.10}$$

One choice, which leads to Daubechies's wavelets (section 3.8), is given by

$$|m_0(\xi)|^2 = 1 - c_N \int_0^\xi (\sin t)^{2N-1} dt, \tag{7.11}$$

where

$$c_N \int_0^\pi (\sin t)^{2N-1} dt = 1,$$

but other choices are possible [65].

As a first example take $m_0 = \frac{1}{2}(e^{-i\xi} + 1)$ and $m_1(\xi) = \frac{1}{2}(e^{-i\xi} - 1)$. Condition (7.10) reduces to

$$\cos^2 \frac{\xi}{2} + \sin^2 \frac{\xi}{2} = 1.$$

A second choice is given by

$$\sqrt{2}\, h_0 = \frac{1}{4}(1 + \sqrt{3}), \qquad \sqrt{2}\, h_1 = \frac{1}{4}(3 + \sqrt{3}),$$

$$\sqrt{2}\, h_2 = \frac{1}{4}(3 - \sqrt{3}), \qquad \sqrt{2}\, h_3 = \frac{1}{4}(1 - \sqrt{3}).$$

Having selected the coefficients h_k, we define the wavelet packets w_n by induction on $n = 0, 1, 2, \ldots$ using the two identities

$$w_{2n}(x) = \sqrt{2} \sum_{k=0}^{2N-1} h_k w_n(2x - k), \tag{7.12}$$

$$w_{2n+1}(x) = \sqrt{2} \sum_{k=0}^{2N-1} g_k w_n(2x - k), \tag{7.13}$$

and the condition $w_0 \in L^1(\mathbb{R})$ with $\int_{-\infty}^{\infty} w_0(x)\, dx = 1$.

We explain the roles of these two identities. Identity (7.12), with $n = 0$, is

$$w_0(x) = \sqrt{2} \sum_{k=0}^{2N-1} h_k w_0(2x - k), \tag{7.14}$$

and the function $\varphi = w_0$ is a fixed point of the operator $T : L^1(\mathbb{R}) \to L^1(\mathbb{R})$ that is defined by

$$Tf(x) = \sqrt{2} \sum_{k=0}^{2N-1} h_k f(2x - k). \tag{7.15}$$

This equation becomes

$$(Tf)\hat{\ }(\xi) = m_0(2^{-1}\xi)\hat{f}(2^{-1}\xi) \tag{7.16}$$

by taking the Fourier transform. If f is normalized by $\int_{-\infty}^{\infty} f(x)dx = 1$, then the fixed point is unique, and it is given by

$$\hat{\varphi}(\xi) = \prod_{k=1}^{\infty} m_0(2^{-k}\xi), \tag{7.17}$$

a relation we have now seen several times. On the other hand, the function φ can be constructed directly in "time" space by iterating T. Figure 7.2 illustrates the iterative scheme for constructing φ using the characteristic function of $[0, 1]$ for the initial value f_0. Here, $f_{j+1} = T(f_j)$, and we have drawn the first few functions f_0, f_1, and f_2. The coefficients h_0, h_1, h_2, and h_3 are approximately those given in the second example mentioned above. Then the sequence f_j converges uniformly to the fixed point φ.

Once the function $\varphi = w_0$ is constructed, we use (7.13) with $n = 0$ to obtain $\psi = w_1$. (The function ψ is the "mother" wavelet in the construction of the "ordinary" orthonormal wavelet bases, and φ is the "father" wavelet.) Next, we use (7.12) and (7.13) with $n = 1$ and obtain w_2 and w_3. By repeating this process, we generate, two at a time, all of the wavelet packets. The support of φ is exactly the interval $[0, 2N - 1]$ (see [65]), and it is easy to show that the supports of w_n, $n \in \mathbb{N}$, are included in $[0, 2N - 1]$.

The central result about the basic wavelet packets we have constructed is that the double sequence

$$w_n(x - k), \quad n \in \mathbb{N}, \quad k \in \mathbb{Z}, \tag{7.18}$$

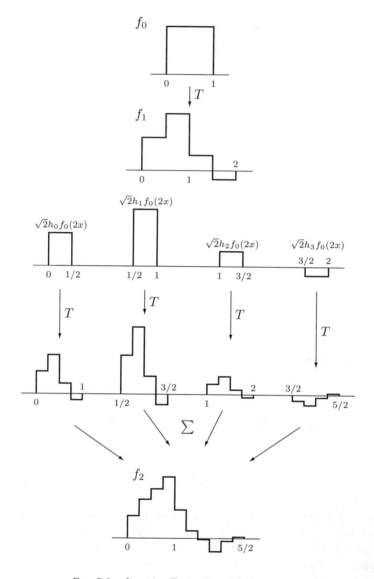

FIG. 7.2. *Iterating T starting with $f_0 = \chi_{[0,1]}$.*

is an orthonormal basis for $L^2(\mathbb{R})$. To be more precise, the subsequence derived from (7.18) by taking $2^j \leq n < 2^{j+1}$ is an orthonormal basis for the orthogonal complement W_j of V_j in V_{j+1}. Recall that, in the language of multiresolution analysis, V_j is the closed subspace of $L^2(\mathbb{R})$ spanned by the orthonormal basis $2^{j/2}\varphi(2^j x - k)$, $k \in \mathbb{Z}$, and similarly, $2^{j/2}\psi(2^j x - k)$, $k \in \mathbb{Z}$, is an orthonormal basis for W_j. Thus, the construction of wavelet packets appears as a change of orthonormal basis inside each W_j.

An interesting observation concerns the case where the filter has length one and $h_0 = h_1 = \frac{1}{\sqrt{2}}$, $g_1 = -g_0 = \frac{1}{\sqrt{2}}$. This brings us back to the Walsh system mentioned at the beginning of the section. We let r denote the one-periodic function that equals 1 on the interval $[0, \frac{1}{2})$ and -1 on the interval $[\frac{1}{2}, 1)$. To define the Walsh system W_n, $n \in \mathbb{N}$, let χ denote the characteristic function of the interval $[0,1)$ and, for $n = \varepsilon_0 + 2\varepsilon_1 + \cdots + 2^j \varepsilon_j$, where $\varepsilon_j = 0$ or 1, write

$$W_n(x) = [r(x)]^{\varepsilon_0}[r(2x)]^{\varepsilon_1}\cdots[r(2^j x)]^{\varepsilon_j}\chi(x).$$

Then it is not difficult to verify that

$$W_{2n}(x) = W_n(2x) + W_n(2x-1)$$

and that

$$W_{2n+1}(x) = W_n(2x) - W_n(2x-1).$$

This shows that, in the case of filters of length one, the construction of wavelet packets leads to the Walsh system. The Walsh system W_n, $n \in \mathbb{N}$, is an orthonormal basis for $L^2[0,1]$, and it follows immediately that the double sequence $W_n(x-k)$, $n \in \mathbb{N}$, $k \in \mathbb{Z}$, is an orthonormal basis for $L^2(\mathbb{R})$. (For more about the Walsh system and its connection with quadrature mirror filters, see [44].)

In the general case of basic wavelet packets (filters longer than one), the supports of $w_n(x-k)$ and $w_{n'}(x-k')$ are not necessarily disjoint when $k \neq k'$, and proving the orthogonality of the double sequence $w_n(x-k)$, $n \in \mathbb{N}$, $k \in \mathbb{Z}$, is more subtle. We will return to this orthogonality issue in section 7.4.

7.3 General wavelet packets

The basic wavelet packets are the functions w_n, $n \in \mathbb{N}$ (which are derived from a filter $\{h_k\}$), and the sequence $w_n(x-k)$, $n \in \mathbb{N}$, $k \in \mathbb{Z}$, is an orthonormal basis for $L^2(\mathbb{R})$. This orthonormal basis is analogous to the Walsh system, but for filters longer than 1, it is more regular. That is, the frequency localization of the functions w_n is better than the frequency localization of the functions in the Walsh system. Nevertheless, this frequency localization does not yield an estimate of the type

$$\inf_{\omega \in \mathbb{R}} \int_{-\infty}^{\infty} (\xi - \omega)^2 |\hat{w}_n(\xi)|^2 d\xi \leq C, \tag{7.19}$$

uniformly in n (see [65]).

The general wavelet packets are the functions

$$2^{j/2} w_n(2^j x - k), \quad n \in \mathbb{N}, \quad j, k \in \mathbb{Z}. \tag{7.20}$$

These are much too numerous to form an orthonormal basis. In fact, we can extract several different orthonormal bases from the collection (7.20). The choice $j = 0$, $n \in \mathbb{N}$, $j, k \in \mathbb{Z}$, leads to the orthonormal basis described in the previous section, while the choice $n = 1$, $j, k \in \mathbb{Z}$, leads to an orthonormal wavelet basis, as described in Chapter 3.

There is another way to select a basis from the function in (7.20). Associate with each of the wavelet packets (7.20) the "frequency interval" $I(j,n)$ defined by $2^j n \leq \xi < 2^j(n+1)$. The following result describes certain sets of wavelet packets that constitute orthonormal bases for $L^2(\mathbb{R})$.

THEOREM 7.1. *Let E be a set of pairs (j,n), $j \in \mathbb{Z}$, $n \in \mathbb{N}$, such that the corresponding frequency intervals $I(j,n)$ constitute a partition of $[0, \infty)$, up to a countable set. Then the subsequence*

$$2^{j/2} w_n(2^j x - k), \quad (j,n) \in E, \quad k \in \mathbb{Z}, \tag{7.21}$$

is an orthonormal basis for $L^2(\mathbb{R})$.

Notice that choosing E is choosing a partition of the frequency axis. This partitioning is "active," whereas the corresponding sampling with respect to the variable x (or t) is passive and is dictated by Shannon's theorem.

Going back to Ville, we see that wavelet packets lead to a signal analysis technique where the process is "first filter different frequency bands; then cut these bands into slices (in time) to study their energy variations." Similarly, we refer to the methodology developed by Liénard: "The proposed analysis process contains the following steps: filtering with a zero-phase filterbank, and modeling the output signals into successive waveforms (channel-to-channel modeling)."

When we have at our disposal a "library" of orthonormal bases, each of which can be used to analyze a given signal of finite energy, we are necessarily faced with the problem of knowing which basis to choose. We settle this problem with the same approach that we used for the Malvar–Wilson wavelets: The optimal choice is given by the entropy criterion that we have already used in the preceding chapter. This entropy criterion provides an adaptive filtering of the given signal.

7.4 Splitting algorithms

Let (α_k) and (β_k), $k \in \mathbb{Z}$, be two sequences of coefficients that satisfy the following conditions: $\sum |\alpha_k|^2 < \infty$, $\sum |\beta_k|^2 < \infty$, and, by defining $m_0(\theta) = \sum \alpha_k e^{-ik\theta}$ and $m_1(\theta) = \sum \beta_k e^{-ik\theta}$, the matrix

$$U(\theta) = \begin{bmatrix} m_0(\theta) & m_1(\theta) \\ m_0(\theta + \pi) & m_1(\theta + \pi) \end{bmatrix} \quad \text{is unitary.}$$

Consider a Hilbert space H with an orthonormal basis $(e_k)_{k \in \mathbb{Z}}$, and define the sequence f_k, $k \in \mathbb{Z}$, of vectors in H by

$$f_{2k} = \sqrt{2} \sum_{-\infty}^{\infty} \alpha_{2k-l} e_l, \qquad f_{2k+1} = \sqrt{2} \sum_{-\infty}^{\infty} \beta_{2k-l} e_l. \qquad (7.22)$$

Then the sequence (f_k), indexed by $k \in \mathbb{Z}$, is also an orthonormal basis for the Hilbert space H.

Next, let H_0 be the closed subspace of H generated by the vectors f_{2k}, which we denote by $e_k^{(0)}$; similarly H_1 will be generated by $f_{2k+1} = e_k^{(1)}$, $k \in \mathbb{Z}$.

Nothing prevents us from repeating on $(H_0, e_k^{(0)})$ the operation we have done on (H, e_k) and from iterating these decompositions while keeping the same coefficients (α_k) and (β_k) at each step.

An elementary example is useful for understanding the nature of this splitting algorithm. The initial Hilbert space is $L^2[0, 2\pi]$ with the usual orthonormal basis $e_k = \frac{1}{\sqrt{2}} e^{ik\theta}$, $k \in \mathbb{Z}$. The (2π-periodic) functions m_0 and m_1 are (when restricted to $[0, 2\pi)$) the characteristic functions of $[0, \pi)$ and $[\pi, 2\pi)$. Then the vectors f_{2k} are $\frac{1}{\sqrt{\pi}} e^{i2k\theta} m_0(\theta)$, and they constitute a Fourier basis for the interval $[0, \pi)$, while the vectors f_{2k+1} constitute a Fourier basis for the interval $[\pi, 2\pi)$. Finally, the subspace H_0 of H is composed of the functions supported on the interval $[0, \pi)$, while H_1 is composed of the functions supported on $[\pi, 2\pi)$.

Iterating the splitting algorithm leads to subspaces that are naturally denoted by $H_{(\varepsilon_1, \ldots, \varepsilon_j)}$, where $\varepsilon_i = 0$ or 1, or even by H_I, where I denotes the dyadic interval of length 2^{-j} and origin $2^{-1}\varepsilon_1 + \cdots + 2^{-j}\varepsilon_j$. In the example we have just studied, H_I is exactly the subspace of $L^2[0, 2\pi)$ consisting of the functions that vanish outside the interval $2\pi I$.

This example has guided the intuition of scientists working in signal processing. Assuming that the signal is sampled on $\mathbb{Z}$, they have considered the situation where (α_k) and (β_k) are two finite sequences and where m_0 resembles the transfer function of a low-pass filter while m_1 resembles that of a high-pass filter. One requires, at least, that $m_0(0) = 1$ and that $m_0(\theta) \neq 0$ for $\theta \in [-\frac{\pi}{3}, \frac{\pi}{3}]$.

By analogy with the preceding example, these scientists were led to believe that the iterative scheme, which we have called the splitting algorithm, would provide a finer and finer frequency definition, as one wanders through the maze of "channels" illustrated in Figure 7.3.

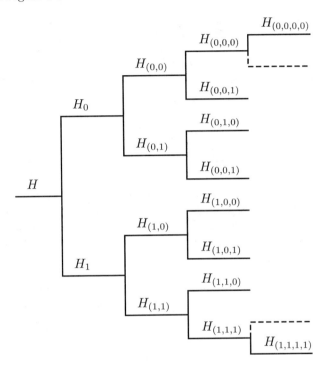

FIG. 7.3. *An illustration of the splitting algorithm.*

The initial Hilbert space H is the direct sum of various combinations of these subspaces. In particular, H is the direct sum of all the subspaces at the same "splitting level": at the first level there are 2 subspaces, at the next level there are 4, then 8, 16, and so on.

To give a better understanding of the construction of wavelet packets and the exact nature of the splitting algorithm, consider the case where the initial Hilbert space is the space V_j, $j \geq 1$—in the language of multiresolution analysis—with the orthonormal basis $2^{j/2}\varphi(2^j x - k)$, $k \in \mathbb{Z}$. Next, suppose that the splitting algorithm has operated j times. Then we arrive exactly at the sequence of functions $w_n(x - k)$, $k \in \mathbb{Z}$, $0 \leq n < 2^j$, and $n = \varepsilon_0 + 2\varepsilon_1 + \cdots + 2^{j-1}\varepsilon_{j-1}$ is the index of the "frequency channel" $H_{(\varepsilon_0, \varepsilon_1, \ldots, \varepsilon_{j-1})}$.

The frequency localization of wavelet packets does not conform to the intuition of the scientists who introduced these algorithms, and the only case where there is a precise relation between the integer n and a frequency in the sense of Fourier analysis is the case where m_0 and m_1 are the transfer functions of "ideal filters."

7.5 Conclusions

It remains for us to indicate how to use wavelet packets. We begin by selecting, for use throughout the discussion, two sequences h_k, g_k, $0 \leq k \leq 2N-1$, that satisfy the conditions for constructing wavelet packets. The choice of these sequences results from a compromise between the length $(2N-1)$ of the filters and the quality of the frequency resolution. Once the filters are selected, we set in motion the algorithm for constructing the wavelet packets. We obtain a huge collection of orthonormal bases for $L^2(\mathbb{R})$ from this process. It is then a question of determining, for a given signal, the optimal basis. And again, the optimal basis is the one (among all those in the wavelet packets) that gives the most compact decomposition of the signal.

We determine this optimal basis by using a "fine-to-coarse" type strategy and the method of merging. We start from the finest frequency channels H_I; these are associated with the dyadic intervals I of length $|I| = 2^{-m}$. The integer m is taken to be as large as necessary to be consistent with the chosen precision. The algorithm proceeds by making the following decision: It combines the left and right halves, I' and I'', of a dyadic interval I whenever the orthonormal basis of H_I yields a more compact representation than that obtained by using the two orthonormal bases of $H_{I'}$ and $H_{I''}$.

The discrete version of wavelet packets also can be used and is immediately available. This is obtained by starting with the Hilbert space $H = l^2(\mathbb{Z})$ of a signal sampled on $\mathbb{Z}$ and the canonical orthonormal basis $(e_k)_{k \in \mathbb{Z}}$, where $e_k(n) = 1$ when $n = k$ and 0 elsewhere. Here there is perfect resolution in position but no resolution in the frequency variable. Next, we systematically apply the splitting algorithm to improve the frequency definition until we reach the spaces H_I associated with the dyadic intervals I of length $|I| = 2^{-m}$. Finally, we apply the algorithm to choose the best basis (section 6.7).

Wavelet packets offer a technique that is dual to the one given by the Malvar–Wilson wavelets. In the case of wavelet packets, we effect an adaptive filtering, whereas the Malvar–Wilson wavelets are associated with an adaptive segmentation of the time (or space) axis.

As was the case with wavelets, wavelet packet orthonormal bases exist in two dimensions, where they have interesting applications to coding efficiently textured images. We quote from [200], where the basic ideas are developed and interesting examples of the compression of textured images are presented. Having pointed out that wavelets provide good compression for smooth images, François Meyer writes:

> Wavelets, however, are ill suited to represent oscillatory patterns. Rapid variations of intensity can only be described by the small scale wavelet coefficients. Long oscillatory patterns thus require many such fine scale coefficients. Unfortunately those small scale coefficients carry very little energy, and are often quantized to zero, even at low compression rates. In order to describe long oscillatory patterns, much larger libraries of waveforms, called wavelet packets, have been developed.

After presenting several examples where the best wavelet packet basis outperforms wavelet coding (both visually and in terms of the quadratic mean), the author offers this criticism:

> We realize that when coding images that contain a mixture of smooth and textured features, the best-basis algorithm is always trying to find a compromise between two conflicting goals: describe the large scale

smooth regions, and describe the local oscillatory patterns. The best basis is chosen in order to minimize the entropy, but such a choice may not always yield "visually pleasant" images. In fact we sometimes notice ringing artifacts on the border of smooth regions, when the basis is mostly composed of oscillatory patterns.

These comments highlight one of the fundamental problems in image processing, which is that no single basis—that we are aware of today—is well suited to compress all images. A recent approach proposed by François Meyer and his collaborators is reminiscent of color separation in the printing industry [201]. An image is separated into several layers, such as the smooth-regions layer and the textures layer. Each layer is coded differently using the transform, or basis, most appropriate for the layer. This is done is such a way that the compressed layers can be restored and put back together to produce a good image. The process is similar to the denoising algorithm used in [34], which we alluded to in sections 5.4 and 6.12. We also note the work of Jacques Froment for another approach to the problem of separating natural images into smooth regions and textured regions [122].

The sparse representation of images is a subtle and controversial issue. Wavelet packets offer an interesting option and perform better than ordinary wavelet expansions when one wants to represent textured features accurately. However, the same remark applies to Malvar–Wilson bases, and one must decide which of these options to use. As if this were not complicated enough, highly textured images are well represented with *brushlets* [202]. Brushlets provide an improvement on wavelet packets by having better frequency localization. In both cases, one is trying to fit the frequency channels to the signal. We have mentioned that wavelet packets do not enjoy the desired frequency localization; in the Fourier domain, their decay is not ideal. Using brushlets amounts to decomposing the Fourier transform of a given signal with an adapted Malvar–Wilson basis.

CHAPTER 8

Computer Vision and Human Vision

We propose to describe and comment on a small part of David Marr's work. We limit our discussion to Marr's analysis of the "low-level" processing of luminous information by the retinal cells. Marr suggested that the coding of this luminous information was based on the zero-crossings of an operator that is now called a wavelet transform. This hypothesis leads us to state the famous "Marr conjecture" and then to state its precise form as conjectured by Mallat. This precise form yields a remarkably effective algorithm. We will see, however, that Mallat's conjecture is not generally correct, and this poses some fascinating new problems.

8.1 Marr's program

Marr's book, *Vision, A Computational Investigation into the Human Representation and Processing of Visual Information* [198], appeared in 1982. Stylistically it is reminiscent of Descartes's *Discours de la méthode*. Exactly as Descartes did, Marr takes us into his confidence and speaks to us as if we were a friend or colleague from his laboratory. Marr confides in us his intellectual progress and tells us about his doubts, his hopes, and his enthusiasms. He gives a lively description of the theories he has struggled with and rejected, and he explains his own research with an infectious enthusiasm.

We recall that the goal of Marvin Minsky's group at the Massachusetts Institute of Technology (MIT) artificial intelligence laboratory was to solve the problem of artificial vision for robots. The challenge was to construct a robot endowed with a perception of its environment that enabled it to perform specific tasks. It turned out that the first attempts to construct a robot capable of understanding its surroundings were completely unsuccessful.

These surprising setbacks showed that the problem of artificial vision was much more difficult than it seemed. The idea then occurred to imitate, within the limits imposed by robot technology, certain solutions found in nature. Marr, who was an expert on the human visual system, was invited to leave Cambridge, England, for Cambridge, Massachusetts, to join the MIT group. According to Marr, the disappointments of the robot scientists were due to having skipped a step. They had tried to go directly from the statement of the problem to its solution without having at hand the basic scientific understanding that is necessary to construct effective algorithms.

Marr's first premise is that there exists a science of vision, that it must be developed, and that once there has been sufficient progress, the problems posed by vision for robots can be solved.

Marr's second premise is that the science of human vision is no different from the science of robot (or computer) vision.

Marr's third premise is that it is as vain to imitate nature in the case of vision as it would have been to construct an airplane by imitating the form of birds and the structure of their feathers. On the other hand, he notes that the laws of aerodynamics explain the flight of birds and enable us to build airplanes.

Thus it is important, as much for human vision as for computer vision, to establish scientific foundations rather than blindly to seek solutions. To develop this basic science, one must carefully define the scope of inquiry. In the case of human vision, one must clearly exclude everything that depends on training, culture, taste, and similar "conditioning." For instance, the ability to distinguish the canvas of a master from that of an imitator has nothing to do with the science of basic human vision. One retains only the mechanical or involuntary aspects of vision, that is, those aspects that enable us to move around, to drive a car, and so on. Thus we limit the following discussion to low-level vision. This is the aspect of vision that enables us to re-create the three-dimensional organization of the physical world around us from the luminous excitations that stimulate the retina.

The notion that low-level vision functions according to universal scientific algorithms seemed to be an implausible idea to some scientists, and it encountered two kinds of opposition. In the first place, neurophysiologists had discovered certain cells having specific visual functions. But Marr was opposed to this reductionist approach to the problems of vision, and he offered two criticisms on this subject:

(a) After several very stimulating discoveries, neurophysiologists had not made sufficient progress to enable them to explain the action of the human visual system based on a collection of ad hoc cells.

(b) It would be absurd to look for the cell that lets you immediately recognize your grandmother.

On another front, Marr was opposed to attempts by psychologists to relate the performance of the human visual system to a learning process. Roughly, the idea is that we recognize the familiar objects of our environment by dint of having seen and touched them simultaneously. In fact, Bela Julesz made a fundamental discovery that eliminated this as a working hypothesis.

Julesz made a systematic study of the response of the human visual system when it was presented with completely artificial images (synthetic images having no significance) that were computer-generated, random-dot stereograms. If these synthetic images presented a certain "formal structure" that stimulated stereovision reflexes, the eye deduced, in several milliseconds and without the slightest hesitation, a three-dimensional organization of the image. This organization "in relief" is clearly only a mirage in which the mechanism of stereovision finds itself trapped. This mechanism acts with the same speed, the same quality, and the same precision as if it were a matter of recognizing familiar objects. The conclusion is that familiarity with the objects one sees plays no role in the primary mechanisms of vision. Marr set out to understand the algorithmic architecture of these low-level mechanisms.

This venture can be likened to that of the seventeenth-century physiologists who studied the human body by comparing it to a complex and subtle machine—an assembly of bones, joints, and nerves whose functioning could be explained, calculated, and predicted by the same laws that applied to winches and pulleys. A century and a half later, Claude Bernard made a similar connection between the organic functioning of the human body and results from the nascent field of organic chemistry. The synthesis of urea (Wöhler, 1828) again reduced the gap between the chemistry of life and organic chemistry.

In their scientific approach, these researchers relied on solid, well-founded knowledge that came either from mechanics or from chemistry. They then tried to effect a technology transfer and to apply results acquired in the study of matter to the life sciences. But what Marr set out to do was much more difficult because the relevant knowledge base, namely, an understanding of robots, was too tenuous to serve as the nucleus for an explanation of the human visual system.

Marr asserted that the problems posed by human vision or by computer vision are of the same kind and that they are part of a coherent and rigorous theory, of an articulate and logical doctrine.

It is necessary, at the outset, to set aside any consideration of whether the results will ultimately be implemented with copper wires or nerve cells and to limit the investigation to the following four properties of human vision that we wish to imitate or reproduce in robots:

(a) The recognition of contours of objects. These are the contours that delimit objects and structure the environment into distinct objects.

(b) The sense of the third dimension from two-dimensional retinal images and the ability to arrive at a three-dimensional organization of physical space.

(c) The extraction of relief from shadows.

(d) The perception of movement in an animated scene.

The fundamental questions posed by Marr are the following:

(a) How is it scientifically possible to define the contours of objects from the variations of their light intensity?

(b) How is it possible to sense depth?

(c) How is movement sensed? How do we recognize that an object has moved by examining a succession of images?

Marr opened a very active area of contemporary scientific research by giving each of these problems a precise algorithmic formulation and by furnishing parts of the solution in the form of algorithms.

Marr's working hypothesis is that human vision and computer vision face the same problems. Thus, the algorithmic solutions can and must be tested within the framework of robot technology and artificial vision. In case of success, it is necessary to investigate whether these algorithms are physiologically realistic. For example, Marr did not believe that human neuronal circuits used iterative loops, which are an essential aspect of the existing algorithms.

This discussion raises the basic problem of knowing the nature of the *representation* on which the algorithms act. Marr used a simple comparison to help us understand the implications of a representation. If the problem at hand was adding integers, then the representation of the integers could be given in the Roman system, in the decimal system, or in the binary system. These three systems provide three representations of the integers. But the algorithms used for addition will be different in the three cases, and they will vary greatly in difficulty. This shows that the choice of this or that representation involves significant consequences. (See the quotation on page 12.)

8.2 The theory of zero-crossings

Marr felt that image processing in the human visual system has a complex hierarchical structure, involving several layers of processing. The "low-level processing" furnishes a representation that is used by later stages of visual information processing. Based on a very precise analysis of the functioning of the ganglion cells, Marr was led to this hypothesis: The basic representation ("the raw primal sketch") furnished by the retinal system is a succession of sketches at different scales and these scales are in geometric progression. These sketches are made with lines, and these lines are the zero-crossings that Marr uses in the following argument [198, p. 54]:

> The first of the three stages described above concerns the detection of intensity changes. The two ideas underlying their detection are (1) that intensity changes occur at different scales in an image, and so their optimal detection requires the use of operators of different sizes; and (2) that a sudden intensity change will give rise to a peak or trough in the first derivative or, equivalently, to a *zero-crossing* in the second derivative ....
>
> These ideas suggest that in order to detect intensity changes efficiently, one should search for a filter that has two salient characteristics. First and foremost, it should be a differential operator, taking either a first or second derivative of the image. Second, it should be capable of being tuned to act at any desired scale, so that large filters can be used to detect blurry shadow edges, and small ones to detect sharply focused fine details in the image.

Marr and Hildreth [199] argued that the most satisfactory operator fulfilling those conditions is the filter ΔG, where Δ is the Laplacian operator $(\partial^2/\partial x^2 + \partial^2/\partial y^2)$ and G stands for the two-dimensional Gaussian distribution $G(x,y) = \frac{1}{2\pi\sigma^2} e^{-(x^2+y^2)/2\sigma^2}$, which has standard deviation σ. ΔG is a circularly symmetric Mexican-hat-shaped operator whose distribution in two dimensions may be expressed in terms of the radial distance r from the origin by the formula[5]

$$\Delta G(r) = \frac{-1}{\pi\sigma^4} \left(1 - \frac{r^2}{2\sigma^2}\right) e^{-r^2/2\sigma^2}.$$

Marr is computing the two-dimensional wavelet transform of the image using the wavelet ψ, which is the Laplacian of the Gaussian G. Today, ψ is known as *Marr's wavelet*. If a black and white image is defined by the gray levels $f(x,y)$, the zero-crossings of Marr's theory are the lines defined by the equation $(f * \psi_\sigma)(x,y) = 0$. Since the function ψ is even, the values of convolution product $f * \psi_\sigma$ are (up to a proportionality factor) the wavelet coefficients of f, analyzed with the wavelet ψ. Hence, the zero-crossings are defined by the vanishing of the wavelet coefficients.

The values of σ remain to be specified. The values used in human vision are in geometric progression, and they were discovered by Campbell, Robson, Wilson, Giese, and Bergen, based on neurophysiological experiments. These experiments led to the values $\sigma_j = (1.75)^j \sigma_0$.

Marr's conjecture is that the original image f is completely determined by the sequence of lines defined by $(f * \psi_{\sigma_j})(x,y) = 0$. Interest in this representation of an

---

[5] We have changed the notation and corrected typos.

image stems from its invariance under translations, rotations, and dilations. Here are some of Marr's thoughts about this representation [198, p. 67]:

> Zero-crossings provide a natural way of moving from an analog or continuous representation like the two-dimensional image intensity values $I(x,y)$ to a discrete, symbolic representation. A fascinating thing about this representation is that it probably incurs no loss of information. The arguments supporting this are not yet secure.

In the following pages, we propose to study Marr's conjecture. We will show first of all that it is incorrect for periodic images covering an unbounded area. In particular, we will construct a whole family of periodic functions that have the same zero-crossings. However, since our counterexample is unbounded, it does not exclude the possibility that the conjecture is true for images having finite extent.

We will then examine Mallat's conjecture, which is a version of Marr's conjecture. Mallat's conjecture leads to an explicit algorithm for reconstructing the image. This algorithm works very well in spite of the fact that Mallat's conjecture is in general false. Although the algorithm is not widely used in practice, there is continuing research interest in this technique. The counterexample that we construct is, in a certain sense, more realistic than the one we present in the case of Marr's conjecture.

8.3 A counterexample to Marr's conjecture

We begin with a counterexample in one dimension. It will then be easy to transform it into a two-dimensional counterexample. This counterexample has the property of being periodic in x (or in x and y in the two-dimensional case). We do not know how to construct other counterexamples.

Consider all the functions f of the real variable x, having real values, and given by the series

$$f(x) = \sin x + \sum_{k=2}^{\infty} \alpha_k \sin kx, \qquad (8.1)$$

where we require that

$$\sum_{k=2}^{\infty} k^3 |\alpha_k| < 1. \qquad (8.2)$$

We are going to show that all choices of the coefficients α_k lead to the same zero-crossings. For example, $\sin x$ and $\sin x + \frac{1}{9}\sin 2x$ have the same zero-crossings. We prove this assertion by applying the following simple observation: If u and v are two continuous functions of x, and if, for some constant $r \in [0,1)$, $|v(x)| \leq r|u(x)|$ for all x, then $u(x) + v(x) = 0$ is equivalent to $u(x) = 0$.

Returning to (8.1), define $g_\delta(x) = \frac{1}{\delta\sqrt{2\pi}} e^{-x^2/2\delta^2}$. Then

$$f * g_\delta(x) = e^{-\delta^2/2} \sin x + \sum_{k=2}^{\infty} \alpha_k e^{-k^2\delta^2/2} \sin kx.$$

It follows from this that

$$-\frac{d^2}{dx^2}(f * g_\delta)(x) = e^{-\delta^2/2} \sin x + \sum_{k=2}^{\infty} k^2 \alpha_k e^{-k^2\delta^2/2} \sin kx$$

$$= u(x) + v(x).$$

Since $|\sin kx| \leq k|\sin x|$, $|v(x)| \leq r|u(x)|$, where $r = \sum_2^\infty k^3|\alpha_k| < 1$. Thus the zero-crossings of all the functions f are $x = m\pi$, $m \in \mathbb{Z}$.

If we wish to have $0 \leq f(x) \leq 1$, it is sufficient to add a suitable constant to $f(x)$ (defined by (8.1)) and then to renormalize the result by multiplication with a suitable positive constant. These two operations do not change the positions of the zero-crossings.

A nontrivial two-dimensional counterexample is given by the function

$$f(x,y) = \sin x \sin y + \sum_{k=2}^\infty \alpha_k \sin kx \sin ky,$$

where we now require that

$$2 \sum_{k=2}^\infty k^4 |\alpha_k| < 1.$$

8.4 Mallat's conjecture

The existence of these counterexamples and several remarks Marr made in his book led Stéphane Mallat to a more precise version of Marr's conjecture. Mallat observed that numerical image processing using certain kinds of pyramid algorithms (quadrature mirror filters) and Marr's approach represented two particular examples of wavelet analysis of an image. In fact, one has $\Delta(f * g_\delta) = \delta^{-2} f * \psi_\delta$, where

$$\psi(x,y) = -\frac{1}{\pi}\left(1 - \frac{x^2+y^2}{2}\right)\exp\left(-\frac{x^2+y^2}{2}\right)$$

is Marr's wavelet. With this in mind, Mallat took up a promising approach: The idea was to give Marr's conjecture a precise numerical and algorithmic formulation by taking advantage of the progress that had been made in image processing in the early 1980s using pyramid algorithms.

We start with the one-dimensional case. Mallat replaced the Gaussian $\frac{1}{\sqrt{2\pi}} e^{-x^2/2}$ with the basic cubic spline θ, whose support is the interval $[-2,2]$. Recall that $\theta = T * T$, where T is the triangle function whose value is $1 - |x|$ if $|x| \leq 1$ and 0 if $|x| > 1$.

Let f be the function we wish to analyze by the method of zero-crossings and write $\theta_\delta(x) = \delta^{-1}\theta(\delta^{-1}x)$. Then the zero-crossings are the values of x where the second derivative $\frac{d^2}{dx^2}(f * \theta_\delta)$ is zero and changes sign. To use the pyramid algorithms, Mallat assumes that $\delta = 2^{-j}$, $j \in \mathbb{Z}$. He then proposes to code the signal f with the double sequence $(x_{q,j}, y_{q,j})$, where

(a) $x = x_{q,j}$ is (for $\delta = 2^{-j}$) a zero of $\frac{d^2}{dx^2}(f * \theta_\delta)$ where this function changes sign, and

(b) $y_{q,j} = \frac{d}{dx}(f * \theta_\delta)(x_{q,j})$.

In other words, Mallat considers the values of $x = x_{q,j}$ where $\frac{d}{dx}(f * \theta_\delta)$ has an extremum, and he keeps the values of these local extrema in memory.

Certain of these local extrema are related to points where the signal f changes rapidly; this is the case for the points x_1 and x_2 in Figure 8.1. Other extrema are related to points where the function changes very little. Mallat had the idea

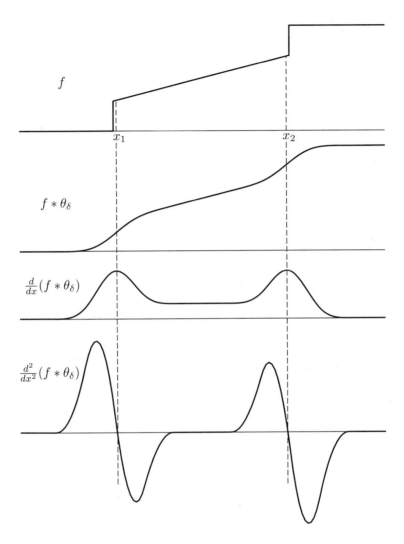

FIG. 8.1. *A schematic of Mallat's process.*

to consider only the first of these and thus to retain only the local maxima of $|\frac{d}{dx}(f * \theta_\delta)|$. This will not change the critical analysis that follows.

Coding f with the double sequence $(x_{q,j}, y_{q,j})$ meets two objectives: It is invariant under translation, and it corresponds to a precise form of Marr's conjecture. Here is what Marr wrote [198, p. 68]:

> On the other hand, we do have extra information, namely, the values of the slopes of the curves as they cross zero, since this corresponds roughly to the contrast of the underlying edge in the image. An analytic approach to the problem seems to be difficult, but in an empirical investigation, Nishihara (1981) found encouraging evidence supporting the view that a two-dimensional filtered image can be reconstructed from its zero-crossings and their slopes.

We are going to show that this conjecture is in general incorrect. However, this assertion must be tempered, since our counterexample depends on a specific choice for the function θ. If θ is the cubic spline, then we have a counterexample. If, on the other hand, the cubic spline θ is replaced with the function that is equal to $1+\cos x$ if $|x| \leq \pi$ and to 0 if $|x| > \pi$ (which is the Tukey window), then, for all signals f with compact support, reconstruction is theoretically possible but unstable (see Appendix C). In this case, it comes down to determining a function with compact support from the knowledge of its Fourier transform in the neighborhood of zero, and this is an unstable process.

Appendix C contains a complete description of our counterexample, but for those who wish to skip the details, we provide an outline here in the main text. We begin by making a change of scale so that the values of δ are $2\pi 2^{-j}$ rather than 2^{-j}, $j \in \mathbb{Z}$. (This is a convenience rather than an essential point.) We then define $f_0(x) = 1 + \cos x$ if $-\pi \leq x \leq \pi$ and $f_0(x) = 0$ if $|x| \geq \pi$. The first step consists in finding the zeros of $\frac{d^2}{dx^2}(f_0 * \theta_\delta)$, which are the inflection points of $f_0 * \theta_\delta$. We note that $f_0 * \theta_\delta(x) = 0$ whenever $|x| \geq \pi + 2\delta$, and thus we search for the other zeros. Since $(f_0 * \theta_\delta)'' = f_0'' * \theta_\delta$, θ is even, and $\cos(\frac{\pi}{2} + x) = -\cos(\frac{\pi}{2} - x)$, it is clear just by examining the integral that $(f_0 * \theta_\delta)''(\pm \frac{\pi}{2}) = 0$ for all $\delta \leq \frac{\pi}{2}$. When δ is large, the roles played by f_0 and θ_δ are interchanged, and we write $(f_0 * \theta_\delta)''(x) = f_0 * \theta_\delta''(x)$. Then when $\delta \geq 3\pi$, we see that $f_0 * \theta_\delta''(\pm \frac{2}{3}\delta) = 0$, again by just examining the integral.

We introduce a perturbation R that belongs to $C^\infty(\mathbb{R})$, is even, and is supported by $\frac{\pi}{8} \leq |x| \leq \frac{\pi}{4}$. We also require that the first three moments of R vanish. Having fixed such an R, the perturbation of f_0 is $f = f_0 + \varepsilon R$, where $\varepsilon > 0$ is small. We then prove—and this is all in Appendix C—that $(f_0 * \theta_\delta)''(x) = 0$ implies $(R * \theta_\delta)''(x) = 0$ and $(R * \theta_\delta)'(x) = 0$. A stronger statement is actually needed: There exists a constant C such that

$$\left| \frac{d^2}{dx^2}(R * \theta_\delta)(x) \right| \leq C \left| \frac{d^2}{dx^2}(f_0 * \theta_\delta)(x) \right|$$

uniformly for all $\delta = 2\pi 2^{-j}$. Once this is proved, it is clear from the argument given above for the Marr counterexample that f and f_0 have the same zero-crossings. The fact that $(f * \theta_\delta)'(x) = (f_0 * \theta_\delta)'(x)$ at these zero-crossings follows from the definition of R.

If the function f that we wish to analyze by Mallat's algorithm is a step function (with an arbitrarily large number of discontinuities), then Mallat's conjecture is correct. In fact, thanks to the symmetry of the function θ, the zero-crossings occur (for sufficiently small $\delta > 0$) at the points of discontinuity, while the values of the first derivatives of the smoothed signal furnish the jumps in the signal at these discontinuities. In this case, we have perfect reconstruction of the signal.

All this explains, without doubt, why Mallat's algorithm works in practice with such excellent precision, no matter which signals are treated. The signals in question have more in common with step functions than with the subtle functions described in the counterexamples.

8.5 The two-dimensional version of Mallat's algorithm

We start with a two-dimensional image g. From this we create the increasingly blurred versions at scales $\delta = 2^{-j}$, $j \in \mathbb{Z}$, by taking the various convolution products

$g * \theta_\delta$, where, in two dimensions, $\theta_\delta(x,y) = \theta_\delta(x)\theta_\delta(y)$. The function θ is the basic cubic spline used in one dimension.

Next we consider the local maxima of the modulus of the gradient of $g * \theta_\delta$. We keep in memory the positions of these local maxima as well as the gradients at these points. The conjecture is that this data, computed for $\delta = 2^{-j}$, characterizes the image whose gray levels are given by $g(x,y)$.

We will show that this conjecture is incorrect in this general form. This does not exclude the possibility of its being true if (1) more restrictive assumptions are made about the function g or (2) the definition of the smoothing operator is changed.

The counterexample in two dimensions will not be compactly supported. Finding a counterexample whose support is a square is an unsolved problem. Our counterexample will be $g(x,y) = f(x) + f(y)$, where f is the counterexample in one dimension. Then

$$g(x,y) * \theta_\delta(x)\theta_\delta(y) = f(x) * \theta_\delta(x) + f(y) * \theta_\delta(y),$$

and the gradient of this function is the vector

$$\left(\frac{d}{dx}(f * \theta_\delta)(x), \frac{d}{dy}(f * \theta_\delta)(y)\right).$$

Its length is $(|\frac{d}{dx}(f * \theta_\delta)|^2 + |\frac{d}{dy}(f * \theta_\delta)|^2)^{1/2}$, and it has a maximum if and only if $|\frac{d}{dx}(f * \theta_\delta)|$ and $|\frac{d}{dy}(f * \theta_\delta)|$ are at a maximum. But the set of functions f has been constructed so that the positions of the maxima of $|\frac{d}{dx}(f * \theta_\delta)|$ are independent of the choice of f and the same is true for the values of $\frac{d}{dx}(f * \theta_\delta)$ at these points when $\delta = 2\pi 2^{-j}$, $j \in \mathbb{Z}$.

8.6 Conclusions

All of this shows that Marr's conjecture is doubtful. Nevertheless, the underlying heuristics are playing a key role in signal processing. A successful example is the signal analysis being done by Alain Arneodo and his group to reveal the complex nature of certain signals, particularly velocity signals from fully developed turbulence. This processing has been used on other signals, including signals derived from DNA and financial time series, with impressive results. (The application to turbulence will be described in detail in the next chapter.) We note that the problem of reconstruction is irrelevant for these applications: One wishes to extract some meaningful characteristics of the signal, but one is not interested in reconstructing the signal. Thus, even if Marr's conjecture is doubtful, its spirit is alive.

Regarding Mallat's conjecture, one must distinguish between the problem of unique representation and that of stable reconstruction. In our opinion, the reconstruction is never stable (unless the class of images to which the algorithm is applied is seriously limited). But it is, in certain cases, a representation that defines the image uniquely (see [166] and Appendix C).

CHAPTER 9
Wavelets and Turbulence

9.1 Introduction

Studying turbulence with wavelets is a controversial scientific program. This is not surprising, since we are attacking one of the most difficult and itself controversial problems in science with a rather simple tool. Criticism arose originally when a few scientists announced that spectacular results had been obtained by wavelet methods. It was highly unlikely, however, that one of the oldest fundamental problems of classical physics—a problem whose solution has eluded some of the outstanding scientists of the twentieth century—would suddenly be resolved by the mere introduction of a new tool.

Similar criticisms arose when wavelet methods were first applied to image processing. Today we have a much better understanding of how low-level processing can benefit from wavelet methods. We also understand that some aspects of image processing, such as pattern recognition, are not directly accessible through wavelet methods. In our report on wavelets and turbulence, we hope to draw similar balanced conclusions by indicating what is working and what is not.

Wavelets have been applied to at least three problems in fluid dynamics during the past 15 years. The first one concerns a line of research that was introduced by Benoît Mandelbrot and developed by Uriel Frisch and Giorgio Parisi; it is the program that has led to the recent results by Alain Arneodo and his coworkers in Bordeaux, France. These programs seek to unravel the intricate fine-scale geometrical structure of fully developed turbulence by analyzing time series obtained from wind tunnel experiments. One wishes to know if a fractal or multifractal model is appropriate. A second problem concerns the detection and modeling of the coherent structures that are found in turbulent flows. The third problem mentioned in this chapter deals with the mathematical and numerical treatment of the Navier–Stokes equations. The question here is, Do wavelet-based algorithms perform better than conventional numerical schemes?

Turbulence was studied, described, and modeled long before wavelets existed. As in image processing—and in many other scientific fields—the most readily available and widely used tool was the Fourier transform. We indicate the successes and the limitations of this methodology in the next section. The bulk of the chapter is devoted to discussing the multifractal formalism and the role of wavelets in the continuing development of this approach. Next, we will indicate how wavelets have been used for studying the coherent structures in turbulence. As the last application, we will indicate how wavelets are being used to study the Navier–Stokes equations. (We suggest the book [118] by Uriel Frisch as a general reference on turbulence that discusses the concepts introduced in this chapter.)

9.2 The statistical theory of turbulence and Fourier analysis

The purpose of statistical modeling is to provide useful descriptions of large data sets that originate from complex phenomena. Statistical modeling contrasts sharply with the nineteenth-century approach to science, which culminated with accomplishments like the work of Albert Einstein and the axiomatization of quantum mechanics. Physicists were looking for a few fundamental equations (indeed, partial differential equations) that would describe all the laws of the universe. They sought beauty and simplicity, and measured by technological accomplishments, this approach has been remarkably successful. This very success has led us in the late twentieth century to attack increasingly complex problems, which for one reason or another have resisted a purely deterministic approach. Statistical modeling is often an appropriate intellectual approach when faced with large data sets that are generated by a specific procedure and that present similarities that must be accurately described and understood. This approach is appropriate whenever a purely deterministic attack on a problem is impossible or impractical.

The study of fluid dynamics is an intermediate case. The mathematical equations that govern the evolution of fluid flows have been known for a century; it is a system of nonlinear partial differential equations known as the Navier–Stokes equations. In principle, everything is written in these equations, but in practice we are faced with three monumental problems, which are surely related: We do not understand the mathematics of the Navier–Stokes equations, we cannot efficiently compute the solutions of these equations, and it is difficult to access experimentally the full space–time complexity of fully developed turbulence. In some practical situations, simplifications can be introduced that lead to tractable numerical simulations: They are of great importance in aerodynamics, for instance, where they replace costly wind-tunnel experiments. In the nontractable situations, stochastic modeling is often used. Once this is accepted another quandary must be faced: Should the model be determined by a data-fitting procedure or should it be based on plausible assumptions compatible with physical principles? Initially, the second approach was taken in the case of fully developed turbulence. However, data-fitting has gained importance as instrumentation and experimental techniques have improved. The theoretical successes of the mid twentieth century have been followed by the challenge of creating more sophisticated models to fit the accurate experimental data of the latter part of the century.

The statistical theory of turbulence was introduced more or less simultaneously by Kolmogorov in 1941 [167], [168], Obukhov in 1941 [220], Onsager in 1945 [221], Heisenberg in 1948 [141], and von Weizsäcker in 1948 [255]. This work involved applying the statistical tools used for studying stationary process to understand the partition of energy at different scales in the solutions of the Navier–Stokes equations.

According to Leray, this statistical point of view could be justified by the loss of stability and uniqueness of the solutions for very large Reynolds numbers and for large values of time [172]. More recently, with the advent of computers, we have become even more aware of how sensitive the Navier–Stokes equations are to small errors (such as the inevitable computer round-off errors), to the point that computing deterministic solutions at high Reynolds numbers does not make sense. An example of this problem in the context of weather prediction is known as the "butterfly effect," the idea that a butterfly's passage can change the prediction. This implies that only statistical averages are relevant in many situations.

For modeling fully developed turbulence, we need to distinguish three, roughly defined, scalar regions. The intermediate scales (the inertial zone) lie between the smallest scales (where, through viscosity, the dynamic energy is dissipated in heat) and the largest scales (where exterior forces supply the energy). In this inertial zone, the theory of Kolmogorov stipulates that energy is neither produced nor dissipated but only transferred from one scale to another at a constant rate ε. The statistical modeling of turbulence applies only to the inertial zone.

Other assumptions are that turbulence is statistically homogeneous (invariant under translation), isotropic (invariant under rotation), and self-similar (invariant under dilations when considering scales in the inertial zone). The velocity components are treated as random variables, and the statistical description is derived from the corresponding correlation functions. In view of the space homogeneity, the Fourier transform is the mathematical tool adapted to this statistical approach. Kolmogorov and Obukhov used dimensional analysis to show that the average spectral distribution of energy must scale like $\varepsilon^{2/3}|k|^{-5/3}$, where k is the vector variable of the Fourier transform of the three-dimensional velocity. This means that a log-log plot of the energy versus $|k|$ has a slope equal to $-5/3$. This scaling law is very well verified experimentally for a large range (roughly three decades) of $|k|$.

Perhaps the simplest statistical process that has the same power spectrum is fractional Brownian motion (fBm) with scaling exponent $H = 1/3$. This process is denoted by $B_H(t)$ and is defined by the following three properties: $B_H(t)$ is Gaussian, $B_H(t)$ has stationary increments, and $B_H(t)$ satisfies the scaling law $\lambda^{-H} B_H(\lambda t) \sim B_H(t)$ for $\lambda > 0$. The second requirement means that for each increment h, $B_H(t+h) - B_H(t)$ is a stationary process, and the last one means that $\lambda^{-H} B_H(\lambda t)$ and $B_H(t)$ have the same statistics. The structure functions of fractional Brownian motion are $\mathbb{E}[\,|B_H(t+\tau) - B_H(t)|^p\,]$, where $\mathbb{E}$ is the expectation, and they satisfy the identity

$$\mathbb{E}[\,|B_H(t+\tau) - B_H(t)|^p\,] = c_p |\tau|^{Hp}, \tag{9.1}$$

where $0 < p < \infty$ and c_p is a constant. Structure functions are used as a classification tool. However, in many cases, such as turbulence, we do not have access to the expectation or ensemble average. We are then forced to make an ergodic assumption and to replace this expectation with an integral with respect to the space variable.

Another problem makes life even more complicated. As will be explained later, the experimental data is the velocity of the flow measured at a given point x_0 as a function of time t. The $|k|^{-5/3}$ law concerns the Fourier transform of the velocity at a given time as a function of the space variable. An assumption, called the Taylor hypothesis, is needed to obtain $u(x_1, x_2, x_3, t)$ as a function of the longitudinal variable x_1. We will never have access to the full space–time information. This issue will be discussed further in the next section.

We end this section by noting that fractional Brownian motion had to be abandoned as a model of turbulence. Precise wind-tunnel measurements have shown that the exponent in the structure functions of fully developed turbulence (9.3) is not a linear function of p. We will be looking at this nonlinear behavior later in the chapter. It is often interpreted as the signature of *intermittency*, which is an informal term meaning that certain quantities, particularly energy dissipation, vary greatly in time and space.

9.3 Multifractal probability measures and turbulent flows

We will describe in section 9.4 the multifractal signal processing that has been proposed by Frisch and Parisi, but first we wish to mention the pioneering work of Mandelbrot. Mandelbrot wished to model the rate of energy dissipation in a turbulent flow. This dissipation rate is defined as

$$\varepsilon(x,t) = \frac{\nu}{2} \sum_{i,j} \left(\frac{\partial u_i}{\partial x_j} + \frac{\partial u_j}{\partial x_i} \right)^2,$$

where u_1, u_2, u_3 are the three components of the velocity field. Before describing Mandelbrot's ideas for modeling $\varepsilon(x,t)$, we indicate how $\varepsilon(x,t)$ is measured. Our knowledge of the small-scale structure of a turbulent flow is derived from wind-tunnel experiments. A small wire is placed in a tunnel where the flow is turbulent, and the wire is heated at some point. The thermal decay is related to the fluid velocity $u(x_0, t)$ at this point, as a function of time. For computing the energy dissipation rate, we need instead $u(x, t_0)$ as a function of the space variable x. The Taylor hypothesis, which says that the time variations are equivalent to the space variations, applies to wind-tunnel experiments, and thus $\varepsilon(x,t)$ is computed as $c\bigl(\frac{\partial u}{\partial t}(x_0, t)\bigr)^2$, where c is a constant. Various wind-tunnel experiments, going back to Batchelor and Townsend in the mid-1940s, have suggested that the energy dissipation at the smallest scales is not uniformly distributed [26]. More recently, Alain Arneodo and his group [15] have analyzed very accurate data that were obtained by Gagne and Hopfinger and colleagues in a large wind tunnel in Modane, France [5]. These wind-tunnel measurements confirm that the energy dissipation rate associated with the small scales of a turbulent flow is spatially intermittent. Observations like these led Mandelbrot to propose a random multiplicative model for the energy dissipation $\varepsilon(x,t)$ [190, 191].

Mandelbrot's model and ones that have followed describe how energy cascades from large scales to the smallest scales, where it is dissipated as heat. Thus, speaking informally, there are two aspects to these models: the rule that governs how the energy is partitioned from scale to scale and the set on which the dissipation occurs. To gain some intuition about these concepts, we describe the construction of the multifractal Bernoulli measure μ_p. This is perhaps the simplest mathematical construct that models these ideas.

This probability measure depends on a parameter $p \in (0,1)$. Let $q = 1 - p$, and inductively define $\mu = \mu_p$ on dyadic subintervals I of $[0,1)$. In other words, $I = [k2^{-j}, (k+1)2^{-j})$ with $0 \leq k < 2^j$, $j \geq 0$. If $I' = [k2^{-j}, (k+\frac{1}{2})2^{-j})$ and $I'' = [(k+\frac{1}{2})2^{-j}, k2^{-j})$ are the "sons" of I, then the three measures $\mu(I)$, $\mu(I')$, and $\mu(I'')$ of these intervals are related by $\mu(I') = p\mu(I)$ and $\mu(I'') = q\mu(I)$. When $p = q = \frac{1}{2}$, μ is the ordinary Lebesgue measure on $[0,1]$, and when $0 < p < q < 1$, μ is a singular measure. The function $F(x) = \mu([0,x])$ is an example of a devil's staircase. Indeed, F is a continuous strictly increasing function whose derivative F' vanishes almost everywhere. The velocity vanishes almost everywhere, but we are still moving! The construction of these measures serves as the model for what we call a *multiplicative cascade*, and the analysis of these measures serves as a model for the *multifractal formalism*.

If μ is a given probability measure on $[0,1]$, its local scaling exponents $\alpha(x_0)$ are defined by comparing $\mu(I)$ with $|I|^\alpha$, where $x_0 \in I$, as its length $|I|$ tends to zero. Mathematicians ask if $\mu(I) \leq C|I|^\alpha$ when $x_0 \in I$, and if so define $\alpha(x_0)$ to be the upper bound of these exponents α. Physicists are more optimistic and find $\alpha(x_0)$ by making a log-log plot of the mass $\mu(I)$ carried by I versus the length of I. If the log-log plot is close to a straight line with slope α, they write $\mu(I) \sim |I|^\alpha$. (The notation $\mu(I) \sim |I|^\alpha$ means that $\frac{\log \mu(I)}{\log |I|^\alpha} \to 1$ as $|I| \to 0$. Elsewhere, we write expressions like $A(x) \sim B(x)$ to mean that $\frac{\log A(x)}{\log B(x)} \to 1$ as $x \to 0$.)

Returning to the special case of Bernoulli measures, $\alpha(x_0)$ can be computed explicitly, and it depends only on the average number of 0's in the dyadic expansion of x_0. This is the reason that $\alpha(x_0)$ is highly unstable as a function of x_0: It is discontinuous at every x_0. In general, it is impossible to compute $\alpha(x_0)$, and if one wants information about μ, it is necessary to look elsewhere. As is often the case, another function, such as an average, is more useful than the function itself. Here is what is done for measures: For a given exponent h in $[0,1]$, denote by $E(h)$ the set of x_0 such that $\alpha(x_0) = h$ and denote by $D(h)$ its Hausdorff dimension. Then $D(h)$ measures how likely it is that the scaling exponent is h. This function $D(h)$ is called the spectrum of singularities of μ; it has a characteristic concave shape, and it plays an important role in what follows. (The notion of Hausdorff dimension has become an indispensable tool for characterizing fractal sets, and it is used throughout this chapter and the next. For completeness, we include the definition and a brief discussion in section 9.10.)

Mandelbrot generalized the construction of the Bernoulli measures to provide a better model for the rate of energy dissipation in fully developed turbulence. In this new construction, p is replaced by a random variable $p(I,\omega)$ that depends on the dyadic interval I, which will be subdivided into two subintervals I' and I'', as in the Bernoulli construction. These random variables all belong to $[0,1]$ and are independent and identically distributed as I runs over dyadic intervals. The existence and properties of these random measures μ are discussed by J.-P. Kahane and J. Peyrière in [165]. This model has been extensively studied by mathematicians, and its generalizations form an active area of research ([22] and [212] contain some of the latest results). An obvious drawback of this model is the dyadic partitioning, which has no reason to appear in fluid mechanics. Continuous-scale cascade models introduced by B. Castaing and his coworkers avoid this problem, since these models favor no particular scale [47].

9.4 Multifractal modeling of the velocity field

The goal of this research is to study the three-dimensional velocity field of fully developed turbulence, but as we have emphasized above, our knowledge about fully developed turbulence comes to us from a one-dimensional velocity signal. This signal is not a probability measure nor is there any reason to believe that it should be modeled as the primitive of a probability measure. The analysis outlined for measures does not apply, so if we want a similar analysis that applies to functions, it is necessary to extend the ideas. This step was taken by Parisi and Frisch [223]. The starting point for this analysis is the following observation: $\mu(I) \sim |I|^\alpha$ as I contains x_0 and as its length $|I|$ tends to zero can be rewritten as $|F(x) - F(x_0)| \sim |x - x_0|^\alpha$, where F is a primitive (indefinite integral) of μ. This suggests computing the pointwise Hölder exponents.

Recall the definition given in Chapter 2: If f is continuous at x_0 and if $0 < \alpha \leq 1$, then one writes $f \in C^\alpha(x_0)$ when $|f(x) - f(x_0)| \leq C|x - x_0|^\alpha$ for some constant C. If $\alpha > 1$, then $f(x) - f(x_0)$ is replaced by the error term in the Taylor expansion of f at x_0. The scaling exponent $\alpha(x_0)$ is defined as the supremum of those α for which f belongs to $C^\alpha(x_0)$. Mathematicians then write

$$\alpha(x_0) = \liminf \frac{\log|f(x) - f(x_0)|}{\log|x - x_0|}. \tag{9.2}$$

Physicists might expect this lim inf to be a limit, in which case $\alpha(x_0)$ can be measured by a log-log plot.

Once $\alpha(x_0)$ is defined, $E(h)$ is defined to be the set of all x_0 for which $\alpha(x_0) = h$, $h \in [0, 1]$. Finally, $D(h)$ is the Hausdorff dimension of $E(h)$, and the multifractal spectrum of singularities of f is precisely this function $D(h)$, which is defined on $[0, 1]$.

Note that this procedure can be applied to any function f of the real variable x. (In fact, it can be generalized to functions of several variables.) Note also that $D(h)$ can be defined on the whole real line if Hölder exponents $h > 1$ and $h < 0$ are used. In the case $h < 0$, one uses the *weak scaling exponent* defined in [208]. We will discuss the computational problems of determining $D(h)$ later in this section.

Multifractal signal processing consists in computing the spectrum of singularities $D(h)$ and using it as a classification tool. We will use two examples from mathematics to illustrate the power of this tool. It allows us to distinguish between regularly irregular functions and irregularly irregular functions. In the former case, irregularity can be anticipated, while in the latter case, this wild behavior cannot be anticipated at all. Strong, unexpected transients are responsible for this erratic behavior. The first situation is illustrated by the Weierstrass function $\mathcal{W}(t) = \sum_{n=0}^{\infty} B^n \cos(A^n t)$, where $0 < B < 1$ and $AB \geq 1$. In this case $\alpha(x_0) = -\frac{\log B}{\log A} = h_0$ everywhere, and the singularity spectrum $D(h)$ is trivial: $D(h) = 0$ if $h \neq h_0$, while $D(h_0) = 1$. This situation is similar to that of fractional Brownian motion ($B_H(t)$ in (9.1)), where $\alpha(x_0) = H$ for all x_0.

The second case is illustrated by the function $\mathcal{R}(x) = \sum_{n=1}^{\infty} n^{-2} \sin(\pi n^2 x)$, which is attributed to Bernhard Riemann, who is said to have suggested it as an example of a continuous, nowhere-differentiable function. Its complexity eluded mathematicians for more than a century, for $\mathcal{R}$ is truly an erratic function. Its pointwise Hölder exponent $\alpha(x_0)$ is everywhere discontinuous. However, the spectrum of singularities $D(h)$ of the Riemann function is amazingly simple: $D(h) = 0$ if $h \leq \frac{1}{2}$ or $h \geq \frac{3}{4}$, and $D(h) = 4h - 2$ if $\frac{1}{2} \leq h \leq \frac{3}{4}$ [153]. (All of this is explained in more detail in Chapter 10.) The spectrum of singularities of other "special" functions can be computed explicitly, and it is amazing to see how many of these functions have nontrivial spectra (see [155] and [159]).

We are now going to look at what the physicists have been doing. They make measurements; they measure the velocity of a turbulent flow or of other complicated functions. They then wish to know if the complexity of an experimental function can be explained by a multiplicative cascade or by some other hidden dynamical system. For example, in the case of the Bernoulli measure μ_p, or of the corresponding devil's staircase F, one would like to recover p and rules for constructing μ_p from the knowledge of F. This is an inverse problem. Looking for a hidden multiplicative cascade that would generate a given signal is a great scientific challenge. This challenge is called *multiscale system theory* by Albert Benveniste and Alan S. Willsky. The goal is to recognize and analyze phenomena occurring at different scales. One wants to build "multiscale autoregressive processes" that will play the same role when one zooms across scales as ARMA processes do when one moves across time. The desire is to have an algorithm for detecting transients across scales. These would be the unexpected events that appear as one zooms across scales. Neither algorithms nor software are yet available, and multifractal signal processing can be viewed as a limited attempt to achieve a multiscale system theory. However, some promising results on multiscale autoregressive models can be found, for example, in [23], [24], [25], and [70].

An initial piece of information that would help the search for the multiscale autoregressive system or some other multiscale dynamic that explains the data set is the spectrum of singularities. Conversely, finding the spectrum of singularities becomes much easier if we know a priori that the signal has some simple multiscale structure. This is indeed fortunate, since determining $D(h)$ directly from the definition clearly requires an infinite amount of computing: The pointwise scaling exponent $\alpha(x_0)$ must be computed at every point, which is obviously impossible. Furthermore, computing Hausdorff dimensions is also not feasible in practice, since it involves all possible coverings of the set being analyzed. The way around this impasse that Frisch and Parisi proposed involves the use of the structure functions

$$I(y,p) = \int_{-\infty}^{+\infty} |f(x+y) - f(x)|^p \, dx,$$

and it is based on the following heuristic reasoning.

To speak of a multifractal structure means that for every positive exponent h, there is a set of singular points with Hausdorff dimension $D(h)$ on which the increment $|f(x+y) - f(x)|$ acts like $|y|^h$. The contribution of these "singularities of exponent h" to $I(y,p) = \int_{-\infty}^{+\infty} |f(x+y) - f(x)|^p \, dx$ is the order of magnitude of the product $|y|^{ph}|y|^{1-D(h)}$, where the second factor is the probability that an interval of length $|y|$ intersects a fractal set of dimension $D(h)$. As y tends to zero, the dominant term in $I(y,p)$ is the one with the smallest possible exponent. This leads to the relation

$$I(y,p) \sim |y|^{\zeta(p)}, \tag{9.3}$$

where

$$\zeta(p) = \inf_{h>0} \{ph + 1 - D(h)\}. \tag{9.4}$$

The exponent $\zeta(p)$ is thus given by the Legendre transform of the codimension $1 - D(h)$, where $D(h)$ is the Hausdorff dimension of the exceptional points x where $|f(x+y)-f(x)|$ behaves like $|y|^h$. If (9.4) is valid and if D is a concave function, then the spectrum of singularities can be recovered by the Legendre inversion formula

$$D(h) = \inf_p \{ph + 1 - \zeta(p)\}. \tag{9.5}$$

We say that the multifractal formalism applies to a given function f if (9.5) holds. This is not true in general. In fact, it was proved in [154] that any continuous function D that is defined on $[0,1]$ with values in $[0,1]$ is the spectrum of singularities of some function f. The function D is not necessarily concave, so it cannot always be computed by (9.5). One might expect that (9.5) yields the convex hull of the spectrum of singularities. Unfortunately, this more conservative result is again too optimistic, since there are other reasons that cause the multifractal formalism to fail. One of these is the presence of chirps in the signal. This will be discussed in the next section.

Alain Arneodo and his team modified the definition of the structure functions [16]. They replaced the crude increments $f(x+y) - f(x)$ with a smooth average of these increments, namely, with the wavelet coefficients

$$W(f; y, x) = \int_{-\infty}^{\infty} f(x + yt)\psi(t)\, dt = \frac{1}{y}\int_{-\infty}^{\infty} f(u)\psi\left(\frac{u-x}{y}\right) du. \quad (9.6)$$

As observed in [154], writing

$$\int_{-\infty}^{\infty} |f(x+y) - f(x)|^p\, dx \leq C|y|^{\alpha p}, \quad (9.7)$$

where $0 < \alpha < 1$ and $1 \leq p \leq \infty$, is equivalent to

$$\int |W(f; y, x)|^p\, dx \leq C'|y|^{\alpha p}, \quad (9.8)$$

or to f being in the Besov space $B_p^{\alpha,\infty}$. (The relation (9.8) is a possible definition of this Besov space. Besov spaces will appear again in Chapter 11. See Appendix D for a definition and discussion of Besov spaces.)

It is actually possible to derive (9.3) and (9.4) from (9.8). Indeed, if f is in $C^\alpha(x_0)$, then we will see in Chapter 10 that the wavelet transform of f near x_0 satisfies the relation

$$|W(f; a, b)| \leq Ca^\alpha \left(1 + \frac{|b - x_0|}{a}\right)^\alpha. \quad (9.9)$$

Thus, in the cones $|b - x_0| \leq C'a$ we expect that

$$|W(f; a, b)| \sim a^h, \quad (9.10)$$

where h is the Hölder exponent of f at x_0. The derivation of (9.3) and (9.4) then follows the same argument that was used to derive these relations from the definition of $I(y, p)$.

The relation (9.8) makes perfectly good sense when α is either negative or greater than one, whereas the structure functions $\int_{-\infty}^{\infty} |f(x+y) - f(x)|^p\, dx$ do not offer the same flexibility. When the given function f is corrupted by noise, (9.8) can still be used on a range of scales, while the ordinary structure functions do not have a meaning. At this point, the program proposed by Frisch and Parisi needs to be reformulated. One starts with an arbitrary function f of a real variable and defines

$$\tau(p) = \sup\{s \mid f \in B_p^{s/p,\infty}\}. \quad (9.11)$$

One then asks if

$$D(h) = \inf_p \{hp + 1 - \tau(p)\}. \quad (9.12)$$

But (9.12) is not always true. Even if D is a concave function, this concave function is not generally given by an infimum of affine functions with positive slopes. If $D(h)$ is bell shaped, negative values of p also are needed to obtain the decreasing part (to the right of the maximum) of $D(h)$ in (9.12). The best result, which is given in [154], is this: If $f \in C^\varepsilon(\mathbb{R})$ for some $\varepsilon > 0$, then

$$D(h) \leq \inf_{p > p_c} \{hp + 1 - \tau(p)\}, \quad (9.13)$$

where p_c is the only value of p for which $\tau(p) = 1$.

Since p is the slope of the $D(h)$ curve, we cannot expect to reconstruct the decreasing part of the curve without using negative values of p. Negative p's are needed, but Besov spaces with negative p's do not make sense. Indeed, the defining relation (9.8) does not make sense for negative p's: Estimating the integral in (9.8) for negative p's is clearly a totally unstable calculation because the integral will diverge whenever the wavelet transform $W(f; a, b)$ vanishes.

To proceed with the program, it is necessary to "renormalize" this divergence by eliminating most locations where $W(f; a, b)$ is very small. To do this cleverly, one must keep in mind the point of the computations: If the spectrum of singularities has a bell shape, we want to obtain the decreasing part of $D(h)$. This part corresponds to the (relatively) small sets of points where the function has a large Hölder exponent. Equation (9.9) shows that $|W(f; a, b)|$ must be *uniformly* small on the cones above x_0 if $\alpha(x_0)$ is large, and, conversely, (9.9) shows that if $|W(f; a, b)|$ is small at some places and large at others in these cones, then f will not be smooth at x_0. This observation provides the clue to the needed renormalization. The idea is to eliminate from the computation of the integral in (9.8) small values of $|W(f; a, b)|$ that are surrounded by large ones and to keep only those values for which the wavelet transform is uniformly smaller around x_0. In other words, one takes into account only the local maxima of the wavelet transform.

This is what Arneodo and his group do. They implement Mallat's idea of using only the local maxima of the wavelet transform.[6] The technique is known as the wavelet transform modulus maxima (WTMM) algorithm, and this is how they apply it: Let f be the function to be analyzed and let

$$W(f; a, b) = \int f(b + at)\psi(t)\, dt$$

be its wavelet transform, where $a > 0$ and $b \in \mathbb{R}$. The first step is to find, for a fixed scale a, the values of b where $|W(f; a, b)|$ attains a local maximum. These values are denoted by $b_j(a)$, $j = 0, 1, 2, \ldots$, but one retains only those arguments b for which the points $(b_j(a), a)$ belong to a continuous curve that eventually reaches the horizontal axis, $a = 0$. These curves often bifurcate as a tends to zero. It is believed that the pattern of these branchings is a symbolic representation of a multiplicative cascade. This means that we are looking for a multiscale autoregressive process that would yield the data set.

The next step consists of computing the partition functions

$$\mathcal{Z}(p, a) = \sum_j |W(a, b_j(a))|^p, \qquad (9.14)$$

where the sum runs over the skeleton of the function f, that is, over those connected lines of local maxima. Finally, one hopes to find a power-law behavior that reads

$$\mathcal{Z}(p, a) \sim a^{\tau(p)}. \qquad (9.15)$$

This optimistic search is made by looking at a plot of $\log \mathcal{Z}(p, a)$ versus $\log a$. The anticipated result is a linear plot whose slope $\tau(p)$ does not depend on the choice of analyzing wavelet ψ that is used to compute $W(f; a, b)$.

---

[6] See [7] and [16] for a more complete account of Arneodo's work. More about the WTMM and Mallat's work can be found in [181], [182], and [184].

The use of partition functions such as (9.14) leads to intractable mathematical difficulties (see [154] for a discussion), and although the technique yields sharp numerical results, the numerical algorithms must be handled very carefully. There is, however, a variant of (9.14) that is mathematically robust. Again, the idea is to eliminate small values of the wavelet transform that carry no information. This idea was discussed above, but here the recipe is slightly different. We replace $|W(f; y, x)|$ in (9.8) by

$$d(a,b) = \sup_{(a',b')} |W(f; a', b')|,$$

where the supremum is taken over the box $[0, a] \times [b - a, b + a]$. It can be shown that even for $p < 0$, the new exponent $\eta(p)$ that we obtain from the relation

$$\int d^p(a,b)\, db \sim a^{\eta(p)} \tag{9.16}$$

is independent of the wavelet ψ. The exponent $\eta(p)$ is not altered by the addition of a smooth perturbation, and using $\eta(p)$ in (9.5) instead of $\zeta(p)$ actually yields the correct part of the decreasing spectrum for many mathematical functions on which we can test the validity of these formulas (see [157]). We can reformulate these results by saying that the condition

$$\int d^p(a,b)\, db \leq a^{sp}$$

defines a natural extension of the Besov spaces $B_p^{s,\infty}$ for negative values of p.

We now return to fully developed turbulence. The very accurate data obtained from experiments made at the Modane wind tunnel led Frisch and Parisi to make the conjecture that $D(h)$ is a "universal" function, which means that it is independent of the specific medium, the boundaries, and other details of a fully developed turbulent flow. If this were true, the determination of $D(h)$ would yield important information about the nature of turbulence.

It became clear to Arneodo and his colleagues that $\zeta(p)$ cannot be defined precisely using (9.15) [11]. The log-log plot of $\mathcal{Z}(p, a)$ versus a was not a straight line over the full range of scales that represent the inertial range. Something even worse happened: When the scales are restricted to a range where the behavior is approximately linear, the slope depends on the analyzing wavelet. These considerations have led some investigators to question the definition of the inertial range. This consists of the scales that lie between two extremes: the largest scale where the flow is created and the smallest scale where energy is dissipated as heat due to the viscosity. Fitting data to a modified form of Castaing's continuous-scale models led Arneodo to suggest that the scale at which dissipation occurs might fluctuate throughout the signal.

There is both good news and bad news. The bad news is that it is not possible to apply the multifractal formalism in the strict sense given by (9.15) to turbulent flows. The good news is that this failure has opened a new line of research whose objective is to gain a better understanding of the inertial range in fully developed turbulence. The dissipation scale is not constant but instead depends on the dynamical properties of the flow (see [233]).

We mentioned in Chapter 1 and again in Chapter 8 that wavelet techniques have been used to analyze DNA sequences, and this is another application of the

WTMM algorithm. One first associates a sequence of real numbers x_n with a DNA sequence consisting of the four nucleotides A, C, G, and T as follows: Select four real numbers v_a, v_c, v_g, and v_t to represent A, C, G, and T respectively; if the ith nucleotide is $j(i)$, where $j(i) \in \{A, C, G, T\}$, define

$$x_n = \sum_{i=1}^{n} v_{j(i)}. \tag{9.17}$$

One expects that the statistical properties of this sequence will yield pertinent information about the corresponding genome, and indeed this is the case. Arneodo and his coworkers have successfully applied the WTMM algorithm to compare this sequence with an fBm [13]. This technique allowed them to distinguish coding sequences (where (9.17) is statistically similar to regular Brownian motion) from noncoding sequences (where (9.17) is statistically similar to an fBm with index different from $\frac{1}{2}$). A more involved analysis of the long-range correlations in (A.1) and of its multifractal properties can be found in [12] and [8].

Finally, we note that the maxima lines of the wavelet transform are being used by Nicolleau and Vassilicos to characterize the intermittency in turbulence [219].

9.5 Coherent structures

Anyone who has seen experiments done in a water tunnel or seen one of the educational films about turbulence has surely noticed that the flow is not "completely chaotic" and that it exhibits a sort of organization at large scales, at least at relatively low Reynolds numbers. The objects we see are called coherent structures, and technically they represent local condensations of the vorticity field that last longer than other characteristic times associated with the flow [105]. Unfortunately, this is about all we can say, and an initial problem is that there is not a precise mathematical definition for the objects called coherent structures.

Coherent structures cannot be observed directly in wind-tunnel experiments at high Reynolds numbers. However, if one accepts the calculation of $D(h)$ performed by Arneodo and his team on the Modane signal, then one has a first hint of the existence of coherent structures in fully developed turbulence. An examination of the data shows the existence of negative Hölder exponents, and this implies the existence of extremely strong velocity gradients. If one does not dismiss these negative exponents $\alpha(x_0)$ as artifacts, then one interpretation of them is the rare passage of a strong vortex filament past the probe. These vortex filaments are coherent structures; they are rare and elusive, but they are thought by many experts to be one of the keys to understanding turbulence. While coherent structures are not visible in high Reynolds number wind-tunnel experiments, they are accessible in two-dimensional simulations, and this will bring us to another application of wavelets in the field of turbulence. We have already seen how wavelets are used to analyze experimental turbulence data. In section 9.7, we will describe how they are used to analyze simulated turbulence.

The detection of coherent structures and the multifractal analysis of fully developed turbulence are two completely different wavelet-based investigations. First, there is a great difference in scales: Multifractal analysis is concerned with the smallest scales, while the coherent structures that are studied are, for the most part, large-scale objects. Second, the analyses are performed on different signals: Multifractal analysis is done on the very precise one-dimensional signals obtained in wind-tunnel experiments, whereas wavelet analysis "looks for" coherent structures that are generated by two- and three-dimensional numerical simulation. In fact, numerical simulations are not capable of producing useful small-scale data, and this will probably be the case for some time to come. On the other hand, trying to understand coherent structures by analyzing one-dimensional wind-tunnel data may be like trying to understand a symphony by hearing only one instrument.

The use of digital computers to simulate turbulent flows was anticipated at the time the first stored-memory computers were being built. This is what Herman H. Goldstine and John von Neumann wrote in 1946 [131, p. 4]:

> The phenomenon of turbulence was discovered physically and is still largely unexplored by mathematical techniques. At the same time, it is noteworthy that the physical experimentation which leads to these and similar discoveries is a quite peculiar form of experimentation; it is very different from what is characteristic in other parts of physics. Indeed, to a great extent, experimentation in fluid dynamics is carried out under conditions where the underlying physical principles are not in doubt, where the quantities to be observed are completely determined by known equations. The purpose of the experiment is not to verify a proposed theory but to replace a computation from an unquestioned theory by direct measurements. Thus wind tunnels are, for example, used at present, at least in large part, as computing devices of the so-called analogy type (or, to use a less widely used, but more suggestive, expression proposed by Wiener and Caldwell: of the measurement type) to integrate the nonlinear partial differential equations of fluid dynamics.
>
> Thus it was to a considerable extent a somewhat recondite form of computation which provided, and is still providing, the decisive mathematical ideas in the field of fluid dynamics. It is an analogy (i.e., measurement) method, to be sure. It seems clear, however, that digital (in the Wiener–Caldwell terminology: counting) devices have more flexibility and more accuracy, and could be made much faster under present conditions. We believe, therefore, it is now time to concentrate on effecting the transition to such devices, and that this will increase the power of the approach in question to an unprecedented extent.

In spite of the enormous progress made since these comments, the best supercomputers cannot solve three-dimensional Navier–Stokes equations with enough resolution to yield accurately enough scales in the velocity field to verify the multifractal hypothesis. What supercomputers can do is give a good sketch of the solution at (relatively) low Reynolds numbers.

One of the chief advocates for studying turbulence in physical space and, in particular, for studying coherent structures has been Norman Zabusky, who is perhaps best known for his discovery (in collaboration with Kruskal) of solitons. In 1977, he wrote (quoted from [264, p. 41]):

> In the last decade we have experienced a conceptual shift in our view of turbulence. For flows with strong velocity shear... or other organizing characteristics, many now feel that the spectral or wavenumber-space description has inhibited fundamental progress. The next "El Dorado" lies in the mathematical understanding of coherent structures in weakly

dissipative fluids: the formation, evolution and interaction of metastable vortex-like solutions of nonlinear partial differential equations.

The study of coherent structures is pursued both experimentally and computationally, but before we describe some of the latter work we have a few more comments on coherent structures themselves. We mentioned that what we call coherent structures are condensations (or concentrations) of vorticity. These structures include vorticity tubes, those thin, swirling miniature tornadoes that are seen in water tunnel experiments. (For a precise description of vorticity tubes, the reader is referred to [52].) Recall that vorticity is defined as the curl of the velocity, and hence vorticity concentrations corresponding to coherent structures are sources of low pressure. This fact is crucial in the experiments of Yves Couder, which are described in the next section. But before going on, we return for a moment to the negative Hölder exponents found by Arneodo and his team. They could be dismissed as artifacts, but to do so in science can lead to "missing the gold ring." We prefer to regard them as rare but important events as suggested in [15]: "One tentative interpretation could be the occasional passage near the probe of slender vortex filaments of the sort observed in numerical simulations .... At first sight it seems that this interpretation should be rejected on the ground that the probe is measuring along a line and such a line has almost surely an empty intersection with the vortex filaments which, on inertial-range scales, appear as one dimensional objects."

9.6 Couder's experiments

If seeing is believing, then the very clever experiments done by Couder and his collaborators provide convincing evidence for the existence of vortex filaments. We will describe these experiments, but first we need to return to the Navier–Stokes equations and relate the pressure to the viscosity and to

$$\sigma^2 = \frac{1}{2} \sum_{i,j} \left(\frac{\partial u_i}{\partial x_j} + \frac{\partial u_j}{\partial x_i} \right)^2.$$

Note that the energy dissipation rate was $\nu\sigma^2 = \varepsilon$. Now we have

$$\Delta p = \mu(|\omega|^2 - \sigma^2), \qquad (9.18)$$

which explains why large values of the length $|\omega|$ of the vorticity ω correspond to minima of the pressure. Here, $\mu = \frac{\rho}{2}$, where ρ is the fluid density. The identity tells us that p will be recovered as a Coulomb potential of $|\omega|^2 - \sigma^2$. This is an averaged quantity, and this implies more regularity and stability. It means that the regions where the pressure is low are well defined and easier to detect than the places where $|\omega|$ is large.

In the experiments, Couder measures the pressure $p(x,t)$ at a given space location $x = x_0$, as a function of the time variable, and he also images the regions of low pressure, which according to (9.18) correspond to vortex tubes. The pressure is measured by a piezo-electric probe. The low-pressure regions are imaged by injecting microbubbles into the flow. These microbubbles migrate toward the regions of low pressure and accumulate in regions of strong vorticity. The low-pressure vortex filaments can thus be visualized. The pressure is recorded as a function of time, as is the image, and the visualization of the depressions can be correlated with the low peaks in the recorded pressure signal.

Similar experiments have been done by S. Fauve and C. Laroch [106], and the data from these experiments have been analyzed by Patrice Abry [1] using wavelet techniques. Wavelet analysis is shown to be particularly useful in detecting and analyzing the low-pressure peaks; wavelets are used to split the signal into two components, where the first consists of these well-defined low-pressure peaks and the remainder is treated as noise. Abry and his collaborators use a statistical modeling of the "background noise." An abrupt change from this statistical model shows that a vorticity filament has been detected. They then make a statistical decision on the wavelet coefficients to determine the coefficients that are due to the vorticity filament and those that are background. They are then able to do a cascade-type analysis on the "cleaned background." Once the coefficients have been separated, Abry and his collaborators do an analysis similar to Arneodo's (see, for example, [16] and [10]). This original approach borrows ideas from two quite different points of view: cascade models and coherent structures. This kind of processing will be met again in the next section and in a more systematic way when we discuss Donoho's work in Chapter 11.

9.7 Marie Farge's numerical experiments

Marie Farge wished to detect and extract coherent structures in two-dimensional simulated turbulence. Farge explained how and why she was led to use wavelet analysis in her study of numerical simulations of two-dimensional fully developed turbulence [103, p. 289]:

> It is important to realize that the wavelet transform is not being used to study turbulence simply because it is currently fashionable; but rather because we have been searching for a long time for a technique capable of decomposing turbulent flows in both space *and* scale simultaneously. If, under the influence of the statistical theory of turbulence, we had lost in the past the habit of considering the flow evolution in physical space, we have now recovered it thanks to the advent of supercomputers and their associated means of visualization. They have revealed to us a menagerie of turbulent flow patterns, namely, the existence of coherent structures and their elementary interactions ... for which the present statistical theory is not adequate.

What Farge asks of wavelet analysis (or of any other form of time-frequency analysis) is to decouple the dynamics of the coherent structures from the residual flow. The residual flow would play only a passive role in an action whose protagonists would be the coherent structures; these "protagonists" clash or join forces according to their "sign."

One should keep in mind, however, that the Navier–Stokes equations are nonlinear, and the interactions between the coherent structures and the residual flow is one of the main difficulties in Farge's program. The decoupling is a first approximation, which is believed to be valid for only a short time. In particular, it should be noted that, unlike solitons, when two coherent structures meet, there can be a strong interaction that leads to their fusion into a new coherent structure.

Farge, after having tried several methods to extract the coherent structures from the residual flow, decided to use Victor Wickerhauser's algorithm (discussed in Chapter 7), which provides a decomposition in a basis adapted to the signal. The results are surprisingly good and are discussed in [104]. However, this methodology relies heavily on the algorithm being used. The problem here is similar to that in image processing. We are claiming—if we accept the use of the best-basis search—that a compression-oriented method can detect patterns. Coherent structures in turbulence are intricate features, and it seems remarkable that they could be extracted by using such a general tool as a best-basis search. The appropriate scientific explanation of this phenomenon remains to be found.

9.8 Modeling and detecting chirps in turbulent flows

Chirps have been mentioned in several places and contexts. In Chapter 5, we defined a chirp to be a function f of the form $f(t) = A(t)e^{i\varphi(t)}$, where A and φ satisfy the conditions

$$\left|\frac{A'(t)}{A(t)\varphi'(t)}\right| \ll 1 \quad \text{and} \quad \left|\frac{\varphi''(t)}{(\varphi'(t))^2}\right| \ll 1. \tag{9.19}$$

Linear chirps and hyperbolic chirps were also defined in Chapter 5 in the discussion of the Wigner–Ville transform. Chirps appeared again in Chapter 6 in the construction of chirplets and chirplet bases, and we mentioned the specific chirp, $f(t) = (t - t_0)^{-1/4} \cos[\omega(t_0 - t)^{5/8} + \theta]$, in a brief discussion of gravitational waves. We are now going to discuss chirps in turbulent flows.

Determining if there are chirps in fully developed turbulence is a current research issue. Superficially, is seems to be related to the problem of understanding coherent structures. In fact, it is possible to imagine coherent structures as being two- and three-dimensional chirps. Thus, being able to say something about the existence and distribution of chirps might have implications about coherent structures and their distribution. Again we emphasize that our "window on turbulence" is a one-dimensional velocity signal, so the immediate question is whether or not this signal contains chirps.

We mentioned in section 9.4 that chirps in the signal can cause the multifractal formalism to fail. This problem is another motivation for studying chirps in the context of wavelets and the multifractal formalism. Indeed, to establish the multifractal formalism, we made the assumption that the wavelet transform satisfies

$$|W(a,b)| \sim a^h$$

in the cones $|b - t_0| \leq Ca$ above a point where the Hölder exponent is h. This contains implicitly the very specific assumption that there exist only "vertical" ridges in the signal, and such ridges correspond to a cusp singularity like $|t - t_0|^h$. This means that if we wish to detect chirps with a multifractal formalism, then the standard multifractal formalism must be modified. This has been done, and we will describe this "grand canonical" multifractal formalism after some preliminary comments.

It is unrealistic to expect to find an algorithm to detect the general (nonparametric) behavior described by (9.19). If we wish to detect chirps, then it is necessary to be more specific about the objects we are trying to detect. What we want is a parametric model of a chirp. Several mathematical models have been proposed. Perhaps the simplest way to make (9.19) more specific is to require power-law behaviors as $t \to t_0$, and the simplest way to do this is to model chirps by the functions $f_{h,\beta}$ defined by

$$f_{h,\beta}(t) = |t - t_0|^h e^{i|t-t_0|^{-\beta}}, \tag{9.20}$$

where h is the usual Hölder exponent and $\beta > 0$ is called the *oscillation exponent*. We wish to find an algorithm that yields h and β when $f_{h,\beta}$ is the analyzed function, but we also want the algorithm to yield h and β if the analyzed function "looks like" $f_{h,\beta}$ near t_0. This is similar to the situation where an algorithm yields the Hölder exponent when the function $f(t) = |t - t_0|^h$ is analyzed, but it also detects any function whose Hölder exponent is h. Thus, we see that there is the problem of defining the class of functions that "look like" (9.20) at t_0. Defining functions whose Hölder exponent is h at t_0 was relatively simple (recall (2.4) and (2.5)); saying what it means for a function to have a chirp at t_0 is more subtle because it must involve the oscillation exponent β, and there is not yet a universally accepted definition.

Yves Meyer observed that if one integrates (9.20) n times, then the Hölder exponent of f at t_0 becomes $h+n(1+\beta)$. (This is easily checked by repeated integration by parts.) This means that the oscillation exponent β causes the Hölder exponent to increase by $1 + \beta$ after each integration rather than by 1, as might be expected. This led to the following definition.

DEFINITION 9.1. *f has a chirp of type (h, β) at t_0 if f is $C^h(t_0)$ and if the iterated primitives $f^{(-n)}$ are $C^{h+n(1+\beta)}(t_0)$.*

Note that this definition cannot be compared with (9.19) because there is no differentiability assumption in a neighborhood of t_0. This definition is, however, consistent with the "ridge heuristic": Meyer proved that this behavior can be characterized by precise decay estimates of $W(a,b)$ as (a,b) moves away from the ridge $a = |b - t_0|^{1+\beta}$ (see [160]). We will see in the next chapter that Riemann's function $\sum \frac{1}{n^2} \sin \pi n^2 t$ has a chirp at $t_0 = 1$ according to this definition. Although the definition is well adapted to the study of mathematical examples like Riemann's function, it is not well adapted to practical signal analysis. Indeed, a minimal requirement for a definition to be useful for real data processing is invariance under the superposition of smooth noise of small amplitude. This is not the case here as shown by the following example. Let

$$f(t) = t^h \sin \frac{1}{t^\beta} + A \sum 2^{-Hj} \sin(2^j t), \qquad (9.21)$$

where $H \gg h$ and A is small. This models a chirp (the first term) plus smooth noise of small amplitude (the second term). For $A = 0$, f has a chirp of type (h, β) at $t = 0$, but as soon as $A \neq 0$, the Hölder exponent of $f^{(-n)}$ at 0 is $\inf\{h + n(1+\beta), H + n\}$. Thus if n is large enough, the Hölder exponent increases by one at each integration and the oscillation exponent of (9.21) is not β as it should be by Definition 9.1, but it is 0.

An alternative definition, which is robust with respect to the addition of smooth noise, was proposed by Arneodo, Bacry, Jaffard, and Muzy [9]. It is based on the experimental observation that one can see the oscillations of the chirp in the graph of (9.21) as long as $H > h$ and on the belief that in this case, the oscillation exponent should reflect these oscillations. The next definition meets this requirement.

DEFINITION 9.2. *Let $(I - \frac{d^2}{dt^2})^{-s/2}$ denote the fractional integral of order s and let $h_s(t)$ be the Hölder exponent of $(I - \frac{d^2}{dt^2})^{-s/2} f$. The oscillation exponent of f at t_0 is defined to be $\lim_{s \to 0} \frac{h_s(t_0) - h_0(t_0)}{s} - 1$, where $h_0(t_0) = h(t_0)$ is the ordinary Hölder exponent of f at t_0.*

(Note that the operator $(I - \frac{d^2}{dt^2})^{-s/2}$ amounts to multiplying the Fourier transform of the signal by $(1 + \xi^2)^{-s/2}$.)

To apply Definition 9.2 one needs only to check how much the Hölder exponent increases under fractional integration of infinitesimal order, while Definition 9.1 requires one to check infinitely many integrations. Although Definition 9.2 is even further from the original definition of a chirp (9.19), it also has a characterization in terms of decay away from the ridge; however, in this case, the decay is much slower than for Definition 9.1.

Having a robust definition is only the first step to detecting chirps in turbulence, but it leaves unsolved the fundamental problem: It is not at all clear what elusive chirps look like. Whatever they are, we expect them to be three-dimensional objects; in fact, we expect them to be coherent structures, perhaps like very fine vortex threads. The available signals are one-dimensional cuts, and it is not clear that one-dimensional cuts of "three-dimensional chirps" (whatever they are) are chirps as we have defined them. This problem has been studied by J.-M. Aubry in [14], and the situation is far from clear. Aubry developed a menagerie of three-dimensional chirps, and for some of them, almost every cut is a chirp. Nevertheless, as things now stand, we have no theoretical or numerical reason to believe that the possible chirps in turbulence belong to one category or another.

If one is determined to go on a snark hunt, then it is necessary to have a bag. To look for chirps in turbulence, this means that we must have an algorithm that can be applied to the one-dimensional data. There are two approaches: If one expects to detect a few isolated spiral-like structures in a noisy environment—like the problem of detecting gravitational waves—then ridge identification could be considered (see section 6.11 and [148]). However, if one believes that chirps are so pervasive that only a statistical approach makes sense, then it is reasonable to look for an extension of the multifractal formalism. Such an extension has been proposed by Stéphane Jaffard [156], and it is currently being implemented as a numerical algorithm by Alain Arneodo and his group in Bordeaux. The goal is much more ambitious than in the classical multifractal formalism. We now wish to determine the Hausdorff dimension of the set of points where there is a chirp with Hölder exponent h and oscillation exponent β. We denote this dimension by $D(h, \beta)$. This function is called the spectrum of oscillating singularities.

We are going to describe this extension, but first we wish to show by an example why the classical formalism fails. We mentioned that the derivation of the standard multifractal formalism makes the assumption that all Hölder singularities are cusp-like. An even more radical way to see the limitation of this multifractal formalism is to compare its behavior on the devil's staircase, which we call F, and on the chirp

$$f(t) = t^h \sin \frac{1}{t^\beta}. \tag{9.22}$$

We are going to compare the behavior of $W(F; a, b)$ and $W(f; a, b)$ at a small fixed scale a.

For such a fixed a, there are positive constants C_1 and C_2 such that

$$C_1 a^{\log 2/\log 3} \leq |W(F; a, b)| \leq C_2 a^{\log 2/\log 3} \tag{9.23}$$

on intervals of length a around the lines where $|W(F; a, b)|$ attains it maxima (as a function of b), and $|W(F; a, b)|$ decays rapidly outside these intervals. There are about $a^{-(\log 2/\log 3)}$ such lines at the scale a, so that the total length of the region where (9.23) holds is about $a^{1-\log 2/\log 3}$.

Similarly, for the chirp, $C'_1 a^{\frac{h}{1+\beta}} \leq |W(a,b)| \leq C'_2 a^{\frac{h}{1+\beta}}$ on intervals of length $a^{\frac{1}{1+\beta}}$ around the ridges defined by $a = |b|^{1+\beta}$. This means that the statistics based on the wavelet coefficients at a given scale will be exactly the same for the devil's staircase F and the chirp (9.22) if we choose

$$\frac{1}{1+\beta} = 1 - \frac{\log 2}{\log 3} \quad \text{and} \quad h = \frac{\log 2/\log 3}{1-\log 2/\log 3},$$

which amounts to having $h = \beta = \frac{\log 2}{\log 3 - \log 2}$.

The message from this example is this: To differentiate two such behaviors using a multifractal formalism, it is necessary to capture more information than is carried by the wavelet transform on the lines $a = $ a constant. This touches on a recurrent theme associated with the use of the wavelet transform for analyzing the local behavior of functions: It is necessary to use all of the information about the wavelet transform contained in a full neighborhood of the points t_0. This is the heuristic: One must examine "some function" of $W(a,b)$ in neighborhoods $(0,a] \times [b-a, b+a]$ as $a \to 0$. The "function" varies depending on the task. Recall that when faced with the problem of making sense of $\int |W(a,b)|^p\, db$ for $p < 0$, we were led to consider the function

$$d(a,b) = \sup |W(f; a', b')|,$$

where the supremum is taken over $(a', b') \in [0, a] \times [b-a, b+a]$. To extend the multifractal formalism to one that will detect chirps, we use a slight variation of this idea. We consider the function

$$d_s(a,b) = \sup |(a')^s W(a', b')|,$$

where the supremum is taken over $(a', b') \in (0, a] \times [b-a, b+a]$, and we define $\eta(p, s)$ by

$$\int d_{s/p}^p(a,b)\, db \sim a^{\eta(p,s)}. \tag{9.24}$$

We now use a heuristic argument similar to the one used to derive (9.3) and (9.5) to derive $D(h, \beta)$ from (9.24). We begin by estimating the contribution of the chirps with Hölder exponent h and oscillation exponent β to $d_s(a,b)$. If the box $(0, a] \times [b-a, b+a]$ is centered above such a chirp, the ridge intersects the box at $a' = a^{1+\beta}$, and at this point $|W(a', b')| \sim a^h$. Thus, $d_s(a,b) \sim a^{(1+\beta)s} a^h$ and $d_{s/p}^p(a,b) \sim a^{(1+\beta)s+ph}$. We now follow exactly the argument presented in section 9.4. The total contribution of these chirps to the left-hand side of (9.24) is thus

$$a^{(1+\beta)s+ph+1-D(h,\beta)},$$

so that

$$\eta(p,s) = \inf_{h,\beta}\{(1+\beta)s + ph + 1 - D(h, \beta)\}.$$

$\eta(p,s)$ is obtained from $1 - D(h, \beta)$ by a two-dimensional Legendre transformation. Thus, if $D(h, \beta)$ is concave, then

$$1 - D(h, \beta) = \inf\{(1+\beta)s + ph + \eta(p, s)\}.$$

The validity of this formula has been successfully tested on functions that display fractal sets of chirps [156], and as indicated above, its numerical implementation is being now undertaken by Arneodo and his team.

9.9 Wavelets, paraproducts, and Navier–Stokes equations

This section is devoted to discussing the use of wavelets for solving the Navier–Stokes equations numerically. Divergence-free orthonormal wavelet bases were first found by Battle and Federbush [30], and the construction was later improved by P. G. Lemarié-Rieusset [171]. It is natural to try using these bases in place of finite element methods in numerical schemes. By doing so, the Galerkin method is consistent with the invariance of the Navier–Stokes equations under certain transformations. We begin by writing these equations:

$$\begin{cases} \dfrac{\partial u}{\partial t} = \nu \Delta u - (u_1 \partial_1 + u_2 \partial_2 + u_3 \partial_3) u - \nabla p, \\ \partial_1 u_1 + \partial_2 u_2 + \partial_3 u_3 = 0, \\ u(x, 0) = u_0(x), \end{cases} \tag{9.25}$$

where $x = (x_1, x_2, x_3)$ belongs to $\mathbb{R}^3$. In our model problem, there is no boundary, the fluid fills the space $\mathbb{R}^3$, and there are no external forces. The system (9.25) contains four unknown functions u_1, u_2, u_3, and p and consists of four equations (plus an initial condition), so the balance is correct. The transformations we considered are the group actions defined by

$$\begin{cases} u(x,t) \to \lambda u(\lambda x, \lambda^2 t), \quad \lambda > 0, \\ u(x,t) \to u(x - y, t), \quad y \in \mathbb{R}^3. \end{cases}$$

The Battle–Federbush basis provides a Galerkin scheme into which the affine group actions can be incorporated. The Battle–Federbush basis is $2^{3j/2} \vec{\psi}(2^j x - k)$, $j \in \mathbb{Z}$, $k \in \mathbb{Z}^3$, $\vec{\psi} \in A$, where A is a collection of 14 divergence-free wavelets. This basis spans the closed subspace of $L^2(\mathbb{R}^3) \times L^2(\mathbb{R}^3) \times L^2(\mathbb{R}^3)$ defined by $u = (u_1, u_2, u_3)$, $u_j \in L^2(\mathbb{R}^3)$, and $\partial_1 u_1 + \partial_2 u_2 + \partial_3 u_3 = 0$. We have $\vec{\psi} = (\psi_1, \psi_2, \psi_3)$, and these three functions belong to the Schwartz class. Furthermore, in one of the constructions [207], the Fourier transform of each of these functions is supported by the annulus defined by

$$\frac{2}{3}\pi \leq \sup\{|\xi_1|, |\xi_2|, |\xi_3|\} \leq \frac{8}{3}\pi.$$

The numerical scheme that follows is aimed at decoupling the Navier–Stokes into a sequence of equations. This idea was proposed by several authors, including P. Frick and V. Zimin, whom we quote [116, p. 265]:[7]

> Ideas, like the ones used to create wavelet analysis, were proposed by Zimin (1981) for construction of a hierarchical model of turbulence.
> In a paper by Zimin (1981) a special functional basis has been presented. Functions of this basis are related to a hierarchical system of vortices of different sizes. The number of vortices in a unit volume increases with decreasing size and each function is well localized both in

---
[7]References cited here are in the original article.

Fourier and physical spaces. The product of the characteristic scales of localization in Fourier and coordinate spaces satisfies the uncertainty condition.

The cascade equations, written for the quantities A_i, each define the velocity oscillations in the interval of wave numbers and describe the principal characteristics of energy redistribution processes between different scales. The cascade equations minimize the dimensionality of systems, which describe the turbulent flows in a wide range of wave numbers, and have a form

$$d_t A_i = \sum_{j,k} X_{ijk} A_j A_k + Y_i A_i + F_i,$$

where F_i characterize the energy sources in corresponding interval of spectrum ....

The hierarchical model of turbulence is based on the natural assumption that turbulence is an ensemble of vortices of progressively diminishing scales. The hierarchical basis for two-dimensional (2D) turbulence describes the ensemble of the vortices, in which any vortex of the given size consists of four vortices of half size and so on. The ensemble of vortices of the same size forms a "level." The functions of the hierarchical basis are constructed in such a way that Fourier-images of vortices of [a] single level occupy only [a] single octave in the wave-number space and regions of localization of different levels in the Fourier space do not overlap.

The wave-number space is divided at ring zones such that

$$\pi 2^N < |k| < \pi 2^{N+1}.$$

This quotation from Frick and Zimin implies that these authors are using the Shannon wavelet basis. This remark is made explicitly in their paper. The scaling function of the one-dimensional Shannon basis is the sinc function $\varphi(t) = \frac{\sin \pi t}{\pi t}$, while the corresponding mother wavelet is given by

$$\psi(t) = 2\varphi(2t) - \varphi(t).$$

These functions have poor localization in the coordinate space, although they have an ideal localization in the frequency domain. Shannon's wavelets correspond to ideal filters in signal processing, and these ideal filters are unrealistic because their numerical support is infinite. For the same reason, Shannon's wavelets cannot be used in numerical analysis. Indeed, the nonlinear terms that appear in the Galerkin scheme do not have a rapid off-diagonal decay, and the Navier–Stokes equations are not decoupled in the Shannon's basis. The obvious question is, What happens if Shannon's basis is replaced with the Battle–Federbush basis?

The corresponding Galerkin scheme looks like this:

$$u(x,t) = \sum_{\lambda \in \Lambda} \alpha_\lambda(t) \psi_\lambda(x), \tag{9.26}$$

where ψ_λ is a condensed notation for the Battle–Federbush basis, and

$$\frac{d}{dt}\alpha_\lambda(t) = \sum_{\lambda'} \eta(\lambda, \lambda') \alpha_{\lambda'}(t) + \sum_{\lambda', \lambda''} \beta(\lambda, \lambda', \lambda'') \alpha_{\lambda'}(t) \alpha_{\lambda''}(t), \tag{9.27}$$

where $\eta(\lambda,\lambda') = (\Delta\psi_{\lambda'},\psi_\lambda)$, $\beta(\lambda,\lambda',\lambda'') = b(\psi_{\lambda'},\psi_{\lambda''},\psi_\lambda)$, and

$$b(u,v,w) = \sum_{k=1}^{3}\sum_{l=1}^{3}\int u_k(x)(\partial_k v_l)(x)\overline{w}_l(x)\,dx. \qquad (9.28)$$

The Navier–Stokes equations are decoupled by this Galerkin scheme if and only if both $\eta(\lambda,\lambda')$ and $\beta(\lambda,\lambda',\lambda'')$ have a rapid off-diagonal decay.

Concerning $\eta(\lambda,\lambda')$, we observe that $\eta(\lambda,\lambda') = 0$ whenever $|j'-j| \geq 3$. (The wavelet ψ_λ is located at $k2^{-j}$ and the corresponding scale is 2^{-j}.) If $|j'-j| \leq 2$, then these coefficients $\eta(\lambda,\lambda')$ have a rapid off-diagonal decay, since the wavelets ψ_λ belong to the Schwartz class. If the Shannon wavelets were used, this would not be the case, which rules out this basis.

Now for the bad news. Even if the Battle–Federbush basis is used, $\beta(\lambda,\lambda',\lambda'')$ takes large off-diagonal values. Indeed, if $\lambda' = \lambda''$ and $j \to -\infty$, then $\beta(\lambda,\lambda',\lambda'')$ does not decay rapidly. This problem appears whenever one considers the product fg of two functions f and g whose Fourier transforms are supported by the ring $R \leq |\xi| \leq 2R$, where R is large. In this situation, the product fg may generate large low-frequency terms.

This remark serves as an introduction to the so-called paraproduct algorithms that apply to the pointwise product between two nonsmooth functions f and g. Paraproduct algorithms are a way to analyze the application of nonlinear operators on highly oscillating functions. It is a way of rewriting the result as a hierarchy of terms that are easier to analyze. These techniques have been used successfully in the mathematical resolution of fluid dynamics equations (see [45], [50], [51], and [207]).

Taking a broader perspective, we note that the pertinence of wavelet methods to the numerical solution of partial differential equations remains an unclear issue. One significant drawback is the lack of flexibility in constructing wavelets adapted to complicated geometry. Equally significant is the fact that multigrid algorithms, which share many of the desirable properties of wavelet algorithms, attained a mature development before wavelets were introduced. One point where wavelets seem to be competitive is in local refinements. It is easy to add a few new wavelets where "something" seems to be happening, whereas local refinements of meshes of finite elements are more complicated to handle. (We suggest [56] by A. Cohen, W. Dahmen, and R. DeVore, where these questions are discussed.)

9.10 Hausdorff measure and dimension

Hausdorff dimension is a mathematical tool that allows one to quantify the fractal behavior of the functions and measures that have been mentioned in this chapter. It is a key tool in the mathematical development of multifractal analysis, and thus this last section also provides background for the next chapter.

We are mainly interested in Hausdorff dimension, but to get there it is necessary to pass through the definition of Hausdorff measure. (Hausdorff measure appears once in section 11.4.) Our discussion follows that of Falconer in [101], and we recommend this book to anyone who wishes to learn more about these ideas.

For any nonempty subset $U \subset \mathbb{R}^n$, the diameter of U is defined to be

$$|U| = \sup\{|x-y| \mid x,y \in \mathbb{R}^n\}.$$

An ε-cover of a subset $A \subset \mathbb{R}^n$ is any countable (or finite) collection of set $\{U_i\}$ such that $A \subset \cup_{i=1}^{\infty} U_i$ and $0 < |U_i| \leq \varepsilon$.

For any subset $A \subset \mathbb{R}^n$, any $s \geq 0$, and any $\varepsilon > 0$, define

$$\mathcal{H}^s_\varepsilon(A) = \inf \sum_{i=1}^\infty |U_i|^s, \qquad (9.29)$$

where the infimum is taken over all ε-covers of A. If $\varepsilon' < \varepsilon$, then every ε'-cover is an ε-cover, and hence $\mathcal{H}^s_{\varepsilon'}(A) \geq \mathcal{H}^s_\varepsilon(A)$. Thus $\mathcal{H}^s_\varepsilon(A)$ tends to a limit as $s \to 0$, and we write

$$\mathcal{H}^s(A) = \lim_{\varepsilon \to 0} \mathcal{H}^s_\varepsilon(A). \qquad (9.30)$$

$\mathcal{H}^s(A)$, which is often infinite, is called the s-dimensional Hausdorff measure of $A \subset \mathbb{R}^n$. It can be shown that $\mathcal{H}^s$ is a measure, and, in fact, n-dimensional Hausdorff measure is, up to a constant multiple, Lebesgue measure. We are not concerned here with Hausdorff measure, so we will move directly to the definition of Hausdorff dimension.

Observe that if $t > s$ and $\{U_i\}$ is an ε-cover of A, then

$$\sum_{i=1}^\infty |U_i|^t \leq \varepsilon^{t-s} \sum_{i=1}^\infty |U_i|^s.$$

Taking the infimum of both sides shows that $\mathcal{H}^t_\varepsilon(A) \leq \varepsilon^{t-s}\mathcal{H}^s_\varepsilon(A)$. If $\mathcal{H}^s(A) < \infty$, then by taking the limit as $\varepsilon \to 0$ we see that $\mathcal{H}^t(A) = 0$. In short, $t > s$ and $\mathcal{H}^s(A) < \infty$ imply that $\mathcal{H}^t(A) = 0$. A direct consequence is that $\mathcal{H}^t(A) = 0$ for all $A \subset \mathbb{R}^n$ whenever $t > n$: Since $\mathcal{H}^n$ is Lebesgue measure (up to a factor), the $\mathcal{H}^n$ measure of the unit ball in $\mathbb{R}^n$ is finite; it follows that $\mathcal{H}^t(\mathbb{R}^n) = 0$ if $t > n$.

The Hausdorff dimension of A is then defined as

$$\dim_H(A) = \inf\{s \mid \mathcal{H}^s(A) = 0\}. \qquad (9.31)$$

If $\mathcal{H}^0(A) < \infty$, then it follows from the definition that A is finite. Thus with this exception, it is clear from the discussion that

$$\dim_H(A) = \inf\{s \mid \mathcal{H}^s(A) = 0\} = \sup\{s \mid \mathcal{H}^s(A) = \infty\}. \qquad (9.32)$$

Thus the Hausdorff dimension of F is the point where the graph of $\mathcal{H}^s(A)$ as a function of s "jumps" from infinity to zero.

The Hausdorff dimension agrees with the ordinary definition of dimension for smooth objects: A smooth curve in $\mathbb{R}^n$ has Hausdorff dimension one, a smooth surface has Hausdorff dimension two, and, in general, a smooth m-dimensional manifold has Hausdorff dimension m. In particular, the unit sphere in $\mathbb{R}^n$ has Hausdorff dimension $n-1$. On the other hand, the Hausdorff dimension of Cantor's triadic set is $\frac{\log 2}{\log 3}$, as shown by Hausdorff in [140].

CHAPTER 10

Wavelets and Multifractal Functions

10.1 Introduction

We presented the conjecture of Frisch and Parisi concerning the multifractal nature of the velocity of a turbulent fluid in the previous chapter on wavelets and turbulence. They introduced the hypothesis that there is a set of points with Hausdorff dimension $D(h)$ where the velocity increments satisfy

$$|v(x + \Delta x, t) - v(x, t)| \sim |\Delta x|^h,$$

and from this they argued that

$$\int_{\mathbb{R}} |v(x + \Delta x, t) - v(x, t)|^p \, dt \sim |\Delta x|^{\zeta(p)} \tag{10.1}$$

as $|\Delta x| \to 0$, where

$$\zeta(p) = \inf_h \{hp + 1 - D(h)\}. \tag{10.2}$$

We are interested in $D(h)$ because it tells us about the fractal or multifractal nature of fully-developed turbulence, but $\zeta(p)$ is the quantity we can compute numerically. Fortunately, under the assumption that $D(h)$ is concave, it can be recovered from $\zeta(p)$ by a classical Legendre inversion formula

$$D(h) = \inf_p \{hp + 1 - \zeta(p)\}. \tag{10.3}$$

Since Hölder exponents and Hausdorff dimensions cannot be reasonably computed numerically, (10.3) is the only way to obtain the spectrum of singularities of a signal. Unfortunately, our understanding of this formula is quite poor; there are examples and counterexamples of its validity. (See [154] for a discussion.) The good news is that we can test (10.3) on several mathematically defined functions for which both sides of the equality can be computed independently, and this provides some intuition about the range of validity and the limitations of (10.3). We present two examples of functions that are fractal or multifractal, and we show how wavelet methods can be used to compute their Hölder exponents and their spectrums of singularities. The two functions we study are the Weierstrass function

$$\mathcal{W}(t) = \sum_{n=0}^{\infty} B^n \cos(A^n t),$$

where $0 < B < 1$ and $AB \geq 1$, and the Riemann function

$$\mathcal{R}(t) = \sum_{n=1}^{\infty} \frac{1}{n^2} \sin(\pi n^2 t).$$

In addition to exhibiting an example for which (10.3) holds, the analysis of the Riemann function will provide an opportunity to compare the performances of wavelets and Fourier analysis in the context of "multifractal analysis."

This chapter is more technical than the others, in the sense that we have chosen to present the proofs of certain results. The proofs have been selected to illustrate techniques that we feel are basic to wavelet analysis. On the other hand, we warn the reader that this does not imply that the chapter is self-contained. To obtain a balance between telling the story and avoiding too much detail, we refer to other sources for certain key results and proofs. This chapter differs from the others in another respect: The emphasis is on the use of wavelets to analyze the detailed structure of functions, and thus it illustrates the use of wavelets "within mathematics," as alluded to at the end of section 1.7.

10.2 The Weierstrass function

Historians tell us that Karl Weierstrass mentioned the function $\mathcal{R}$ in a talk to the Academy of Sciences in Berlin on 18 July 1872 and indicated that Riemann had introduced this function to warn mathematicians that a continuous function need not have a derivative [97]. This function, which first appeared in print in 1875 in [96], has come to be known as Riemann's function, although there seems to be no written evidence, other than that given by Weierstrass, that connects Riemann directly with this function. (See [43] for a fascinating discussion of the mystery surrounding the origin of $\mathcal{R}$.)

Weierstrass was not able to analyze $\mathcal{R}$. Instead, he introduced the much more lacunary series $\mathcal{W}(t) = \sum_{n=0}^{\infty} B^n \cos(A^n t)$, $0 < B < 1$, and showed that if A is an odd integer and if AB is sufficiently large, then $\mathcal{W}$ is nowhere differentiable. We will see that the result is true if $AB \geq 1$. (Weierstrass's proof can be found in his collected work [256].)

We intend to show that the function $\mathcal{W}(t) = \sum_{j=0}^{\infty} B^j \cos(A^j t)$ is nowhere differentiable and that the same is true for the function $\widetilde{\mathcal{W}}(t) = \sum_{j=0}^{\infty} B^j \sin(A^j t)$. These proofs will use wavelet analysis, which in this example appears in general outline as a form of Littlewood–Paley analysis. The method we follow is due to Geza Freud [115]. The proof is quite simple, but it is based on an important aspect of wavelet analysis: Analyzing wavelets abound, and success follows from a judicious choice.

We begin by defining the wavelet ψ in terms of its Fourier transform $\hat{\psi}$. We first require that $\hat{\psi}$ satisfy the following three conditions:

(a) $\hat{\psi}(\xi) = 0$ if $\xi \leq A^{-1}$, $A > 1$ (in particular on $(-\infty, 0]$).
(b) $\hat{\psi}(\xi) = 0$ if $\xi \geq A$.
(c) $\hat{\psi}(1) = 1$.

Since there is no problem in doing so, we will assume that $\hat{\psi}$ is infinitely differentiable. By construction, $\hat{\psi}^{(k)}(0) = 0$, so $\int t^k \psi(t)\, dt = 0$ for $k \in \mathbb{N}$. Furthermore, since $\hat{\psi}$ is infinitely differentiable and has compact support, ψ is in the Schwartz class, and, in particular, $|t|^k |\psi(t)| \to 0$ as $|t| \to +\infty$ for all $k \in \mathbb{N}$. This is more than is needed for the proof, but it is there for the asking.

Write $\psi_j(t) = A^j\psi(A^j t)$, $j \in \mathbb{N}$, and denote the convolution operators $f \mapsto f * \psi_j$ by Δ_j. These operators constitute a sequence of bandpass filters. The analysis of a real function f using the sequence Δ_j resembles a Littlewood–Paley analysis that would be carried out on the analytic signal whose real part is f. Freud's method is based on the following lemma.

LEMMA 10.1. *Let f be a bounded, continuous function of the real variable t. If f is differentiable at t_0, then*

$$\Delta_j f(t_0) = A^{-j}\varepsilon_j, \quad \text{where } \varepsilon_j \to 0 \text{ as } j \to +\infty.$$

Proof. By definition, $\Delta_j f(t_0) = A^j \int f(t_0 - t)\psi(A^j t)\,dt$. We can write

$$f(t_0 - t) = f(t_0) - tf'(t_0) + t\varepsilon(t),$$

where $\varepsilon(t) \to 0$ as $t \to 0$ and $|\varepsilon(t)| \leq C$ for some $C > 0$. This gives three terms for $\Delta_j f(t_0)$. The first two are zero because $\int \psi(t)\,dt = \int t\psi(t)\,dt = 0$. The third term is $A^j \int t\varepsilon(t)\psi(A^j t)\,dt = A^{-j}\int \varepsilon(A^{-j}t)t\psi(t)\,dt$. But we have $|\varepsilon(A^{-j}t)| \leq C$, $\lim_{j\to +\infty} \varepsilon(A^{-j}t) = 0$ (simple convergence), and $\int |t||\psi(t)|\,dt < \infty$. From this it follows that $\varepsilon_j = \int \varepsilon(A^{-j}t)t\psi(t)\,dt \to 0$ as $j \to +\infty$. □

To prove Weierstrass's result, we apply the operators Δ_j to the two functions

$$\mathcal{W}(t) = \sum_{j=0}^{\infty} B^j \cos(A^j t) \quad \text{and} \quad \widetilde{\mathcal{W}}(t) = \sum_{j=0}^{\infty} B^j \sin(A^j t).$$

By direct computation,

$$(\Delta_j \mathcal{W})(t) = \frac{1}{2}A^{-j}(AB)^j e^{iA^j t} \quad \text{and} \quad (\Delta_j \widetilde{\mathcal{W}})(t) = -i\frac{1}{2}A^{-j}(AB)^j e^{iA^j t}.$$

Lemma 10.1 applies, and since $(AB)^j e^{iA^j t} \not\to 0$ as $j \to +\infty$, the conclusion is that $\mathcal{W}$ and $\widetilde{\mathcal{W}}$ are nowhere differentiable.

We pause to make an observation about the choice of the analyzing wavelet ψ. If we had initially chosen $\hat{\psi}$ to be real-valued and even, with $\hat{\psi}(\xi) = 0$ for $|\xi| \leq A^{-1}$ and for $|\xi| \geq A$, then the analyzing wavelet ψ would have been real-valued and even. This would have led to $(\Delta_j \widetilde{\mathcal{W}})(t) = B^j \sin(A^j t)$, and we could not have concluded from Lemma 10.1 that $\widetilde{\mathcal{W}}$ is not differentiable at $t = pA^{-q}\pi$ whenever A is an integer. From this example we see the merit of choosing an analyzing wavelet that is analytic: The information contained in $\Delta_j f(t)$ is more specific.

This choice of analyzing wavelets loses its importance if we rephrase Lemma 10.1 in the following, more precise form.

LEMMA 10.2. *Let f be a bounded, continuous function of the real variable t. If f is differentiable at t_0, then there exists a function η defined for $x \geq 0$ such that it is increasing, it is continuous at 0 with $\eta(0) = 0$, and*

$$|\Delta_j f(t_1)| \leq |t_1 - t_0|\eta(|t_1 - t_0|) + A^{-j}\eta(A^{-j}) \tag{10.4}$$

for all $j \geq 0$ and all real t_1.

Proof. The proof is similar to that of Lemma 10.1, but here it is necessary to do some tinkering to separate the parameters $t_0 - t_1$ and A^{-j} in the error term. As before, we write

$$f(t) = f(t_0) + (t - t_0)f'(t_0) + (t_0 - t)\varepsilon(t_0 - t),$$

and then

$$\Delta_j f(t_1) = \int (t_0 - t_1 + A^{-j}t)\varepsilon(t_0 - t_1 + A^{-j}t)\psi(t)\, dt. \tag{10.5}$$

Define the function β on $[0, +\infty)$ by $\beta(h) = \sup_{|t| \leq h} |\varepsilon(t)|$. Then β has the following properties:

(i) β is continuous and bounded on $[0, +\infty)$ and $\beta(0) = 0$.
(ii) β is monotonic nondecreasing, that is, $\beta(h_1) \leq \beta(h_2)$ when $h_1 < h_2$.
(iii) $(u+v)\beta(u+v) \leq 2u\beta(2u) + 2v\beta(2v)$ whenever $u \geq 0$ and $v \geq 0$.

Property (iii) follows from (ii). If $u \geq v$, then

$$(u+v)\beta(u+v) \leq 2u\beta(2u) \leq 2u\beta(2u) + 2v\beta(2v).$$

Returning to (10.5), we have

$$|\Delta_j f(t_1)| \leq \int |t_0 - t_1 + A^{-j}t||\varepsilon(t_0 - t_1 + A^{-j}t)||\psi(t)|\, dt$$
$$\leq \int (|t_0 - t_1| + |A^{-j}t|)\, \beta(|t_0 - t_1| + A^{-j}|t|)|\psi(t)|\, dt$$
$$\leq 2|t_0 - t_1|\beta(2|t_0 - t_1|)\int |\psi(t)|\, dt + 2 \cdot A^{-j} \int \beta(2 \cdot A^{-j}|t|)|t||\psi(t)|\, dt.$$

By taking η to be the function defined by

$$\eta(h) = 2\sup\left\{\beta(2h)\int |\psi(t)|\, dt,\ \int \beta(2h|t|)|t||\psi(t)|\, dt\right\},$$

we arrive at the statement of the lemma. □

The proofs of the two lemmas use only the following properties of the wavelet ψ: $\int \psi(t)\, dt = \int t\psi(t)\, dt = 0$ and $t\psi(t) \in L^1(\mathbb{R})$. This leaves plenty of room for choosing a wavelet to fit the task at hand.

To see the advantage of Lemma 10.2 over Lemma 10.1, suppose that we had made the "bad choice" of a real, even wavelet, in which case we ended up with $(\Delta_j \widetilde{\mathcal{W}})(t) = B^j \sin(A^j t)$, and we were not able, using Lemma 10.1, to reach the desired conclusion. The result follows from Lemma 10.2, however. For example, for $t_0 = 0$, take $t_1 = \frac{\pi}{2}A^{-j}$ so that $\sin A^j t_1 = 1$.

The statement of Lemma 10.2 comes close to being a necessary and sufficient condition for differentiability at t_0. (The sharpest results about computing the regularity of a function using the wavelet transform can be found in [160] and [208].)

10.3 Regular points in an irregular background

We now propose to determine the points x_0 where a function, which may be very irregular at other points, has a given Hölder regularity. This form of regularity is expressed by the following condition: For $0 < \alpha < 1$, f is said to be $C^\alpha(x_0)$ if there exists a $C > 0$ such that

$$|f(x) - f(x_0)| \leq C|x - x_0|^\alpha.$$

If there exists a constant C such that this relation holds uniformly for all $x_0 \in \mathbb{R}$, we say that f is in the Hölder space $C^\alpha(\mathbb{R})$ and write $f \in C^\alpha(\mathbb{R})$. (Note that these definitions are consistent with those given in section 2.2.)

We discussed the Grossmann–Morlet analysis of a function f in $L^2(\mathbb{R})$ in section 2.7. There we introduced the notation

$$W(a,b) = \langle f, \psi_{(a,b)} \rangle,$$

where

$$\psi_{(a,b)}(x) = a^{-1/2} \psi\left(\frac{x-b}{a}\right), \quad a > 0, \quad b \in \mathbb{R}.$$

The term $a^{-1/2}$ was chosen so $\|\psi_{(a,b)}\|_2 = \|\psi\|_2$, since we were interested in an L^2 analysis. In the current chapter, we are interested in the analysis of functions in L^∞, and we change the normalizing factor to a^{-1} so $\|\psi_{(a,b)}\|_1 = \|\psi\|_1$. Thus,

$$W(a,b) = \frac{1}{a} \int_\mathbb{R} f(x) \overline{\psi}\left(\frac{x-b}{a}\right) dx. \tag{10.6}$$

This transform make sense if, for example, f is bounded and $\psi \in L^1(\mathbb{R})$.

We also mentioned in Chapter 2 the reconstruction formula

$$f(x) = \int_{a>0} \left(\int_{b \in \mathbb{R}} W(a,b) \psi\left(\frac{x-b}{a}\right) \frac{db}{a} \right) \frac{da}{a}, \tag{10.7}$$

which converges in the sense of $L^2(\mathbb{R})$ when the wavelet ψ satisfies appropriate conditions. If f is bounded, then under suitable conditions on f and ψ, the inversion formula (10.7) holds at all points where f is continuous. (Precise statements and technical details for two inversion theorems are given in Appendix B.)

Wavelet analysis provides a direct and rather easy access to the pointwise behavior of functions and signals. This statement often has been used as an advertisement for wavelet analysis. We wish to back up this claim with a precise mathematical formulation when the pointwise behavior is measured by the Hölder exponent $\alpha(x_0)$. This goal will be reached with Theorem 10.1.

The first result we will prove is a simple generalization of Lemma 10.2. It states that if $f \in C^\alpha(x_0)$, then $|W(a,b)| \leq C(a^\alpha + |b - x_0|^\alpha)$, where a^α has replaced $A^{-j}\eta(A^{-j})$ and $|b - x_0|^\alpha$ has replaced $|t_1 - t_0|\eta(|t_1 - t_0|)$. (Note that t_0 is now x_0, t_1 is b, and A^{-j} is a.)

Here, and elsewhere in this chapter, we require the analyzing wavelet ψ to satisfy at least the condition $|\psi(x)| \leq C(1 + |x|^2)^{-1}$. Of course, we require the usual condition $\int \psi(x)\, dx = 0$. Other conditions will be added as needed. As shown in section 10.2, there are plenty of wavelets with these properties. We are going to state and prove the next few results under the assumption that $\alpha \leq 1$. This hypothesis is not essential, and the results extend to $\alpha > 1$ (see [151]).

LEMMA 10.3. *If f is bounded and in $C^\alpha(x_0)$, then its wavelet transform satisfies this condition: There exists a constant $C > 0$ such that, if $a \leq 1$ and $|b - x_0| \leq 1$, then*

$$|W(a,b)| \leq C(a^\alpha + |b - x_0|^\alpha). \tag{10.8}$$

Proof. The proof is simpler than that of Lemma 10.2. Since $\int \psi(x)\,dx = 0$, we can write

$$W(a,b) = \frac{1}{a} \int [f(x) - f(x_0)] \overline{\psi}\left(\frac{x-b}{a}\right) dx.$$

Then

$$|W(a,b)| \leq \frac{1}{a} \int |f(x) - f(x_0)| \left|\psi\left(\frac{x-b}{a}\right)\right| dx$$

$$\leq \frac{C}{a} \int |x - x_0|^\alpha \left|\psi\left(\frac{x-b}{a}\right)\right| dx,$$

and by making the change of variable $u = \frac{x-b}{a}$, we have

$$|W(a,b)| \leq C \int |au + b - x_0|^\alpha |\psi(u)|\,du$$

$$\leq C a^\alpha \int |u|^\alpha |\psi(u)|\,du + C|b - x_0|^\alpha \int |\psi(u)|\,du.$$

The result follows from the observation that

$$\int |u|^\alpha |\psi(u)|\,du \leq C \int \frac{|u|^\alpha}{1 + |u|^2}\,du < +\infty,$$

since $0 < \alpha < 1$. $\square$

Note that if $f \in C^\alpha(\mathbb{R})$, then the lemma implies that $|W(a,b)| \leq Ca^\alpha$. We will state and prove a converse to Lemma 10.3, but first we wish to comment on this estimate.

The "cone of influence," $\Gamma(x_0)$ of x_0, is defined by $a \geq |b - x_0|$. If $(b,a) \in \Gamma(x_0)$, then (10.8) becomes $|W(a,b)| \leq 2Ca^\alpha$; however, if (b,a) is not in $\Gamma(x_0)$, then $|W(a,b)| \leq 2C|b-x_0|^\alpha$. Some scientists thought at first that the Hölder exponent $\alpha(x_0)$ of f at x_0 could be computed by estimating $|W(a,b)|$ inside the cone of influence $\Gamma(x_0)$. This belief is based on the following reasoning: Assume that the support of ψ is contained in $[-1,1]$ and assume that we compute $W(a,b)$ when $(b,a) \notin \Gamma(x_0)$. Then $\varepsilon = |b - x_0| - a > 0$ and

$$W(a,b) = \int f(x) \psi_{(a,b)}(x)\,dx = \int_{|x-x_0| > \varepsilon} f(x) \psi_{(a,b)}(x)\,dx.$$

This computation led some to believe that the behavior of f near x_0 did not influence the wavelet coefficients of f outside the cone of influence of x_0.

An example that supports this idea is given by the function $f(x) = |x - x_0|^\alpha$. In this case, $W(a,b) = a^\alpha \psi_\alpha\left(\frac{b-x_0}{a}\right)$, where $\hat{\psi}_\alpha(\xi) = c(\alpha)|\xi|^{-1-\alpha}\hat{\psi}(\xi)$. If $\psi_\alpha(\lambda) \neq 0$, then it suffices to read $W(a,b)$ on the half-line $a = \lambda^{-1}(b - x_0) > 0$ to recover the exponent α.

A counterexample is the chirp $f(x) = |x - x_0|^\alpha \exp(i(x - x_0)^{-1})$. Integration by parts shows that $|W(a,b)| \leq C_N a^N$ when $a \geq \beta|x - x_0|$, $\beta > 0$. If estimates inside the cone of influence were sufficient for determining the Hölder regularity, then f would belong to $C^N(x_0)$ for every integer N. But this is not the case, and thus, this counterexample shows that examining the wavelet coefficients inside the cone of influence is not sufficient for determining the Hölder regularity of a function at a given point. Furthermore, the inequality $|W(a,b)| \leq C(a^\alpha + |b-x_0|^\alpha)$ is not sufficient either, and Lemma 10.3 does not yield a necessary and sufficient condition for $f \in C^\alpha(x_0)$. Nevertheless, the sufficient condition is only an epsilon away from (10.8), and the following theorem comes close to being the converse of Lemma 10.3.

THEOREM 10.1. *Assume that $0 < \alpha' < \alpha < 1$. If the wavelet transform of $W(a,b)$ of a bounded function f satisfies*

$$|W(a,b)| \leq Ca^\alpha \left(1 + \frac{|b-x_0|}{a}\right)^{\alpha'} \tag{10.9}$$

in some neighborhood $0 < a < a_0$, $|b-x_0| < b_0$, then f belongs to $C^\alpha(x_0)$.

We will prove this theorem, but, before doing so, we wish to comment on some of its implications. The first observation is that the estimate (10.9) implies that $|W(a,b)| \leq Ca^{\alpha-\alpha'}$, and this implies (by Theorem 10.1) that $f \in C^{\alpha-\alpha'}(\mathbb{R})$. Applying Theorem 10.1 means looking for points x_0 where f is more regular than its "average" regularity. The global regularity is given by $\alpha - \alpha'$, and we are looking for points where the regularity is given by α. The next observation is that this theorem yields an algorithm for computing pointwise Hölder exponents.

THEOREM 10.2. *Assume that f is a bounded function that belongs to the Hölder space $C^\beta(\mathbb{R})$ for some β, $0 < \beta < 1$. Then for every point $x_0 \in \mathbb{R}$, the Hölder exponent $\alpha(f, x_0)$ is given by*

$$\alpha(f, x_0) = \liminf_{\substack{a \to 0 \\ b \to x_0}} \frac{\log|W(a,b)|}{\log(a + |b-x_0|)}. \tag{10.10}$$

Proof. Recall that $\alpha(f, x_0) = \sup\{\alpha \mid f \in C^\alpha(x_0)\}$ and write (10.8) as $|W(a,b)| \leq C(a + |b-x_0|)^\alpha$. Then it follows from Lemma 10.3 that

$$\alpha(f, x_0) \leq \liminf_{\substack{a \to 0 \\ b \to x_0}} \frac{\log|W(a,b)|}{\log(a + |b-x_0|)}. \tag{10.11}$$

To prove the result in the other direction, suppose that (10.11) is not an equality. Then there is an α such that

$$\alpha(f, x_0) < \alpha < \liminf_{\substack{a \to 0 \\ b \to x_0}} \frac{\log|W(a,b)|}{\log(a + |b-x_0|)},$$

and we have $|W(a,b)| \leq C(a + |b-x_0|)^\alpha$. The assumption that $f \in C^\beta(\mathbb{R})$ implies that $\beta \leq \alpha(f, x_0) < \alpha$ and (by Lemma 10.3) that $|W(a,b)| \leq Ca^\beta$. By interpolating between these two estimates, we obtain $|W(a,b)| \leq Ca^\gamma \left(1 + \frac{|b-x_0|}{a}\right)^\eta$, where $\eta = \theta\alpha$ and $\gamma = \theta\alpha + (1-\theta)\beta$, $0 < \theta < 1$. By applying Theorem 10.1, we see that $f \in C^\gamma(x_0)$. Since γ is any real number in (β, α), we conclude that $\alpha(f, x_0) \geq \alpha$. Thus (10.11) is an equality, which proves the result. □

A nice example where Theorem 10.2 applies is given by the Riemann function $\mathcal{R}(x) = \sum_{n=1}^\infty \frac{1}{n^2} \sin(\pi n^2 x)$. To see that this is true, we show that $\mathcal{R}$ belongs to the global Hölder space $C^{1/2}(\mathbb{R})$. We write $\mathcal{R}(x) = \sum_{j=0}^\infty R_j(x)$, where

$$R_j(x) = \sum_{2^j \leq n < 2^{j+1}} \frac{1}{n^2} \sin(\pi n^2 x).$$

We immediately have $\|R_j\|_\infty \leq 2^{-j}$ and $\|R'_j\|_\infty \leq \pi 2^j$. Hence,

$$|\mathcal{R}(x+h) - \mathcal{R}(x)| \leq \pi \sum_{j=0}^{N} |h| \|R'_j\|_\infty + 2 \sum_{j=N+1}^{\infty} \|R_j\|_\infty$$
$$\leq 2\pi |h| 2^N + 2 \cdot 2^{-N}.$$

The optimal choice of N is determined by $|h| \leq 4^{-N} < 4|h|$. For this N we have $|\mathcal{R}(x+h) - \mathcal{R}(x)| \leq C|h|^{1/2}$, which means that $\mathcal{R} \in C^{1/2}(\mathbb{R})$.

The proof of Theorem 10.1 will follow a similar strategy. We will give the full details, but first we need to set up some notation and prove another lemma. Since we will be estimating separately the contributions of each scale a in (10.7), we introduce the following notation:

$$(\Delta_a f)(x) = \int_{b \in \mathbb{R}} W(a,b) \psi\Big(\frac{x-b}{a}\Big) \frac{db}{a}.$$

LEMMA 10.4. *Assume that $|\psi(x)| + |\psi'(x)| \leq C(1+|x|^2)^{-1}$. If the wavelet transform of f satisfies the inequality*

$$|W(a,b)| \leq Ca^\alpha \Big(1 + \frac{|b-x_0|}{a}\Big)^{\alpha'} \tag{10.12}$$

for some $C > 0$ and some $\alpha' < \alpha$, then

$$|(\Delta_a f)(x)| \leq Ca^\alpha \Big(1 + \Big|\frac{x-x_0}{a}\Big|^{\alpha'}\Big), \tag{10.13}$$

and

$$|(\Delta_a f)'(x)| \leq Ca^{\alpha-1}\Big(1 + \Big|\frac{x-x_0}{a}\Big|^{\alpha'}\Big). \tag{10.14}$$

Proof. Using (10.12) and the localization of ψ, we see that

$$|(\Delta_a f)(x)| \leq Ca^\alpha \int \frac{1+|\frac{b-x_0}{a}|^{\alpha'}}{1+|\frac{x-b}{a}|^2} \frac{db}{a}.$$

By introducing the new variable $u = \frac{b-x}{a}$ and noting that $|x+y|^{\alpha'} \leq |x|^{\alpha'} + |y|^{\alpha'}$, we have

$$|(\Delta_a f)(x)| \leq Ca^\alpha \Big[\int \frac{1+|u|^{\alpha'}}{1+u^2} du + \Big|\frac{x-x_0}{a}\Big|^{\alpha'} \int \frac{du}{1+u^2}\Big].$$

Since $\alpha' < 1$, (10.13) follows immediately. The proof of (10.14) is similar since

$$(\Delta_a f)'(x) = \int_{b \in \mathbb{R}} W(a,b) \psi'\Big(\frac{x-b}{a}\Big) \frac{db}{a^2}. \qquad \square$$

Proof of Theorem 10.1. We use (10.7) to write

$$f(x) - f(x_0) = \int_{a>0} \big[(\Delta_a f)(x) - (\Delta_a f)(x_0)\big] \frac{da}{a}.$$

For $a \geq |x - x_0|$, using the mean value theorem and (10.14), we have

$$\left| \int_{a \geq |x-x_0|} [(\Delta_a f)(x) - (\Delta_a f)(x_0)] \frac{da}{a} \right| \leq C|x - x_0| \int_{a \geq |x-x_0|} a^{\alpha-1} \frac{da}{a}$$
$$\leq C|x - x_0|^\alpha.$$

For $a < |x - x_0|$, we estimate $(\Delta_a f)(x)$ and $(\Delta_a f)(x_0)$ separately using (10.13) so that

$$\left| \int_{a < |x-x_0|} [(\Delta_a f)(x) - (\Delta_a f)(x_0)] \frac{da}{a} \right| \leq C \int_{a < |x-x_0|} |x - x_0|^{\alpha'} a^{\alpha-\alpha'} \frac{da}{a}$$
$$\leq C|x - x_0|^\alpha.$$

Note that this is the point in the proof where it is crucial to have $\alpha' < \alpha$. □

Observe that (10.12) is stronger than (10.8) since $\alpha' < \alpha$. It often happens, however, that the large wavelet coefficients that determine the regularity at x_0 are in a cone $\left|\frac{b-x_0}{a}\right| \leq C$, in which case the right-hand sides of (10.8) and (10.12) are of the same order of magnitude, and the wavelet criterion is sharp. This is true for the Weierstrass function, as will be shown below.

Having established Theorems 10.1 and 10.2, we can easily prove a result mentioned in section 9.4, namely, that the Hölder exponent of the Weierstrass function $\mathcal{W}(t) = \sum_{n=0}^{\infty} B^n \cos(A^n x)$, $0 < B < 1$, $AB \geq 1$, is $-\frac{\log B}{\log A}$ everywhere. The first step is to compute the wavelet transform of $\mathcal{W}$ with the wavelet that was used in Lemma 10.1. This is a straightforward computation, and we have

$$W_{\mathcal{W}}(a,b) = \frac{1}{2} \sum_{n=0}^{\infty} B^n e^{iA^n b} \overline{\hat{\psi}(A^n a)}. \quad (10.15)$$

Since the support of $\hat{\psi}$ is contained in $[A^{-1}, A]$, the only nonzero terms in the sum occur when $-1 - \frac{\log a}{\log A} < n < 1 - \frac{\log a}{\log A}$, and from this it follows that

$$|W_{\mathcal{W}}(a,b)| \leq Ca^{-\frac{\log B}{\log A}}.$$

This proves that $\mathcal{W} \in C^{-\frac{\log B}{\log A}}(\mathbb{R})$ (by Theorem 10.1) and that Theorem 10.2 applies. Define $a_n = A^{-n}$ for $n \geq 1$. Then for $a = a_n$ there is only one term in the right-hand side of (10.15), and $W(\mathcal{W}; a_n, b) = \frac{1}{2} a_n^{-\frac{\log B}{\log A}} e^{ia_n^{-1} b}$. Using Theorem 10.2, we have

$$\alpha(x_0) = \liminf_{\substack{a \to 0 \\ b \to x_0}} \frac{\log |W_{\mathcal{W}}(a,b)|}{\log(a + |b - x_0|)} \leq \lim_{n \to \infty} \frac{\log |W_{\mathcal{W}}(a_n, b)|}{\log a_n} = -\frac{\log B}{\log A}.$$

Since $\mathcal{W} \in C^{-\frac{\log B}{\log A}}(\mathbb{R})$, this proves that $\alpha(x_0) = -\frac{\log B}{\log A}$ everywhere.

10.4 The Riemann function

In 1916, G. H. Hardy proved that the Riemann function

$$\mathcal{R}(x) = \sum_{n=1}^{\infty} \frac{1}{n^2} \sin(\pi n^2 x)$$

is at best $C^{3/4}(x_0)$ in the following three cases [136]:[8]

---

[8]This proof uses results from a paper Hardy wrote with J. E. Littlewood in 1914 [137].

(a) x_0 is irrational.

(b) $x_0 = \frac{p}{q}$ with $p \equiv 0 \pmod{2}$ and $q \equiv 1 \pmod{4}$.

(c) $x_0 = \frac{p}{q}$ with $p \equiv 1 \pmod{2}$ and $q \equiv 2 \pmod{4}$.

Hardy's proof is a precursor of Lemma 10.3. To obtain the irregularity of $\mathcal{R}$ at a given point, Hardy showed that a wavelet transform of $\mathcal{R}$ is "large" near that point. Of course, Hardy did not use wavelet language, but the "ancestor" of the wavelet transform he used is a perfectly good one, namely, the derivative of the Poisson kernel.

Two problems remained open after Hardy's work: the question of differentiability at the rationals $\frac{p}{q}$ where p and q are odd and the determination of the exact Hölder exponents for all x.

Serge Lang suggested the first of these problems to an undergraduate class in December 1967, and to the general surprise of the mathematics world, Joseph L. Gerver, one of Lang's sophomore students, resolved the problem by proving the following unexpected result: If $x_0 = \frac{p}{q}$ where p and q are odd, then $\mathcal{R}$ is differentiable at x_0 and $\mathcal{R}'(x_0) = -\frac{1}{2}$. He then showed that $\mathcal{R}$ is differentiable at no other points, and the problem of the differentiability of the Riemann function was completely settled (see [127] and [128]).

We will follow Itatsu [149] and give a direct proof—which is based on Fourier analysis—of Gerver's result. This method will actually give us a very precise description of the oscillating behavior of $\mathcal{R}$ near these rationals.

For the irrationals, we will reformulate Hardy's method and, following Duistermaat [97], obtain the best possible "irregularity" at the irrationals. Hardy's method cannot yield information about the "regularity" at those points since this necessitates Theorem 10.1, which was first proved in 1988 [151]. But we will see that Hardy's method and Theorem 10.1 give the exact Hölder exponent at every point.

10.4.1 Hölder regularity at irrationals

Following a variant of Hardy's method, we use the wavelet analysis proposed by Lusin (section 2.6). Thus we take $\psi(x) = \frac{1}{\pi(x+i)^2}$ to be our analyzing wavelet. It is easy to check that $\psi \in L^1(\mathbb{R})$ with $\int |\psi(x)|\, dx = 1$, that $\int \psi(x)\, dx = 0$, and that ψ satisfies the conditions of Theorem 10.1.

We begin by computing the wavelet transform $W_{\mathcal{R}}(a,b) = \langle \mathcal{R}, \psi_{(a,b)} \rangle$. For this, we define the function

$$\mathcal{T}(x) = \mathcal{R}(x) - i\mathcal{S}(x),$$

where $\mathcal{S}(x) = \sum_{n=1}^{\infty} \frac{1}{n^2} \cos(\pi n^2 x)$. $\mathcal{T}$ has an analytic extension

$$\mathcal{T}(z) = \sum_{n=1}^{\infty} \frac{1}{in^2} e^{i\pi n^2 z}$$

in the upper half-plane $z = x + iy$, $y > 0$, where it is uniformly bounded by $\sum_{n=1}^{\infty} \frac{1}{n^2}$. Furthermore,

$$\mathcal{R}(x) = \frac{1}{2}(\mathcal{T}(x) + \widetilde{\mathcal{T}}(x)),$$

where $\widetilde{\mathcal{T}}(x) = -\mathcal{T}(-x)$. Thus,

$$W_{\mathcal{R}} = \frac{1}{2}(W_{\mathcal{T}} + W_{\widetilde{\mathcal{T}}}).$$

It is particularly easy to compute the wavelet transform of $\mathcal{T}$. In fact,

$$W_\mathcal{T}(a,b) = \frac{a}{\pi} \int_{-\infty}^{\infty} \frac{\mathcal{T}(x)\,dx}{(x-(b+ia))^2} = 2ia\mathcal{T}'(b+ia), \qquad a > 0.$$

This is just the form of Cauchy's theorem that says

$$f'(\zeta) = \frac{1}{2\pi i} \int_{-\infty}^{\infty} \frac{f(z)\,dz}{(z-\zeta)^2}$$

whenever f is holomorphic and bounded in the upper half-plane and $\operatorname{Im}\zeta > 0$. A similar argument shows that

$$W_{\widetilde{\mathcal{T}}}(a,b) = \frac{-a}{\pi} \int_{-\infty}^{\infty} \frac{\mathcal{T}(x)\,dx}{(x+(b+ia))^2} = 0, \qquad a > 0.$$

Thus we have

$$W_\mathcal{R}(a,b) = \frac{1}{2} W_\mathcal{T}(a,b) = ia\mathcal{T}'(b+ia).$$

Term-by-term differentiation shows that $\mathcal{T}'(z) = \pi \sum_{n=1}^{\infty} e^{i\pi n^2 z}$, so we have

$$W_\mathcal{R}(a,b) = i\pi a \mathcal{T}'(b+ia)$$
$$= \frac{i\pi a}{2}(\theta(b+ia) - 1), \tag{10.16}$$

where θ is Jacobi's Theta function defined by

$$\theta(z) = \sum_{n \in \mathbb{Z}} e^{i\pi n^2 z}, \qquad \operatorname{Im} z > 0. \tag{10.17}$$

We know from Lemma 10.3, Theorem 10.1, and Theorem 10.2 that one way to determine the regularity of $\mathcal{R}$ at x_0 is to investigate the behavior of $\theta(z)$ in a neighborhood of x_0. To carry out this program, it is necessary to understand how $\theta(z)$ is transformed under a group of transformations $z \mapsto \gamma(z)$ known as the *theta modular group*. This group is defined by

$$\gamma(z) = \frac{rz+s}{qz-p}, \tag{10.18}$$

where $rp + sq = -1$, r, s, p, q are integers, and the matrix

$$\begin{pmatrix} r & s \\ q & p \end{pmatrix} \text{ is of the form } \begin{pmatrix} \text{even} & \text{odd} \\ \text{odd} & \text{even} \end{pmatrix} \text{ or } \begin{pmatrix} \text{odd} & \text{even} \\ \text{even} & \text{odd} \end{pmatrix}.$$

A discussion of the theta modular group and its action on the Jacobi Theta function would be too much of a detour from our main objective. We will quote the needed results and refer the reader to the paper [97] by Duistermaat for a complete development.

It is easy to see that the $\gamma : \mathbb{C} \to \mathbb{C}$ maps the upper half-plane into itself. It is slightly more involved to establish the following result: When γ belongs to the theta modular group, θ is transformed as follows:

$$\theta(z) = \theta(\gamma(z))\, e^{im\pi/4}\, q^{-1/2} \left(z - \frac{p}{q}\right)^{-1/2}, \tag{10.19}$$

where m is an integer that is a rather complicated function of $\frac{p}{q}$. This formula, which is the cornerstone for the study of the Theta function, can be proved by first showing that the theta modular group is generated by the translation $z \mapsto z+2$ and by the inversion $z \mapsto -\frac{1}{z}$. With this established, it is only necessary to verify (10.19) for these two transforms. The first transformation just expresses the periodicity of θ. The second can be obtained by applying Poisson's summation formula

$$\sum_{n\in\mathbb{Z}} f(n) = \sum_{n\in\mathbb{Z}} \hat{f}(2\pi n)$$

(which holds at least for all f in the Schwartz class) to the Gaussian $x \mapsto e^{-\pi y x^2}$, $y > 0$. This yields

$$\sqrt{y}\sum_{n\in\mathbb{Z}} e^{-\pi n^2 y} = \sum_{n\in\mathbb{Z}} e^{-\pi n^2/y}.$$

By extending this relation analytically to all of the upper half-plane, $z = x + iy$, $y > 0$, we have

$$\sqrt{-iz}\,\theta(z) = \theta\left(-\frac{1}{z}\right).$$

We will be using the fact that $\theta(z) \to 1$ as $y \to +\infty$ uniformly in x, where $z = x + iy$. In fact,

$$|\theta(z) - 1| \leq 2\sum_{n=1}^{\infty} e^{-\pi n^2 y} \leq 2\sum_{n=1}^{\infty} e^{-\pi n y} = 2\frac{e^{-\pi y}}{1 - e^{-\pi y}}. \tag{10.20}$$

Our first result is that $\alpha(\mathcal{R}, x_0) = \frac{1}{2}$ when $x_0 = \frac{p}{q}$ and p and q are not both odd. In this case, it is easy to show that there is a γ in the theta modular group that maps x_0 to infinity. Take $b + ia = \frac{p}{q} + ia$. We are going to examine the behavior of $|W_\mathcal{R}(a,b)|$ as $a \to 0$. From equations (10.16) and (10.19), we see that

$$|W_\mathcal{R}(a,b)| = a^{1/2}\frac{\pi}{2}\left|\theta\left(\frac{r}{q} + \frac{i}{q^2 a}\right)e^{im\pi/4}(iq)^{-1/2} - a^{1/2}\right|. \tag{10.21}$$

The estimate (10.20) implies that $\frac{\pi}{2}\left|\theta\left(\frac{r}{q} + \frac{i}{q^2 a}\right)e^{im\pi/4}(iq)^{-1/2} - a^{1/2}\right| \to \frac{\pi}{2\sqrt{q}}$ as $a \to 0$, and this implies that

$$\lim_{a\to 0}\frac{\log|W_\mathcal{R}(a,b)|}{\log a} = \frac{1}{2}.$$

The result follows from Theorem 10.2: We proved before that $\mathcal{R} \in C^{1/2}(\mathbb{R})$ and, hence, that Theorem 10.2 applies. We have just shown that

$$\liminf_{\substack{a\to 0 \\ b\to x_0}} \frac{\log|W_\mathcal{R}(a,b)|}{\log(a + |b - x_0|)} \leq \frac{1}{2},$$

so from Theorem 10.2, $\alpha(\mathcal{R}, x_0) \leq \frac{1}{2}$. On the other hand, $\mathcal{R} \in C^{1/2}(\mathbb{R})$ implies that $\alpha(\mathcal{R}, x_0) \geq \frac{1}{2}$. Thus, $\alpha(\mathcal{R}, x_0) = \frac{1}{2}$ whenever $x_0 = \frac{p}{q}$ and p and q are not both odd. The case where p and q are both odd will be treated separately, but first we are going to determine $\alpha(\mathcal{R}, x_0)$ when x_0 is irrational.

If x_0 is irrational, then it cannot be mapped to infinity with an element of the theta modular group. However, it is possible to choose a rational $\frac{p}{q}$ very close to x_0 and map it to infinity. This will provide an estimate of $\theta(z)$ for points z near $\frac{p}{q}$ and hence an estimate of $\theta(z)$ for points near x_0.

But what do we mean by rationals "very close to x_0"? In this case, we mean the rationals given by the continued fraction[9] expansion of x_0. This is a sequence of rationals $\frac{p_n}{q_n}$ such that

$$\left|x_0 - \frac{p_n}{q_n}\right| \leq \frac{1}{2q_n^2}. \tag{10.22}$$

A result from the theory of continued fractions states that no rationals other than those in this sequence approximate x_0 better. However, some irrational numbers are much better approximated by their continued fraction expansion than is indicated by (10.22). The exponent 2 of q_n in (10.22) is the "worst possible." A degree of approximation to x_0 by rationals can be defined by considering the set

$$\mathrm{T}(x_0) = \left\{\tau \,\Big|\, \left|x_0 - \frac{p_n}{q_n}\right| \leq \frac{1}{q_n^\tau}\right\}, \tag{10.23}$$

where the inequality in (10.23) must hold for infinitely many n such that p_n and q_n are not both odd. (We are only interested in these $\frac{p_n}{q_n}$, since they are the ones that can be mapped to infinity.) Then $\tau(x_0)$ is defined by

$$\tau(x_0) = \sup_{\tau \in \mathrm{T}(x_0)} \tau.$$

Note that $\tau(x_0)$ can be $+\infty$. This is the case, for example, when $x_0 = \sum_{n=1}^\infty 2^{-n!}$. On the other hand, $\tau(x_0) \geq 2$. (The reference for this and other results about continued fractions will always be [138].)

We are now going to show that

$$\alpha(\mathcal{R}, x_0) \leq \frac{1}{2} + \frac{1}{2\tau(x_0)} \tag{10.24}$$

whenever x_0 is irrational. With what we have already shown, this proves that $\frac{1}{2} \leq \alpha(\mathcal{R}, x_0) \leq \frac{3}{4}$. The proof is similar to the one given for $x_0 = \frac{p}{q}$, p and q not both odd.

The first step is to choose a γ_n in the theta modular group that maps $\frac{p_n}{q_n}$ to infinity when p_n and q_n are not both odd. (In what follows we are only considering $\frac{p_n}{q_n}$ where p_n and q_n are not both odd.) A simple computation using the fact that $rp + sp = -1$ shows that this is always possible. Now define

$$z_n = b_n + ia_n = \frac{p_n}{q_n} + i\left|x_0 - \frac{p_n}{q_n}\right|.$$

We are going to examine the behavior of $|W_\mathcal{R}(a,b)|$ at the points $z_n = b_n + ia_n$. For this, it is convenient to define τ_n by

$$\left|x_0 - \frac{p_n}{q_n}\right| = \frac{1}{q_n^{\tau_n}}.$$

---

[9]For information about continued fractions we recommend *An Introduction to the Theory of Numbers* by G. H. Hardy and E. M. Wright [138].

Since $|x_0 - \frac{p_n}{q_n}| < \frac{1}{q_n^2}$ for all n, it is clear that $\tau_n > 2$ for all n. Armed with this notation and equation (10.21), we have

$$|W_\mathcal{R}(a_n, b_n)| = a_n \frac{\pi}{2} \left| \theta\left(\frac{r_n}{q_n} + \frac{i}{q_n^2 a_n}\right) e^{im_n\pi/4} (iq_n a_n)^{-1/2} - 1 \right|$$

$$= q_n^{-\frac{1+\tau_n}{2}} \frac{\pi}{2} \left| \theta\left(\frac{r_n}{q_n} + iq_n^{\tau_n - 2}\right) e^{im_n\pi/4} (i)^{-1/2} - iq_n^{\frac{1-\tau_n}{2}} \right|.$$

It follows from (10.20) ($|\theta(z) - 1| \leq \frac{1}{2}$ if $\text{Im } z \geq 1$) and the fact that $\tau_n > 2$ that

$$\frac{1}{4} \leq \left| \theta\left(\frac{r_n}{q_n} + iq_n^{\tau_n - 2}\right) e^{im_n\pi/4} (i)^{-1/2} - iq_n^{\frac{1-\tau_n}{2}} \right| \leq \frac{3}{4}$$

for all sufficiently large n. We now wish to estimate

$$\liminf_{\substack{a_n \to 0 \\ b_n \to x_0}} \frac{\log |W_\mathcal{R}(a_n, b_n)|}{\log(a_n + |b_n - x_0|)}.$$

In our notation, $\log(a_n + |b_n - x_0|) = \log 2q_n^{-\tau_n}$, so we have

$$\frac{\log |W_\mathcal{R}(a_n, b_n)|}{\log(a_n + |b_n - x_0|)} = \left(\frac{1}{2} + \frac{1}{2\tau_n}\right)\left(1 + \frac{\log 2}{\log 2q_n^{-\tau_n}}\right)^{-1}$$

$$+ \frac{\log \frac{\pi}{2} \left| \theta\left(\frac{r_n}{q_n} + iq_n^{\tau_n - 2}\right) e^{im_n\pi/4} (i)^{-1/2} - iq_n^{\frac{1-\tau_n}{2}} \right|}{\log 2q_n^{-\tau_n}}.$$

The second term of the right-hand side of this equation tends to zero as $n \to \infty$, and we conclude that

$$\liminf_{n \to \infty} \frac{\log |W_\mathcal{R}(a_n, b_n)|}{\log(a_n + |b_n - x_0|)} \leq \frac{1}{2} + \frac{1}{2\tau(x_0)}.$$

Theorem 10.2 applies, and since

$$\liminf_{\substack{a \to 0 \\ b \to x_0}} \frac{\log |W_\mathcal{R}(a, b)|}{\log(a + |b - x_0|)} \leq \liminf_{n \to \infty} \frac{\log |W_\mathcal{R}(a_n, b_n)|}{\log(a_n + |b_n - x_0|)},$$

we have

$$\frac{1}{2} \leq \alpha(\mathcal{R}, x_0) \leq \frac{1}{2} + \frac{1}{2\tau(x_0)} \leq \frac{3}{4}, \tag{10.25}$$

which is what we wished to prove.

Thus $\mathcal{R}$ is certainly not smoother than $\frac{1}{2} + \frac{1}{2\tau(x_0)}$ at an irrational x_0. It takes more work, involving the investigation of several cases, but it can be shown using Theorem 10.1 that $\alpha(\mathcal{R}, x_0) = \frac{1}{2} + \frac{1}{2\tau(x_0)}$. This was proved by Stéphane Jaffard, and the details can be found in [152].

We are going to investigate in the next section the behavior of $\mathcal{R}(x)$ at the rationals $x = \frac{p}{q}$ where both p and q are odd. But before doing so, we wish to return to the multifractal formalism, which was mentioned at the beginning of this chapter and elsewhere. A classical result from number theory known as Jarník's theorem (see [101] for a proof) gives the exact Hausdorff dimension of the points having a given order τ of approximation by rationals, namely,

$$d(\tau) = \frac{2}{\tau}.$$

Thus, the spectrum of singularities of $\mathcal{R}$ is

$$d(h) = 4h - 2 \quad \text{if} \quad h \in \left[\frac{1}{2}, \frac{3}{4}\right],$$
$$d\left(\frac{3}{2}\right) = 0. \tag{10.26}$$

For all other values of h, the set is empty, and hence its Hausdorff dimension is zero. The exponent $\frac{3}{2}$ corresponds to the rationals where $\mathcal{R}$ is differentiable. The Hölder exponent is actually $\frac{3}{2}$ at these points, as will be seen below.

The increasing part of the spectrum (corresponding to $h \in [\frac{1}{2}, \frac{3}{4}]$) can be recovered by the multifractal formalism, which is thus valid for Riemann's function [153]. This is significant, since $\mathcal{R}$ contains chirps.

10.4.2 Riemann's function near $x_0 = 1$

The last task in this chapter is to study Riemann's function near the points that were not discussed in the previous section, namely, the rationals $\frac{p}{q} = \frac{\text{odd}}{\text{odd}}$. Recall that Gerver was the first to show that $\mathcal{R}'(\frac{p}{q}) = -\frac{1}{2}$ at these points. To simplify the notation, we will discuss only the case $\frac{p}{q} = 1$. The study near the other rationals can be related to this case by mapping $\frac{p}{q}$ onto 1 with a member of the theta modular group. Also, instead of $\mathcal{R}$, we work with

$$S(x) = \sum_{n=1}^{\infty} \frac{e^{in^2 x}}{n^2}.$$

We can take the imaginary part later, and to simplify notation we have dropped the factor π. Thus we are going to study $S(x)$ at $x = \pi$.

This function satisfies the following recursion relation:

$$\begin{aligned} S(x + \pi) &= \sum_{n=1}^{\infty} \frac{(-1)^n e^{in^2 x}}{n^2} \\ &= \sum_{n \text{ even}} \frac{e^{in^2 x}}{n^2} - \sum_{n \text{ odd}} \frac{e^{in^2 x}}{n^2} \\ &= 2 \sum_{n \text{ even}} \frac{e^{in^2 x}}{n^2} - \sum_{n=1}^{\infty} \frac{e^{in^2 x}}{n^2} \\ &= \frac{1}{2} S(4x) - S(x). \end{aligned} \tag{10.27}$$

We are now going to obtain an asymptotic expansion of $S(x)$ as $x \to 0$. We can restrict the values of x to $x > 0$, since $S(-x) = \overline{S(x)}$.

$$\begin{aligned} S(x) &= \sum_{n=1}^{\infty} \frac{e^{in^2 x} - 1}{n^2} + \sum_{n=1}^{\infty} \frac{1}{n^2} \\ &= \frac{1}{2} \sum_{-\infty}^{\infty} \frac{e^{in^2 x} - 1}{n^2} - \frac{ix}{2} + \sum_{n=1}^{\infty} \frac{1}{n^2}. \end{aligned} \tag{10.28}$$

Let $v(x) = \sum_{-\infty}^{\infty} \frac{e^{in^2 x} - 1}{n^2}$. The Fourier transform of $f(x) = \frac{e^{ix^2} - 1}{x^2}$ has the following asymptotic expansion at infinity: For each fixed $K \geq 1$,

$$\hat{f}(\xi) = e^{-i\xi^2/4} \left(\frac{c_1}{\xi^2} + \cdots + \frac{c_K}{\xi^{2K}} + \xi^{-2K} \varepsilon_K(\xi) \right),$$

where $\varepsilon_K(\xi)$ is bounded and $\varepsilon_K(\xi) \to 0$ as $\xi \to \infty$.

Using Poisson's summation formula, we have

$$v(x) = \sqrt{x} \sum_{n \in \mathbb{Z}} f(n\sqrt{x})$$

$$= \sqrt{x} \sum_{n \in \mathbb{Z}} \hat{f}\left(\frac{2\pi}{\sqrt{x}} n\right) \qquad (10.29)$$

$$= \sqrt{x}\hat{f}(0) + \sqrt{x} \sum_{n \neq 0} e^{-i\pi^2 n^2/x} \left(\sum_{k=1}^{K} \frac{c_k x^k}{(2\pi n)^{2k}} + \frac{x^K}{(2\pi n)^{2K}} \varepsilon_K\left(\frac{2\pi n}{\sqrt{x}}\right) \right),$$

and this is valid for each $K \geq 1$. We write

$$g_k(x) = \frac{c_k}{2} \sum_{n \neq 0} \frac{4^k e^{-i\pi^2 n^2 x/4} - e^{-i\pi^2 n^2 x}}{(2\pi n)^{2k}}.$$

By using equations (10.27), (10.28), and (10.29), we obtain the following asymptotic expansion for $S(\pi + x)$ as $x \to 0$:

$$S(\pi + x) = -\frac{\pi^2}{12} - \frac{ix}{2} + \sqrt{x} \sum_{k=1}^{K} x^k g_k\left(\frac{1}{x}\right) + o(x^{K+1/2}). \qquad (10.30)$$

This proves that $\mathcal{R}'(1) = -\frac{1}{2}$, which was first proved by Gerver. But the technique used here tells us more. It is clear from (10.30) that $\mathcal{R} \in C^{3/2}(1)$ and, in fact, that the Hölder exponent of $\mathcal{R}$ at $x = 1$ is exactly $\frac{3}{2}$. This technique also yields precise information about the oscillatory behavior of $\mathcal{R}$ near 1. Equation (10.30) shows that near $x = 1$, $\mathcal{R}$ "looks like" the chirp $x^{3/2} \sin \frac{1}{x}$ superimposed on a straight line with slope $-\frac{1}{2}$.

The g_k have several interesting properties: They are $\frac{8}{\pi}$-periodic functions that belong to $C^{k-1/2}(\mathbb{R})$, and $\int_0^{8/\pi} g_k(x)\,dx = 0$. Perhaps more remarkable is their direct relation to Riemann's function. For example, for $k = 1$,

$$g_1(x) = \frac{c_1}{4\pi^2}\left[4S\left(-\frac{\pi^2 x}{4}\right) - S(-\pi^2 x) \right].$$

The g_k for $k > 1$ are similarly related to primitives of S.

We have recently received a paper from Joseph L. Gerver wherein he studies the differentiability of the function $\sum_{n=1}^{\infty} \frac{\sin n^3 x}{n^3}$ and its chirp behavior at rational points [129]. Gerver's technique is similar to Itatsu's. We refer to it as a Fourier-type technique because it uses the Poisson formula. In fact, it is a variant of the Poisson formula that was found by Hardy and Littlewood.

10.5 Conclusions and comments

Our analysis of $\mathcal{R}$ for $x = 1$ is a direct Fourier method inspired by the paper [149] by Seiichi Itatsu. This work leads us to the following comparison of wavelet and Fourier methods: The wavelet transform gives a general method to estimate pointwise Hölder regularity, but, in specific cases, a direct Fourier method may be more efficient and provide more information.

We mention, moreover, a general setting where wavelet methods fail: Condition (10.9) implies that f has a positive uniform regularity in a neighborhood of

x_0. This excludes all instances of functions that have a dense set of discontinuities. Such functions are not just curiosities; they include a large and important class of stochastic processes, namely, the *Lévy processes*. These are processes with stationary, independent increments. They are multifractal and they satisfy the multifractal formalism, but wavelets offer no help for their analysis. In this case, one must return to a direct classical method (see [158]).

We indicated above that the multifractal formalism is valid for Riemann's function [153]. However, it was not easy to prove this result, and this example underlines a problem in this area: The derivation of the spectrum of singularities for a signal using the multifractal formalism will never be completely satisfactory because it is necessary to verify that this formalism is valid for the signal or class of signals being analyzed. For Riemann's function and for a handful of other functions, it is possible to compute the spectrum of singularities directly. In the case of turbulence, one can dream of deriving the spectrum of singularities mathematically from the Navier–Stokes equations, but as anyone slightly familiar with the field knows, we have very few results about general solutions of these equations, so it seems that we are very far from being able to reach this goal.

A more modest and realistic program is to investigate the fractal nature of solutions of nonlinear partial differential equations that are mathematically simpler than the Navier–Stokes equation but that are "related" to these equations. Again results are scarce; however, there is at least one notable exception: The one-dimensional Burgers equation

$$\frac{\partial u}{\partial t} + \frac{\partial}{\partial x}\left(\frac{u^2}{2}\right) = 0$$

($u(x,t) : \mathbb{R} \times \mathbb{R}^+ \to \mathbb{R}$) had been suggested as a greatly simplified model of the Navier–Stokes equations in one dimension. J. Bertoin proved that if the initial condition $u(x,0)$ is a Brownian motion, then the solution at time t is a Lévy process, which, as noted above, is multifractal [35]. Thus, we have an example of a nonlinear partial differential equation that can develop multifractal solutions starting with a monofractal initial condition.

This is the only example of this kind that we know of, and so the degree of generality of this phenomenon is not at all clear. We believe that it would be very instructive to generalize this result to three dimensions, since the Burgers equation in three dimensions is used to model the evolution of matter in the universe. Indeed, if one could prove that solutions of the three-dimensional Burgers equation are "generically" multifractal, this would provide a theoretical foundation for the many discussions about the multifractal nature of the distribution of matter in the universe (see, for example, [252]).

Finally, we note that only time-scale wavelets have been used in this chapter. This remark also applies to the analysis of function spaces. It is remarkable that, though many different kinds of wavelet expansions are available and used in signal and image processing, only the time-scale wavelets have the "right" mathematical properties that allow their use for applications inside mathematics, namely, the characterization of function spaces and the analysis of multifractal functions.

CHAPTER 11

Data Compression and Restoration of Noisy Images

11.1 Introduction

Wavelets have often been promoted as being the correct tool for processing nonstationary signals having strong transients. In contrast, Fourier analysis is the appropriate tool for studying stationary Gaussian processes. However, as Patrick Flandrin pointed out [110], being nonstationary is a negatively defined concept, and it is too broad to be mathematically useful; it is a jungle, a terra incognita waiting for proper exploration and clarification. Does this mean that our advertisement about wavelets and nonstationary signals belongs to the collection of unfulfilled claims made by the pioneers of the wavelet saga? Should we be pessimistic and conclude that wavelets have nothing to do with nonstationary signals? Not at all. Thanks to the work of a group at the University of South Carolina, this debate was settled when they found the following result: There exist well-defined classes of signals that are characterized by the fact that their wavelet expansions are sparse. If the wavelet expansion of a signal is sparse, then an efficient approximation of the signal requires only a few terms of its wavelet expansion. This paves the way to an efficient compression and transmission. Moreover, these classes are also characterized by optimal nonlinear rational approximation. Ronald DeVore, Björn Jawerth, P. Petrushev, and V. Popov delimited some precisely defined territories inside the jungle of nonstationary signals. These territories are new function spaces, and they happen to be nicely related to certain Besov spaces. (See Appendix D for the definition of Besov spaces and their characterization in terms of wavelets. See [203] for more complete details.)

This fundamental discovery supports some of our pioneers' claims, and at the same time, DeVore's theorem draws a boundary line, which we illustrate with an example. An otherwise smooth function with isolated singularities of the form $|t-t_0|^\alpha$ has a sparse wavelet expansion. (This example, which is true for arbitrarily small $\alpha > 0$, was used to support the original claim.) However, singularities along a curve in $\mathbb{R}^2$ are forbidden by the two-dimensional version of Theorem 11.1, since a function as simple as $\sup\{1-x_1^2-x_2^2, 0\}$ does not have a sparse wavelet expansion in the strict sense of Theorem 11.1. Furthermore, we note that oscillating singularities such as $t \sin\frac{1}{t}$ are excluded from Theorem 11.1, since they too do not have sparse wavelet expansions in the strict sense.

These remarkable results will be described in the next section. Theorems 11.1, 11.2, and 11.3 characterize functions with sparse wavelet expansions. These characterizations are in terms of the ladder of Besov spaces and depend on several degrees of sparsity. We will see that they provide the background for David Donoho's work (section 11.3), much of which was done in collaboration with Iain Johnstone, Gérard

Kerkyacharian, and Dominique Picard (see [94]). Donoho's work is based on the DeVore model. What we mean is that the object X (function, signal, or image) to be recovered can be modeled efficiently by a function belonging to a specified ball in a Besov space. The problem to be addressed is to recover X from noisy data modeled by $Y = X + \sigma Z$, where $\sigma > 0$ is a small parameter and Z is a standard white noise. This problem leads to a much harder one: The data are given by $Y = AX + \sigma Z$, where A is a compact operator. In image processing, A models the optics of the instrument used to obtain the image. This model is used in astronomy, as we will see in Chapter 12.

An estimator $\hat{X}$ of X is given by a linear or nonlinear functional Φ acting on the data Y: $\hat{X} = \Psi(Y)$. The expected discrepancy between $\hat{X}$ and X will be compared with a power law $C\sigma^\alpha$ as σ tends to zero. The optimal estimator is defined to be the one for which α is largest, irrespective of the constant C.

The fundamental discovery made by Donoho and his coworkers is the following: *Wavelet shrinkage* (to be defined) yields an optimal estimator $\hat{X} = \Psi(Y)$, where optimality is challenged over *all* linear and nonlinear functionals Φ acting on the data. This result relies crucially on the fact that the wavelet series expansion of X is sparse. In this sense, sparsity is responsible for optimal denoising. (Selected references to Donoho's work include [88], [90], [92], and [93].)

Some of the models that are currently used in image processing are discussed in section 11.4. These models amount to writing an image f as a sum $u + v$, where u is supposed to represent the important features of the image, while v is intended to include everything else, such as the noise and textures. But what are these important features? Edges are strong candidates. According to David Marr, evolution shaped the human visual system so that it is very sensitive to edges: We immediately recognize the shape of a shirt, but not necessarily the pattern drawn on the shirt. The human eye needs relatively much longer time to distinguish one texture from another. Marr's scientific program led to the following conjecture: A correctly tuned wavelet shrinkage applied to an image f yields the u component and eliminates the v component. In other words, wavelet shrinkage should be an edge detector. This conjecture will be discussed in section 11.4.

Unfortunately, the class of functions (signals or images) whose wavelet expansions are sparse does not contain images, and this is certainly a limitation on the power of wavelet shrinkage in image processing. The good news is that a new basis called *ridgelets* shows promise of being able to yield representations of cartoon images that are sparser than those given by wavelet representations. A cartoon image is defined to be a piecewise smooth function with possible jump discontinuities across smooth Jordan curves. The construction of ridgelets and this new research are discussed in section 11.5.

11.2 Nonlinear approximation and sparse wavelet expansions

Historically, nonlinear approximation developed from the work of several mathematicians in Central and Eastern Europe on rational approximation. Let f be a function defined on a closed and bounded (compact) interval I. To fix our ideas, assume that f belongs to $L^2(I)$. For each positive integer N, one looks for a rational fraction $g_N(x) = P(x)/Q(x)$ with degree $\leq N$ (defined as the maximum of the degrees of the polynomials P and Q) that gives the best approximation to f in the $L^2(I)$ norm. Thus one seeks, for each value of N, to minimize $\|f - g_N\|_2$ with the constraints $g_N = P/Q$, $\deg P \leq N$, and $\deg Q \leq N$. No hypothesis is made about the position of the poles of g_N. Since the set R_N of rational fractions $g_N = P/Q$ that are examined in seeking the minimum is not a linear subspace of $L^2(I)$, the algorithm defining the best approximation is not linear. Furthermore, the function g_N is not unique; it is, however, unique if the approximation is measured in the uniform (L^∞) norm. (See [226] for a complete discussion of rational approximation.)

The goal is to represent rather complicated functions with only a few numbers, namely, the $2N+1$ coefficients of the polynomials P and Q. For this to make sense, it is necessary to know how to control the approximation error. Thus one tries to estimate $r_N(f) = \|f - g_N\|_2$ as a function of N for large N when g_N provides the best rational approximation of f. Rational approximation will offer an advantage over polynomial approximation only if (for an interesting set of functions) $r_N(f) \to 0$ as $N \to \infty$ much more rapidly in the case of rational approximation than in the case of polynomial approximation. When this is the case, one can represent the function f, with an acceptable error, using very few coefficients. This problem is also studied when the L^2 norm is replaced with other functional norms such as the L^p norm or the uniform norm. Thus, we are concerned with data compression based on a representation adapted to the problem.

In contrast to what happens in polynomial approximation, the sequence of errors $r_N(f) = \|f - g_N\|_p$ can decrease rapidly as $N \to \infty$ without f being regular on I in the usual sense.

We are going to consider an instructive example studied by D. J. Newman in 1964 [218]. If one tries to approximate the function $f(x) = |x|$ on $[-1, 1]$, by a polynomial P_N of degree N, the best possible uniform approximation yields

$$\sup_{x \in [-1,1]} |f(x) - P_N(x)| \leq \frac{\gamma}{N} \tag{11.1}$$

for a $\gamma > 0$. The order of approximation cannot be better because of the angle in the graph of f. Newman made the remarkable observation that if we allow rational fractions P_N/Q_N of degree N, the best order of approximation[10] becomes

$$\sup_{x \in [-1,1]} \left| f(x) - \frac{P_N(x)}{Q_N(x)} \right| \leq C e^{-\pi \sqrt{N}}, \tag{11.2}$$

while the number of parameters is only doubled. Thus, to transmit this very simple signal (the graph of f), rational fractions are much better than polynomials. Approximation by polynomials is linear: Polynomials of degree N form a linear space, and the best approximation of the sum of two functions is the sum of the approximations. Approximation by rational fractions of degree N is not linear: The sum of two rational fractions of degree N usually has degree $2N$. The function $f(x) = |x|$ is an example where rational approximation accelerates convergence; another example, which was mentioned in the introduction, is the function $|x|^\alpha$, $\alpha > 0$. An example where rational approximation offers no decisive advantage is given by the chirp $f(x) = x \sin \frac{1}{x}$.

The proofs and discussion of these striking phenomena have been presented in the work of J. Peetre, V. Peller, A. Pekarskii, P. Petrushev, and V. Popov. Results on nonlinear wavelet approximation were then extended to the multidimensional case by DeVore, Jawerth, and Popov [81].

---

[10]This particular upper bound was obtained by N. S. Vjacheslavov in 1975; D. J. Newman proved (11.2) with an exponent different from π. See [226] for a discussion of these results.

Peller obtained his pioneer results for the periodic case [224]. Let R_N be the collection of all rational functions P_N/Q_N, where $\deg P_N \leq N$, $\deg Q_N \leq N$, and $Q_N(z)$ does not vanish on the unit circle $z = e^{i\theta}$. We also denote by R_N the restrictions of these P_N/Q_N to the unit circle. For a continuous function $f(e^{i\theta})$, we write

$$r_N(f) = \mathrm{dist}_{L^\infty}(f, R_N) = \inf_{g \in R_N} \|f - g\|_\infty. \tag{11.3}$$

Peller's theorem says that $r_N(f) = O(N^{-q})$ for every $q \geq 1$ if and only if f belongs to all the Besov spaces $B^{1/p,p}(L^p)$, $0 < p \leq 1$. (A precise definition of these Besov spaces is given in Appendix D.) Roughly speaking, this condition means that the function f is absolutely continuous, that f' belongs to L^1, that f'' belongs to $L^{1/2}$, that f''' belongs to $L^{1/3}$, and so on. In some sense, f is infinitely differentiable, but its derivatives are measured in weaker and weaker norms. An example of such a function is $f(\theta) = |\sin(\theta - \theta_0)|^\alpha$, where $\alpha > 0$. This is a periodic version of the example mentioned in the introduction.

We are now going to discuss a variant of Peller's result in which one is approximating functions defined on $\mathbb{R}$. For this, we define R_N to be the collection of rational functions P/Q, where $\deg P < \deg Q \leq N$ and $Q(x)$ does not vanish on the real line. We wish to characterize those functions f of the real variable x for which

$$r_N(f) = O(N^{-q}) \tag{11.4}$$

for every $q \geq 1$. This turns out to be equivalent to the wavelet expansion of f being sparse. It is now time to define what we mean by sparse. Since we do not want the smoothness of the analyzing wavelet to be a restriction on the result, we will only consider the specific orthogonal wavelet basis $2^{j/2}\psi(2^j x - k)$, $j, k \in \mathbb{Z}$, where ψ belongs to the Schwartz class. We denote by $\psi_{j,k}(x)$ the function $\psi(2^j x - k)$ and warn the reader that we are using the L^∞ normalization $\|\psi_{j,k}\|_\infty = \|\psi\|_\infty$. (This is the same normalization that is used in Appendix D.) Then the wavelet expansion

$$f(x) = \sum_{j,k} \alpha(j,k) \psi_{j,k}(x) \tag{11.5}$$

of a function f that is continuous on $\mathbb{R}$ and vanishes at infinity is said to be *sparse* if and only if

$$\sum_{j,k} |\alpha(j,k)|^p < \infty \quad \text{for all} \quad 0 < p \leq 1. \tag{11.6}$$

Observe that the smaller the exponent p, the stronger the requirement. In fact, at the limit $p = 0$ (which is not considered) there would be only a finite number of nonzero terms. Note also that (11.6) implies that the series in (11.5) converges uniformly to f.

If condition (11.6) holds, then the absolute values of the wavelet coefficients, when arranged in decreasing order, form a sequence $\{c_n\}_{n \in \mathbb{N}^*}$ that decreases rapidly as $n \to \infty$. Indeed, if $p = \frac{1}{k}$, $k \in \mathbb{N}^*$, the two conditions (11.6) and $c_{n+1} \leq c_n$ imply that $nc_n^{1/k} \leq \sum_{j=1}^n c_j^{1/k} \leq C_k$ for some $C_k > 0$. Thus $c_n \leq \frac{C_k^k}{n^k}$ for all $n \in \mathbb{N}^*$, which means that the sequence decays rapidly.

The first variant of Peller's theorem follows an approach that was proposed by DeVore, Jawerth, and Popov [81].

THEOREM 11.1. *Let f be a continuous function defined on $\mathbb{R}$ and assume that f vanishes at infinity. Then condition (11.4) holds if and only if the wavelet expansion of f is sparse.*

We are going to outline a proof of half of the result, namely, that sparsity implies (11.4). The proof would be trivial if the wavelet ψ were a rational function P/Q, since a rearrangement of the partial sums of (11.5) would yield (11.4). Our tasks are thus (a) to write the wavelet ψ as a series $\psi = \sum \gamma_n \frac{P_n}{Q_n}$, where, for q fixed, $\deg P_n < \deg Q_n \leq q$, and where $\sum |\gamma_n|^p < \infty$ for $p > 0$ and (b) to substitute this expansion of ψ in (11.5). Then we have

$$f(x) = \sum_{n=0}^{\infty} \tilde{\gamma}_n \frac{\tilde{P}_n}{\tilde{Q}_n}, \tag{11.7}$$

where $\deg \tilde{P}_n < \deg \tilde{Q}_n \leq q$, and where $\sum |\tilde{\gamma}_n|^p < \infty$ for $p > 0$. It takes only a moment of reflection to see that (11.7) yields (11.4). Finally, the decomposition of the wavelet is not a difficulty: Take $P_n = 1$ and $Q_n(x) = 1 + (a_n x - b_n)^2$, where $a_n > 0$ and $b_n \in \mathbb{R}$.

This part of the proof can be generalized to any dimension. However, the implication in the other direction is deeper, and it is not true in dimensions greater than one. The converse statement for one dimension relies on some beautiful estimates on rational approximation obtained by A. Pekarskii: For $0 < p \leq 1$, there exists a constant $C(p)$ such that for every pair of polynomials P, Q with $\deg P < \deg Q \leq N$, we have

$$\left\| \frac{P}{Q} \right\|_{\dot{B}_p^{1/p,p}} \leq C(p) N^{1/p} \left\| \frac{P}{Q} \right\|_{\infty}.$$

The reader should observe the similarity with Bernstein's inequalities.[11] Here, it is necessary to use the homogeneous Besov norms (see Appendix D).

Before moving on, we present two examples. If $\alpha_1, \ldots, \alpha_m$ are m positive exponents and $g(x) = \exp(-x^2)$, for example, then the function

$$f(x) = (c_1 |x - x_1|^{\alpha_1} + \cdots + c_m |x - x_m|^{\alpha_m}) g(x)$$

has a sparse wavelet expansion and Theorem 11.1 applies. This function has a finite number of isolated singularities and is smooth elsewhere. These properties are not sufficient to ensure that a function has a sparse expansion. A counterexample is the function $f(x) = x \sin \frac{1}{x} - 1$. Oscillating singularities (chirps) prevent sparse wavelet expansions.

An explanation of the ability of rational functions to mimic strong transients is given by an example. If $q \gg 1$ is large, then $f_q(x) = (1 + iqx)^{-2}$ has a sharp peak at zero and almost vanishes away from the origin. This rational function is quite simple. However, f_q has one strong localized oscillation, just as a wavelet does.

We are now going to change direction slightly and discuss another kind of approximation. Instead of approximating f by elements of the set R_N of rational fractions with degrees less than or equal to N and with poles in arbitrary positions, we can approximate f by splines with free knots. In the simplest case, these are continuous, piecewise linear functions having $N-1$ linear pieces. The N end points

---

[11] We suggest G. G. Lorentz's book [175] for an introduction to Bernstein's results.

of the linear pieces are called *knots*; they are *free* because they can be positioned arbitrarily. Instead of using linear splines (which are only continuous) we can use cubic splines (which will be C^2), or splines of arbitrary regularity. We must assume that the order (of regularity) r of the splines is sufficiently large, given the rate at which we want the error r_N between the signal and the best spline approximation to converge to zero.

As a first step, Petrushev compared the quality of rational approximation with that given by spline approximation using N free knots $t_1 < t_2 < \cdots < t_N$ in the interval I. These N free knots play the role of the N poles of P_N/Q_N. (The norm used for this approximation will be specified a little later when we describe DeVore's algorithm explicitly.) This search for the optimal positions of the N knots $t_1, t_2, \ldots, t_N$ is related to the problem of optimally segmenting a given signal (or function) on the interval I. One wants to determine where there are "natural" changes in a signal: We want to segment a function f into $N-1$ functions $f_1, f_2, \ldots, f_{N-1}$ defined on intervals $I_1, I_2, \ldots, I_{N-1}$ forming a partition of I. Each f_j must be well approximated by a polynomial P_j on I_j, where the degrees of the P_j must be $\leq r+1$. A suitably truncated wavelet expansion gives this kind of approximation, if the wavelets are constructed with splines. The interested reader is invited to consult [86] for a more precise formulation.

If the function to be segmented is strongly oscillating, such as $e^{i\omega x}$ for a large ω, it is clear that the optimal segmentation is a delusion: It amounts to decomposing the sinusoid into a sequence of restrictions to intervals of length $\frac{2\pi}{\omega}$, and this destroys the information given by the periodicity. The same remark is true for a chirp of the form $|x|^\alpha \sin\frac{1}{x}$: It is poorly approximated by rational fractions or by free-knot splines.

This second reading of the best approximation problem allows it to be formulated in any dimension. For instance, in two dimensions, an important problem in numerical analysis is to obtain optimal meshes, say, of the surface of an airplane to study its simulated flight. It is clear that the optimal mesh has to be strongly inhomogeneous. For example, one takes relatively few samples on large flat surfaces. One can, for a given two-dimensional image, look for the optimal triangulation using N vertices and construct the approximation g_N to f that minimizes $\|f - g_N\|_2$ and that is piecewise affine with respect to this triangulation. However, in this case, we do not know how to relate the regularity of f to the rate at which $r_N(f) = \|f - g_N\|_2$ tends to zero, and this difficulty comes from the fact that the eccentricities of the triangles can be arbitrarily large. It is only by limiting a priori the eccentricities of the triangles used in the adaptive triangulation that DeVore and Popov show us how to determine a suboptimal solution.

They went around this problem by proposing a definition of what could be an optimal "segmentation" for a function f of n real variables. Such a segmentation is provided by N disjoint n-cubes $Q_1, \ldots, Q_N$, which play the role of the intervals $[t_j, t_{j+1})$ used by Petrushev. Starting with a fixed "bump" function θ on the unit cube, we let θ_Q be a translated and dilated copy of θ living on Q. Then, for a given function f belonging to $L^p(\mathbb{R}^n)$ and for each integer $N \geq 1$, DeVore, Popov, and Jawerth were looking for an optimal choice of cubes $Q_1, \ldots, Q_N$ and constants $c_1, \ldots, c_N$ such that

$$\rho_N(f) = \|f - c_1 \theta_{Q_1} - \cdots - c_N \theta_{Q_N}\|_p$$

is as small as possible. They observe that a suboptimal solution can be obtained in any dimension by expanding f in a series of wavelets and simply retaining the N

largest coefficients. The corresponding partial sum gives the suboptimal solution. In other words, after having written this series, one determines, in a sense, the histogram of the coefficients, and one uses this histogram to realize an a posteriori compression. We thus return to the point of view expressed by Donoho, which is no longer to respect the natural order given by a particular series development. Rearranging the order of the terms can accelerate convergence of a wavelet series expansion.

We will go into a little more detail about DeVore's algorithm by describing it in two important cases: (1) The error is measured in the L^2 norm (mean-square error), and the functions we wish to approximate are not, a priori, even bounded. (2) The error is measured in the L^∞ norm (uniform approximation), and the functions we wish to approximate have a certain a priori regularity. In the first case, we begin with an irregular function f belonging to $L^2(\mathbb{R}^n)$, although the problem we wish to solve is, in fact, local. We try to estimate, for an arbitrary dimension n, $\rho_N(f) = \inf \|f - f_N\|_2$ in these two cases: (a) f_N belongs to the set S_N of free-knot splines, and (b) $f_N(x) = \sum a_\lambda \psi_\lambda(x)$, where here the index N means that the sum contains at most N terms and the ψ_λ are wavelets. The set S_N is defined by partitioning the domain D where one is working into N dyadic cubes $Q_1, \ldots, Q_N$ and by considering all linear combinations of the basic splines ϕ_{Q_j} fitted to the cubes Q_j. The quality of the approximation is measured in the L^2 norm using a given positive exponent $\beta \in (0, \frac{1}{2})$. We wish to characterize those functions f in $L^2(\mathbb{R}^n)$ for which $\rho_N(f)$ is of the order $N^{-\beta}$. This property should be equivalent to some kind of smoothness property of f. Here is the precise statement of the result.

THEOREM 11.2 (DEVORE, JAWERTH, POPOV [81]). *Assume that β is fixed with $0 < \beta < \frac{1}{2}$. Let q be defined by $\frac{1}{q} - \frac{1}{2} = \beta$ and write $\alpha = n\beta$. Then the following three properties of a function f in $L^2(\mathbb{R}^n)$ are equivalent.*

(1) *The function f belongs to the Besov space $\dot{B}^{\alpha,q}(L^q)$.*
(2) *The wavelet coefficients $\alpha(\lambda)$ of f satisfy the condition $\sum |\alpha(\lambda)|^q < \infty$. (The wavelets are assumed to form an orthonormal basis.)*
(3) *The errors $\rho_N(f)$ in the nonlinear approximation satisfy $\rho_N(f) = N^{-\beta}\varepsilon_N$, where ε_N^q is summable.*

Since $0 < \beta < \frac{1}{2}$, we have $1 < q < 2$, which means that $\sum |\alpha(\lambda)|^q < \infty$ is stronger than the obvious condition $\sum |\alpha(\lambda)|^2 = \|f\|_2^2$. In other words, $\dot{B}^{\alpha,q}(L^q)$ is contained in L^2. Furthermore, the best approximation (using N wavelets) is given by the nonlinear thresholding rule whereby one saves only the N largest wavelet coefficients. This is an approximation scheme where the natural order of the wavelet series is upset, and the terms are rearranged in order of decreasing L^2 norms.

If linear approximation were used, then $\rho_N(f) = N^{-\beta}\varepsilon_N$ and $\varepsilon_N \in l^q$ would imply that f belongs to the Sobolev space H^β. However, the Besov space $\dot{B}^{\alpha,q}(L^q)$ is not contained in H^β; it is only in L^2. Here, nonlinear approximation allows one to "cheat" and pretend that everything works as if β derivatives of f belonged to L^2. (Recall that functions in H^β are smooth; in fact they are $[\beta - \frac{n}{2}]$ times differentiable.)

This brings us to a remark about sparsity. Assuming that $f \in L^2(\mathbb{R}^n)$, the condition $\sum |\alpha(\lambda)|^q < \infty$, where $1 < q < 2$, means that the wavelet expansion $f(x) = \sum \alpha(\lambda)\psi_\lambda(x)$ is "sparser" than we would expect from just knowing that $f \in L^2(\mathbb{R}^n)$. This heuristic is based on the following weak inequalities, which we have already met in a slightly different form. For $0 < \tau < 1$, let N_τ be the number of wavelet coefficients $\alpha(\lambda)$ such that $|\alpha(\lambda)| > \tau$. Then $\sum |\alpha(\lambda)|^q = C_q < \infty$ implies that $N_\tau \leq C_q \tau^{-q}$. This inequality is stronger than $N_\tau \leq \|f\|_{L^2}^2 \tau^{-2}$, which is the best we have knowing only that f is in L^2. Note, however, that this sparsity condition, which we will call q-sparse, is not nearly as strong as the one used in Theorem 11.1.

An interesting application of Theorem 11.2 is given by the characteristic function $f = \chi_\Omega$ of a smooth bounded region Ω in $\mathbb{R}^2$. In this case, f belongs to all of the Besov spaces $\dot{B}^{\alpha,q}(L^q)$ for $\alpha < \frac{1}{q}$ and $1 \leq q \leq \infty$. Thus by Theorem 11.2, the errors in the nonlinear approximation satisfy $\rho_N(f) = O(N^{-\beta})$ for any $0 < \beta < \frac{1}{2}$. In this particular case, a direct check shows that $\rho_N(f) = O(N^{-1/2})$.

We are now going to describe the case where f is "regular" and the error for the nonlinear approximation is measured in the L^∞ norm rather than in the L^2 norm. Since the theorem is mainly used in image processing, we will give only the two-dimensional version of the result.

THEOREM 11.3 (DEVORE, JAWERTH, POPOV [81]). *Assume that* $f \in \dot{B}^{\alpha,q}(L^q)$, *where* $\alpha > \frac{2}{q}$ *and* $1 < q < \infty$. *Then the optimal error* $\varepsilon_N(f) = \inf \|f - f_N\|_\infty$ *measured in the uniform norm satisfies the inequality*

$$\varepsilon_N(f) \leq CN^{-\alpha/2}. \tag{11.8}$$

This is a striking result, since (11.8) would characterize the Hölder space $\dot{C}^\alpha$ if linear approximation were used. But $\dot{B}_q^{\alpha,q}$ is contained in $\dot{C}^\beta$, where $\beta = \alpha - \frac{2}{q}$, and not in $\dot{C}^\alpha$. The weaker assumption about f is compensated by nonlinear approximation to give the decay (11.8).

The function $f(x) = |x| \exp(i|x|^{-1} - |x|^2)$ provides an illustration of Theorem 11.3. This function belongs to the Hölder space $\dot{C}^{1/2}$ but not to $\dot{C}^\alpha$ for $\alpha > \frac{1}{2}$. However, f belongs to $\dot{B}^{\alpha,q}(L^q)$ for $1 \leq q \leq \infty$ and $\alpha < \frac{1}{2} + \frac{1}{q}$. This function is a chirp at zero, and it is better compressed by a wavelet series expansion than by a Fourier series expansion.

Here is a sketch of the proof of Theorem 11.3. To obtain a uniform approximation of f with an error less than or equal to δ, one defines the threshold j_0 to be $\alpha^{-1} \log_2 \delta^{-1}$, and one keeps, in the first place, all of the terms in the orthogonal wavelet decomposition of $f(x,y)$ that correspond to scales $0 \leq j < j_0$. One assumes that the first approximation is at scale one given by a function of V_0 and that one is looking for an approximation on a bounded set. Then this first step amounts to keeping $C2^{2j_0}$ terms. If $j \geq j_0$, one applies at each scale an explicit threshold to the wavelet coefficients: The coefficients satisfying $|c(j,k,l)| < \varepsilon_j = \delta 2^{-\beta(j-j_0)}$ are replaced by zero. The wavelet series is written as $\sum \sum c(j,k,l) \psi_i(2^j x - k, 2^j y - l)$, $\beta = \alpha - \frac{2}{q}$, and the hypothesis is that $c(j,k,l) 2^{j\beta}$ belongs to l^q. The number N_j of coefficients retained at the scale 2^{-j} is estimated by observing that the condition of belonging to l^q implies the corresponding weak inequality. We have $N_j \varepsilon_j^q \leq \sum |c(j,k,l)|^q$ and hence $N < C2^{2j_0} + \sum N_j < C'2^{2j_0}$. The error is no greater than $C\delta$. This approximation, where N terms are sufficient to obtain (in two dimensions) an error less than $N^{-\alpha/2}$, is surprising because the global regularity is given by the Hölder exponent $\beta = \alpha - \frac{2}{q}$. By using a linear algorithm, the error would be $N^{-\beta/2}$, which is significantly larger. We remind the reader that this error is measured in the uniform (L^∞) norm.

The scientific message contained in the preceding proof is more important than the proof itself. It is this: A wavelet thresholding provides a near optimal nonlinear approximation. The same conclusion was reached by T. Lyche and K. Mørken working in Oslo on computer aided design [178]. Their compression algorithm uses a multigrid (fine to coarse) scheme. It mimics the Schauder basis expansion, which is a successive approximation of a continuous function by piecewise affine functions. More precisely, as shown in Chapter 2, one has $f(x) = \sum \sum_{0 \le k < 2^j} \alpha(j,k) \theta(2^j x - k)$, where θ is the triangle function centered at $\frac{1}{2}$ on the interval $[0,1]$ and where $\alpha(j,k) = f(k+\frac{1}{2})2^{-j} - \frac{1}{2}[f(k2^{-j}) + f((k+1)2^{-j})]$. The one-dimensional case of the *Oslo algorithm* would then consist of setting $\alpha(j,k) = 0$ for all $\alpha(j,k)$ that fall below a given threshold. This procedure looks like wavelet thresholding to the extent that the $\alpha(j,k)$ resemble wavelet coefficients. In the two-dimensional case, the Oslo algorithm erases each pixel whose gray level can be computed by averaging the gray levels of the neighboring pixels. This is the reason the algorithm is called *knot removal* in [178]. In [80], DeVore, Jawerth, and Lucier translated the Oslo algorithm into the language of wavelets.

We reiterate that nonlinear approximation is indeed more efficient than linear approximation for many functions. One can with very few terms represent rather irregular functions using nonlinear approximation, while if one wished to obtain the same quality of approximation using a linear scheme, one would be obliged to use significantly more terms in the series (or impose much more regularity on the functions that one seeks to represent). In the context of image processing, the goal of nonlinear approximation is to obtain clean edges while optimizing the bit allocation.

As is often the case, these nonlinear techniques seem, a posteriori, very natural in analysis because they amount to classifying things in order of importance rather than confining oneself to the conventional order, like the order of the terms of a series fixed in advance. These remarks might lead to the optimistic belief that nonlinear approximation would yield a solution to the problem of feature extraction. Since "feature" has not been given a precise meaning, we illustrate this concept with an example. Wavelets have already been used for mammogram segmentation, enhancement, and compression [84] (see also [87] and [85]). The goal is to detect biopsy-proven malignant clusters of calcifications superimposed on ordinary tissues of varying density. These clusters are the features to be enhanced. Moreover, these features should not be degraded by a compression algorithm. Indeed, telediagnostics and teletherapy rely crucially on transmitting medical images, and compression is a key ingredient in efficient transmission. The good news for wavelet enthusiasts is that a wavelet-based algorithm is the only lossy compression algorithm to receive FDA approval for use in medical devices (see http://www.jpg.com). We believe that better suited methods will eventually outperform wavelets. This is based on our belief that proper statistical modeling of the class of images to be compressed will lead to the development of algorithms adapted to the images.

Before moving to the second theme of this chapter, we wish to illustrate with an example the compression issue that concerns wavelet expansions versus Fourier expansions. As already mentioned, Peller's theorem applies to the simple function

$$f(x) = (c_1|x - x_1|^{\alpha_1} + \cdots + c_m|x - x_m|^{\alpha_m})e^{-x^2},$$

where the exponents α_j are positive real numbers. If $\alpha = \inf\{\alpha_1, \ldots, \alpha_m\}$, the Fourier transform of f decays like $O(|\xi|^{-1-\alpha})$. Once f is made 2π-periodic, $N = \varepsilon^{-1/\alpha}$ terms of the Fourier series are needed to ensure an error that is uniformly less than ε. If α is small, then N is a large power of ε^{-1}. If one uses wavelets to expand this particular function, then $O(\alpha^{-1}\log\varepsilon^{-1})$ terms will suffice. This example supported the intuition of the pioneers, but it is only since the work described above that we have had a systematic approach to these compression issues.

We end this section with a more systematic study of the singularities of functions that have sparse wavelet expansions. This problem can be studied in any dimension, but we will focus on the two-dimensional case and applications to image processing. A first step is to extend the definition of a sparse wavelet expansion to functions of n real variables. In $\mathbb{R}^n$, $2^n - 1$ wavelets ψ_i are needed to obtain orthonormal wavelet bases of the form $2^{nj/2}\psi_i(2^j x - k)$, $1 \leq i \leq 2^n - 1$, $j \in \mathbb{Z}$, $k \in \mathbb{Z}^n$. Here again we want these wavelets to belong to the Schwartz class $\mathcal{S}(\mathbb{R}^n)$. As before, we say that the function f has a sparse wavelet expansion

$$f(x) = \sum_{i,j,k} \alpha(i,j,k)\psi_i(2^j x - k) \tag{11.9}$$

if and only if

$$\sum_{i,j,k} |\alpha(i,j,k)|^p < \infty \tag{11.10}$$

for all $0 < p \leq 1$. The following result was recently obtained by adapting an argument due to Stéphane Jaffard in [154].

THEOREM 11.4 (Y. MEYER). *If f has a sparse wavelet expansion, then there exists a set $E \subset \mathbb{R}^n$ with Hausdorff dimension zero such that the pointwise Hölder exponent $\alpha(f,x) = +\infty$ for $x \notin E$.*

As an application of this result, we know immediately that the function f defined by $f(x) = \sup\{0, 1-|x|^2\}$, $x \in \mathbb{R}^n$, does not have a sparse wavelet expansion because it is not smooth across the unit sphere, which has Hausdorff dimension $n-1$. More precisely, the Hölder exponent $\alpha(f,x)$ is 1 for $|x| = 1$.

We will outline the proof of Theorem 11.4, since it has not been published elsewhere. The hypothesis is that the coefficients of f in equation (11.9) satisfy (11.10). Write $\varepsilon = N^{-1}$, $N \in \mathbb{N}^*$, and construct the exceptional set E_N as follows. Let U_j^ε be the union over $k \in \mathbb{Z}^n$ and $1 \leq i \leq 2^n - 1$ of the closed balls $|x - k2^{-j}| \leq |\alpha(i,j,k)|^\varepsilon$ and let E_N be the $\limsup_{j \to +\infty} U_j^\varepsilon$. Finally, let E be the union of all of these E_N. Then the Hausdorff dimension of E_N is zero because $\sum |\alpha(i,j,k)|^{\varepsilon\eta} < \infty$ for every $\eta > 0$, and hence the Hausdorff dimension of E is also zero. If $x \notin E$, then it is not difficult to apply Jaffard's criterion (Theorem 10.1) to prove that $\alpha(f,x) = +\infty$, as announced.

Roughly speaking, this theorem tells us that sparse wavelet expansions model signals with isolated singularities. In the two-dimensional case, images have jump discontinuities over lines, and this is excluded by the theorem. Does this mean that the achievements of DeVore and his collaborators do not apply to images? The situation is more complicated than a "yes" or "no" answer. Indeed, Besov spaces are being used to model images. The Besov space chosen by DeVore and Lucier is $\dot{B}_1^{1,1} = \dot{B}^{1,1}(L^1(\mathbb{R}^2))$. Unfortunately, the characteristic functions of smooth bounded domains do not belong to $\dot{B}_1^{1,1}$. This is why the larger space BV (for "bounded variation") is currently preferred. We are going to say more about the spaces that are used to model images in section 11.4. To prepare for this, we pause here to introduce two concepts that play key roles in current research: the space BV and weak l^p.

The space $BV(\mathbb{R}^2)$ is defined to be those functions f whose partial derivatives $\frac{\partial f}{\partial x_1}$ and $\frac{\partial f}{\partial x_2}$ (taken in the sense of distributions) are Radon measures with finite total mass. The BV norm of f is $\int_{\mathbb{R}^2} |\nabla f|\, dx_1 dx_2$, where $|\nabla f|$ is the length of the gradient of f. The characteristic functions χ_Ω of smooth domains Ω belong to BV, and $\|\chi_\Omega\|_{BV}$ is the length of the boundary $\partial \Omega$ of Ω. Although these characteristic functions do not belong to $\dot{B}_1^{1,1}(\mathbb{R}^2)$, we have the embeddings

$$\dot{B}_1^{1,1}(\mathbb{R}^2) \subset BV(\mathbb{R}^2) \subset \dot{B}_1^{1,\infty}(\mathbb{R}^2),$$

and these embeddings play a key role in Donoho's denoising strategy, which is described in the next section.

The definition of weak l^p is this: A sequence c_n is said to belong to weak l^p if the nonincreasing rearrangement c_n^* of $|c_n|$ satisfies the condition $c_n^* \leq Cn^{-1/p}$ for some constant $C > 0$ and all $n \geq 1$. This condition is implied by $\sum |c_n|^p < \infty$, as was already noted following the definition of "sparse."

There is a remarkable connection between these two concepts that was discovered by A. Cohen, R. DeVore, P. Petrushev, and H. Xu in the case of the Haar system [59] and generalized to other wavelet bases by Y. Meyer.

THEOREM 11.5. *If f belongs to $BV(\mathbb{R}^2)$, then the wavelet coefficients of f, $\alpha(\lambda) = \langle f, \psi_\lambda \rangle$, belong to weak l^1.*

The wavelets ψ_λ are assumed to form an orthonormal basis for $L^2(\mathbb{R}^2)$, and, to be precise about the normalization, the wavelet coefficients are those that appear when f is expanded as

$$f(x) = \sum_{i,j,k} \alpha(i,j,k) 2^j \psi_i(2^j x - k), \quad i = 1, 2, 3, \quad j, k \in \mathbb{Z}^2.$$

This condition is sharp. In fact, if f is the characteristic function of the unit disc and if $|\alpha(\lambda_n)|$ denotes the nonincreasing rearrangement of $|\alpha(\lambda)|$, $\lambda \in \Lambda$, then there is a positive γ such that $|\alpha(\lambda_n)| \geq \gamma n^{-1}$. In spite of this, f contained in weak l^1 does not imply that f is in BV.

Being weak l^1 or, more generally, weak l^p for $0 < p \leq 1$ is a form of sparsity, although it is clearly weaker than having $\sum |\alpha(\lambda)|^p < \infty$ for all $0 < p \leq 1$. These ideas will appear again in section 11.4. For the moment, we simply note that the connection between sparse representations of functions and image processing is an extremely active line of research.

11.3 Denoising

We begin with the simplest example. One wants to recover X from the given data Y, where we assume that $Y = X + \sigma Z$. The term σZ is considered to be noise; typically Z will be standard white noise and $\sigma > 0$ is a small parameter. In Donoho's work, the object X is a function f of a real variable t or an image, in which case we will assume that t belongs to the unit square. To develop an algorithm for recovering X, it is necessary to make some mathematical assumptions about the nature of f. These assumptions should reflect our a priori knowledge about the object X. Making assumptions about f based on our knowledge of what X should be is called modeling, and this issue will be addressed again in section 11.4. For the moment, we are going to follow Donoho, so our modeling of f says that f should be smooth or should belong to some ball B in a given function space. We will argue in the next section that images naturally belong to the space $BV(\mathbb{R}^2)$ of functions of bounded variation in the plane. For convenience of notation, we will use X to denote both the object we wish to recover and the function that models this object.

Our goal is to construct an estimator $\hat{X}$ of X. More precisely, we wish to build a nonlinear mapping Φ that takes Y (which are the data at our disposal) into a good candidate for X. We denote by $\|\cdot\|$ the norm that will be used to measure the error between $\hat{X}$ and X, and we let $\mathbb{E}$ denote the expectation operator taken with respect to the noise. We then consider the average risk $\mathbb{E}[\|\hat{X} - X\|^2]$ and compute it for the worst case. This yields the quantity

$$\sup_{X \in B} \mathbb{E}[\|\hat{X} - X\|^2], \qquad (11.11)$$

where B is the ball containing all the functions we wish to recover. Finally, we would like the estimator to be optimal among all possible linear and nonlinear candidates. This means that we need to solve the following minimax problem:

$$\inf_{\Phi} \sup_{X \in B} \mathbb{E}[\|\hat{X} - X\|^2]. \qquad (11.12)$$

This ambitious program is out of reach in most of the interesting cases, and we must thus be content with a near-optimal (or suboptimal) estimator Φ. Suboptimal is defined as follows: Let α be the largest power of σ such that for every $\varepsilon > 0$ the estimate

$$\inf_{\Phi} \sup_{X \in B} \mathbb{E}[\|\hat{X} - X\|^2] \leq C_\varepsilon \sigma^{\alpha - \varepsilon} \qquad (11.13)$$

is true as $\sigma \to 0$. An estimator $\hat{X}$ is suboptimal if

$$\sup_{X \in B} \mathbb{E}[\|\hat{X} - X\|^2] = O(\sigma^{\alpha - \varepsilon})$$

for all $\varepsilon > 0$ as $\sigma \to 0$, where the exponent α is the same as in the optimal case.

Roughly speaking, Donoho's theorem tells us the following: If the risk is measured in the L^2 norm, then the first thing to do for finding a near-optimal estimator is to construct an orthonormal basis for L^2 in which the functions belonging to B have sparse expansions. Following Donoho, we illustrate this statement with an almost trivial example. In the example, X, Y, and Z will be sequences $\{x_n\}$, $\{y_n\}$, and $\{z_n\}$, $n \geq 1$. When we return to more realistic situations, these sequences will be the coordinates of f and the other objects in some suitable basis. The noise $\{z_n\}$ is not stochastic, but we assume that $|z_n| \leq 1$. In other words, each coordinate x_n is corrupted by an error that does not exceed σ. The error between the sequence $\{x_n\}$ we wish to recover and the estimated sequence $\{\hat{x}_n\}$ will be measured in the $l^2(\mathbb{N})$ norm. We are going to model our a priori knowledge about the solution by the condition $|x_n| \leq C n^{-\beta}$, $n \geq 1$, where C is a given constant and $\beta > \frac{1}{2}$ is a given exponent. We denote this collection of sequences by B.

The estimator we construct for the example is based on a slightly different definition of risk. We do not average over the noise, but instead we focus on the worst case. This leads us to define the risk to be

$$\sup_{X \in B} \sup_{Z} \|\hat{X} - X\|_2^2 \qquad (11.14)$$

and to construct an estimator that minimizes (11.14). Constructing this estimator is an exercise. It is sufficient to do it separately for each coordinate. One first considers the case $Cn^{-\beta} > \sigma$ and then the case $Cn^{-\beta} \leq \sigma$. The resulting decision rule, which constitutes the estimator, depends on C and β.

David Donoho improved this algorithm with a much more intuitive decision rule that does not depend on β. This decision rule is called *thresholding*. It is defined as follows: If $\sigma \geq C$, then it is assumed that the signal is entirely buried in the noise and we set $\hat{X} = 0$. If $0 < \sigma < C$, we first consider those indices n for which $|y_n| < 2\sigma$. For such an n, $|x_n| \leq 3\sigma$, and this coordinate of the signal is considered to be buried in the noise. For these cases we set $\hat{x}_n = 0$. If $|y_n| > 2\sigma$, then $Cn^{-\beta} \geq |x_n| > \sigma$, and we set $\hat{x}_n = y_n - \sigma \operatorname{sign}(y_n)$, which implies that $|\hat{x}_n| \leq |x_n| \leq Cn^{-\beta}$. A simple computation shows that the worst risk is of the order σ^α, where $\alpha = 2 - \frac{1}{\beta}$. Observe that this risk becomes smaller as β increases. This thresholding algorithm is near optimal because it yields the same exponent α as the optimal estimator.

Since the thresholding estimator does not depend on β, the converse problem can be addressed: Given a sequence $\{x_n\}_{n \in \mathbb{N}^*}$, under what condition is the worst risk $\sum_{n=1}^{\infty} |\hat{x}_n - x_n|^2$ of the order σ^α as σ tends to zero? Here, as before, $y_n = x_n + \sigma z_n$, where $|z_n| \leq 1$ and $\hat{x}_n$ is the estimator given by the previously defined thresholding. The answer is that $\sum_{n=1}^{\infty} |\hat{x}_n - x_n|^2 = O(\sigma^\alpha)$ if and only if the nonincreasing rearrangement of $|x_n|$ decays like $O(n^{-\beta})$, where $\alpha = 2 - \frac{1}{\beta}$.

We are now going to leave this simple example and address more realistic situations where the object we wish to recover is modeled by a function f defined on the interval $[0,1]$ and belonging to some ball B in a function space E. The noise is assumed to be standard Gaussian white noise,[12] and we wish to estimate $f(t)$ from the noisy data

$$g(t) = f(t) + \sigma z(t), \qquad 0 \leq t \leq 1. \tag{11.15}$$

We assume that the risk is evaluated in $L^2[0,1]$. With these assumptions and with what we have learned from the simple example, we are naturally led to look for an orthonormal basis $\{e_n\}$ for $L^2[0,1]$ that ensures a fast decay of the coefficients $\langle f, e_n \rangle$, $n \geq 1$, when $f \in B$. More precisely, we are led to search for a "best basis" among all orthonormal bases for $L^2[0,1]$, where "best" means that one for which the decay of $\langle f, e_n \rangle$ is the fastest in the worst case, f running over B. As Paul Lévy pointed out, if $\{e_n\}$ is any orthonormal basis, then the coordinates $z_n = \langle z, e_n \rangle$ of a standard white noise are independent, identically distributed standard Gaussian variables (which we abbreviate by i.i.d. $N(0,1)$).

Before going further with the general theory, it is useful to illustrate these ideas with an example. If, for instance, B is the unit ball of the Hölder space $\dot{C}^\alpha$, $\alpha > 0$, then the Fourier coefficients c_n of $f \in B$ decay like $O(n^{-\alpha})$, which is optimal. The corresponding wavelet coefficients decay like $O(n^{-\alpha - 1/2})$, which is clearly better. Furthermore, the space $\dot{C}^\alpha$ is characterized by this decay of the wavelet coefficients. Hölder spaces are embedded in the larger family of Besov spaces $\dot{B}^{\alpha,q}(L^p)$. These remarks shed light on the deep relations between denoising and finding sparse expansions for certain classes of function spaces.

We are now going to reformulate (11.15). Since the signal we are looking for is a smooth function defined on $[0,1]$, our denoising problem can be restated as follows:

---

[12]See [145] for recent results on wavelet thresholding where the noise is not Gaussian.

The data $y_0, \ldots, y_{N-1}$, $N = 2^q$, that are collected are given by

$$y_k = f\left(\frac{k}{N}\right) + \sigma z_k, \qquad 0 \leq k < N, \tag{11.16}$$

where the z_k are i.i.d. $N(0,1)$. The wavelet transform that is needed yields an isometric isomorphism between $L^2[0,1]$ and $l^2(\mathbb{N})$. Here we are looking for its discrete version. In this discrete version, $L^2[0,1]$ is identified with $l^2\{0, \frac{1}{N}, \ldots, \frac{N-1}{N}\}$, where each point $\frac{k}{N}$ is given the mass $\frac{1}{N}$. With these conventions, the wavelet transform of (11.16) is

$$Y_m = X_m + \frac{\sigma}{\sqrt{N}} Z_m, \qquad 0 \leq m \leq N-1, \tag{11.17}$$

where the X_m are the wavelet coefficients of $f(\frac{k}{N})$ and the Z_m are i.i.d. $N(0,1)$. Here, the index m plays the role of the pair (j,k) that is usually used in the wavelet transform.

If the smoothness assumption about f appears as the condition $|X_m| \leq C m^{-\beta}$, we are not too far from our first example. However, in the case at hand, the Z_m are not uniformly bounded by 1, but rather by $\sqrt{2 \log N}$. This is why the threshold τ used below in Donoho's wavelet shrinkage is not $\frac{\sigma}{\sqrt{N}}$, as our simple example might lead one to believe, but rather $\tau = \frac{\sigma}{\sqrt{N}} \sqrt{2 \log N}$.

One of Donoho's most interesting algorithms has the following remarkable property: Its application does not depend on the exponents α, p, and q of the Besov space $\dot{B}^{\alpha,q}(L^p)$ used to model the data. We consider the case of a noisy image and try to reconstruct $f(t_i)$ from the noisy data $d_i = f(t_i) + \sigma z_i$ where z_i is normalized white noise and where the points $t_i = (\frac{k_{i,1}}{N}, \frac{k_{i,2}}{N})$ belong to the fine grid defining the image. This is how the algorithm works: Starting with the noisy data, one computes the corresponding empirical wavelet coefficients. (We will say something about how these are computed in a moment.) Then one applies the following *wavelet shrinkage* to these empirical coefficients: All the coefficients with modulus less than or equal to $\tau = 2\sigma N^{-1}(\log N)^{1/2}$ are replaced with zero. Those whose modulus is greater than τ are displaced toward zero by an amount equal to τ. In other words, each wavelet coefficient, x, is replaced by $y = \theta(x) = x - \tau \operatorname{sign}(x)$ if $|x| > \tau$ and by $y = 0$ otherwise.

Donoho proved that this estimator has the following properties:

(1) Each time that one has a priori "Besov" knowledge about the signal or image, the algorithm is suboptimal.

(2) The algorithm preserves regularity, that is, the a priori knowledge about the signal.

(3) If the signal is zero, the algorithm returns zero.

We must note, however, that the threshold used in the algorithm depends on the a priori knowledge of the noise level. The suboptimal nature of the algorithm is again defined by the rate at which $\|f - \hat{f}\|_2$ tends to zero as the noise level σ tends to zero. Here, $\hat{f}$ is the estimate of f given by Donoho's algorithm.

In Donoho's algorithm one must compute the wavelet coefficients in a situation that is different from the usual case of a function defined on the whole real line. Here, we have only discrete data defined on an interval. Wavelets tailored to an interval have been constructed by I. Daubechies, A. Cohen, and P. Vial [58]. Roughly speaking, one defines the approximation spaces V_j by first using all of the scaling functions $\varphi(2^j x - k)$ having support in the interval I and by then adjoining other special scaling functions that take care of the ends of I. This is done so as to generate all of the polynomials of degree $\leq N$ (in the case one is using wavelets with N zero moments). Then the construction of the wavelets follows the usual process. Having done this, Daubechies and her collaborators constructed the filters needed to pass from one scale to the next. These are the same filters that are used to process the data in Donoho's algorithm.

This new algorithm is called *soft thresholding*. We will see in Chapter 12 how this technique is used to improve an image reconstruction algorithm used in astronomy. This supports our theme that "specific problems call for tailored solutions."

11.4 Modeling images

Image processing is an important application of Donoho's discoveries. This work concerns *geometric-type images*, and here is what it is about. A real image, like that of a classroom, is composed (approximately) of geometric forms that are outlined by rather simple contours. These geometric forms are "filled in" with variations in the luminous intensity called textures. For example, some students wear pullover sweaters, and a close examination of these sweaters reveals periodic, or almost periodic, patterns that have high spatial frequency with weak intensity. This is to say that the variations in luminous intensity may be very rapid but are weak when compared with the much more pronounced variations at the edges of the students' silhouettes. If we asked a talented draftsperson to make a sketch of this classroom, the lines representing the contours would be very distinct while the textures would be reproduced with much less fidelity. These textures, like those created by hair, would be suggested rather than carefully drawn. This is indeed how an artist works. For example, A. Dürer was famous for being able to create, with a single brush stroke, hair that appeared to be drawn hair by hair.

Such ideas led to the concept of simulating natural textures automatically using algorithmic techniques that imitate Dürer's brush. Currently, two-dimensional versions of fractional Brownian motion can be used to simulate some kinds of textures. These simulations are made by representing a fractional Brownian motion as a series of appropriate wavelets with i.i.d. Gaussian coefficients. This technique was initiated by Fabrice Sellan, and the details can be found in [2]. For more recent work on the synthesis of fractional Brownian motion, see [211].

However, our focus is on the contours, and here we imitate Ingres and his pictorial vision. Marr suggested that the low-level process in the human visual system was based on some kind of wavelet analysis. Indeed, Marr wanted to explain the extraordinary ability of the human visual system to detect edges. Marr's explanations are based on the following model. Consider a piecewise smooth function u with jump discontinuities across the boundaries $\partial D_1, \ldots, \partial D_N$ of the domains $D_1, \ldots, D_N$ in which u is smooth. The given image f is modeled by $f = u + v$, where u is defined as above and v contains the noise and texture. Indeed, if the function u is smooth inside $D_1, \ldots, D_N$ with jump discontinuities across the boundaries $\partial D_1, \ldots, \partial D_N$, then the wavelet coefficients $\int u(x)\psi_\lambda\, dx$ are either rather small or rather large. They are small whenever the support of ψ_λ does not hit one of the boundaries, and they are large when the support of ψ_λ intersects these boundaries. Wavelet thresholding retains only the large wavelet coefficients, and it can thus be interpreted as an edge detector. This is where Donoho's wavelet shrinkage can be presented as an algorithm for finding contours. One can say that Donoho's thinking extends the ideas initiated by Marr.

The basic working hypothesis is that the noise and textures are indistinguishable and that the algorithm should extract the design, that is, the contours, while ignoring the textures and noise. Donoho's algorithm can be compared to the patient and meticulous work of an archaeologist who, faced with broken and weathered fragments of pottery, reconstructs the missing pieces based on thought and experience, and from this deduces the eating habits of a civilization. The kind of information used by the archaeologist is not available to the lay person; it is accessible only to a specialist who is armed with a priori knowledge about what is being sought. Then the piece of broken pottery allows the archaeologist to choose one path among several from a universe of possibilities that has been sufficiently restricted by this a priori knowledge.

It is clear that the paradigm according to which contours and textures are the only components of an image is a simplification. Jean-Michel Morel, for example, describes an image as an ordered collection of level lines. The ordering is defined by the intensity of the gray level. Such a representation is more robust than the one given by contours (see [213]).

The $u + v$ model we have introduced is quite general. We are now going to add some refinements. These new models will come equipped with "denoising algorithms" that are designed to extract the u component from the sum $f = u + v$. This problem of extracting u has been pursued by several authors. We will first describe the approach taken by Mumford and Shah. We then consider work by Osher and Rudin, followed by that of DeVore and Lucier, and finally we return to Donoho's contribution. Most of these authors propose a variational approach: They minimize a functional over a collection of candidates for the cartoon sketch u.

In the Mumford–Shah approach, one is looking for a pair (u, K), where K is a compact set and the function u is smooth on the complement of K. We assume for the sake of simplicity that $f(x)$, $x = (x_1, x_2)$, belongs to $L^2(\Omega)$, where Ω is the unit square. The Mumford–Shah functional J, which is to be minimized, is defined by

$$J(u) = \int_\Omega |f(x) - u(x)|^2 \, dx + \alpha \int_{\Omega \setminus K} |\nabla u(x)|^2 \, dx + \beta \mathcal{H}^1(K),$$

where $\alpha > 0$ and $\beta > 0$ are two parameters that need to be adjusted for the class of images being processed and $\mathcal{H}^1(K)$ is the one-dimensional Hausdorff measure (total length) of K.

We are looking for a cartoon sketch u with jump discontinuities across K. These discontinuities prevent the distributional derivative ∇u from being square integrable, and this is the reason that ∇u is only computed on the complement of K. Note that J is the sum of two terms in competition. The first term measures the quality of the approximation; the second term says how smooth we want u to be outside K; and the third term measures a price or penalty to be paid for this approximation. As indicated above, α and β need to be tuned to the class of images being processed. If β is quite small, this choice might lead to finding too many edges (and objects) in the image. On the other hand, if β is relatively large, some objects will be eliminated along with the additive noise. The optimal value of β depends on the class of images. A similar discussion applies to α.

We now turn to the Osher–Rudin model. The first term, which measures the error, is the same, while the penalty function is $\alpha \int_\Omega |\nabla u(x)| \, dx$. This term is the BV norm of u. We say that a function u defined on $\mathbb{R}^2$ has bounded variation if its gradient, in the sense of distributions, is a signed Radon measure with finite total mass. By an obvious abuse of language, this finite mass is denoted by $\int_\Omega |\nabla u(x)|\, dx$. Observe that $\int_\Omega |\nabla u(x)|\, dx$ is the sum of two terms. Indeed, $\nabla u = f + \mu$ where $f \in L^1(\Omega)$ and where μ is singular with respect to Lebesgue measure. Then $\int_\Omega |f(x)|\, dx$ corresponds to $\int_{\Omega \setminus K} |\nabla u(x)|^2\, dx$ in the Mumford–Shah model, while $\|\mu\|$ corresponds to $H^1(K)$. More precisely, if u is smooth on finitely many domains $D_1, \ldots, D_N$ of Ω with jump discontinuities across their boundaries, then $K = D_1 \cup \cdots \cup D_N$, $\int_\Omega |f(x)|\, dx = \int_{\Omega \setminus K} |\nabla u(x)|\, dx$, and $\|\mu\| = \sum_{j=1}^N \int_{\partial D_j} j(u)\, d\sigma$, where $j(u)$ is the jump discontinuity of u across ∂D_j and $d\sigma$ is the arc length. If $j(u) = 1$ identically, then $\|\mu\| = H^1(K)$. This discussion shows that the Mumford–Shah model and the Osher–Rudin model have much in common.

In the DeVore–Lucier model, the penalty function is further simplified. The BV norm is replaced by a Besov norm, and the functional that is minimized becomes $\|f - u\|_2^2 + \beta \sum |\alpha(\lambda)|$, where $u(x) = \sum \alpha(\lambda)\psi_\lambda(x)$ is an orthonormal wavelet expansion of u. As is mentioned in [83], this optimization problem is trivial in the wavelet domain and leads to wavelet shrinkage. To see this, let $f(x) = \sum \gamma(\lambda)\psi_\lambda(x)$ be the wavelet expansion of f. Then the functional becomes $\sum [(\alpha(\lambda) - \gamma(\lambda))^2 + \beta|\alpha(\lambda)|]$, and this can be minimized by finding the minimum of $(\alpha(\lambda) - \gamma(\lambda))^2 + \beta|\alpha(\lambda)|$ for each λ. Assume that $\gamma(\lambda) > 0$. Then a simple computation shows that the minimum occurs at $\alpha(\lambda) = 0$ if $\gamma(\lambda) \leq \frac{\beta}{2}$ and at $\alpha(\lambda) = \gamma(\lambda) - \frac{\beta}{2}$ if $\gamma(\lambda) > \frac{\beta}{2}$. Similarly, if $\gamma(\lambda) < 0$, the minimum occurs at $\alpha(\lambda) = 0$ if $\gamma(\lambda) \geq \frac{\beta}{2}$ and at $\alpha(\lambda) = \gamma(\lambda) + \frac{\beta}{2}$ if $\gamma(\lambda) < \frac{\beta}{2}$. But this is just wavelet shrinkage with the threshold τ equal to $\frac{\beta}{2}$.

The last model we consider is the one defined by Donoho. The assumptions are slightly different. Donoho wishes to recapture u from $f = u + v$, where v is white noise and where u is subject to an a priori constraint of the form $\|u\|_B \leq C$. Here B is also a Besov space, and this a priori knowledge plays the role of the penalty function.

In the Mumford–Shah model or the DeVore–Lucier model, the decomposition $f = u + v$ is a solution of a variational problem. This appears to be an objective search, but it depends on the small parameters α and β that must be adjusted to the class of images we wish to process. In the DeVore–Lucier model, the decomposition also depends on the specific choice of the wavelet basis. This is due to the fact that $\sum |\alpha(\lambda)|$ is not the Besov norm of $\sum \alpha(\lambda)\psi_\lambda$; it is an equivalent norm. Donoho's algorithm was developed in a stochastic setting. However, wavelet thresholding makes sense for any function f.

One can ask if the u component in $f = u + v$ can be reconstructed from the wavelet coefficients of f that exceed a given threshold. The resulting function u is then the same as the one obtained from the DeVore–Lucier approach. One should also compare the Osher–Rudin model with wavelet shrinkage. As mentioned above, Stan Osher and Leonid Rudin defined the cartoon sketch u of a given image f to be the solution of the variational problem $\inf J(u)$, where

$$J(u) = \|f - u\|_2^2 + \lambda \|u\|_{BV} \tag{11.18}$$

and λ is a small parameter.

In the paper [59] cited in section 11.2, A. Cohen, R. DeVore, P. Petrushev, and H. Xu. addressed the issue of solving the variational problem (11.18) using a wavelet shrinkage algorithm. They proved this result: If instead of a smooth wavelet basis, one uses the Haar system, then wavelet shrinkage yields a cartoon sketch $\tilde{u}$ such that $J(\tilde{u}) \leq C \inf J(u)$. One then says that $\tilde{u}$ is suboptimal. Here C is a fixed constant, and the threshold in the shrinkage must be determined. Theorem 11.5 is a crucial piece of information that is used in this algorithm.

Modeling geometric images with BV functions gives better results than modeling them with Besov spaces, but when Donoho wrote his fundamental papers, nothing better than the embedding $\dot{B}_1^{1,1} \subset BV \subset \dot{B}_1^{1,\infty}$ was known. These embeddings play a key role in Donoho's denoising strategy. The same wavelet shrinkage is suboptimal for both of these Besov spaces, and thus it is suboptimal for $BV(\mathbb{R}^2)$. Furthermore, at that time, information about wavelet coefficients of functions of bounded variation was rather poor, while the characterization of Besov spaces using wavelet coefficients was quite simple. For example, $f \in \dot{B}_1^{1,1}$ if and only if $\sum_{\lambda \in \Lambda} |\alpha(\lambda)| < \infty$. Thus it was natural to use the space $\dot{B}_1^{1,1}$ rather than BV.

11.5 Ridgelets

One can argue that functions of bounded variation do not adequately model images. Indeed, a function of bounded variation is either the characteristic function of a domain whose boundary has a finite length, or it is an average of such functions. This atomic decomposition is provided by the co-area identity. Modeling objects with characteristic functions of domains with finite length boundaries may be inappropriate, since the objects we have in mind are probably not that complicated. Donoho decided to describe an image as a collection of objects delimited by smooth boundaries instead of merely rectifiable ones.

If one wants to efficiently represent (or compress) smooth domains, standard isotropic wavelets are not optimal. A better algorithm relies on an efficient description of the boundary, and this calls for orthonormal bases that can efficiently represent elongated objects, such as the arc of a circle. No one knew how to do this until Donoho constructed a remarkable orthonormal basis that was designed to provide a sparse representation for objects having arbitrary large eccentricities. Donoho's construction improved previous work by E. Candès. We are going to describe this basis, and we begin with one of our main themes.

When constructing a wavelet basis, we should return to the issue raised by Jean Ville: Should we first segment the frequency domain, or should we use some bases that are built on a segmentation of the time (or space) domain? The construction of Donoho's basis uses both strategies.

Let $\xi = (\xi_1, \xi_2)$ be the frequency vector, which will be written in polar coordinates as $\xi = (\rho \cos\theta, \rho \sin\theta)$, $-\infty < \rho < \infty$, $\theta \in [0, 2\pi)$. We let ρ take negative values and identify (ρ, θ) with $(-\rho, \theta + \pi)$. Then $L^2(\mathbb{R}, d\xi)$ is identified with the closed subspace H of $L^2(\mathbb{R} \times [0, 2\pi), |\rho|d\rho d\theta)$ defined by $f \in H$ if and only if

$$f(\rho, \theta) = f(-\rho, \theta + \pi). \tag{11.19}$$

An orthonormal basis for $L^2(\mathbb{R}, d\xi)$ will be written as an orthonormal basis for H through this representation in polar coordinates.

We return to the segmentation issue. The first segmentation is reminiscent of the Littlewood–Paley decomposition. The frequency plane is partitioned into dyadic annuli Γ_j defined by $2^j \leq |\xi| < 2^{j+1}$, $j \in \mathbb{Z}$. To build an orthonormal basis in the ρ variable that is consistent with this segmentation, one uses Malvar–Wilson wavelets. Let $w(\rho)$ be an even function of the real variable ρ with the following properties: $w(\rho)$ is C^∞, $w(\rho) = 0$ if $|\rho| \leq \frac{2}{3}$ or $|\rho| \geq 3$, and $|\rho|^{1/2} w(\rho)$ satisfies the Malvar–Wilson conditions (section 6.3). The orthonormal basis we will use is set of functions $2^j w(2^j \rho) \exp\left[i\pi\left(k + \frac{1}{2}\right)2^j \rho\right]$.

Next, we treat the angular variable θ, and here the segmentation is performed in the frequency domain. We are dealing with 2π-periodic functions, and the corresponding frequencies are integers. The orthonormal basis for $L^2[0, 2\pi)$ that is used is the periodized version of the orthonormal wavelet basis $2^{j/2}\psi(2^j t - k)$, $j \geq 0$, $k \in \mathbb{Z}$, and $\varphi(t - k)$, $k \in \mathbb{Z}$, where both φ and ψ belong to the Schwartz class, $\varphi(-t) = \varphi(t)$, and $\psi(1 - t) = \psi(t)$. This wavelet basis is indexed by the dyadic subintervals I of $[0, 2\pi)$. We write $\psi_I(\theta)$, $I \in \mathcal{I}$.

Finally, the *ridgelets* ρ_λ, $\lambda \in \Lambda$, are defined by their Fourier transforms. By definition,

$$\hat{\rho}_\lambda(\xi, \theta) = 2^j \left[w(2^j |\xi|) \exp\left[i\pi \left(k + \frac{1}{2} \right) 2^j |\xi| \right] \psi_I(\theta) \right. \\ \left. - w(2^j |\xi|) \exp\left[-i\pi \left(k + \frac{1}{2} \right) 2^j |\xi| \right] \psi_I(\theta + \pi) \right]. \tag{11.20}$$

Donoho's original paper on this subject treated the ridgelet expansion of the characteristic function of a half-plane [91]. Since then it has been shown that the ridgelet expansion of a characteristic function of a smooth domain is weak $l^{1/2}$, whereas the best one can do with wavelets is weak l^1. Thus ridgelets provide better compression for this class of images than do wavelets.

11.6 Conclusions

Several problems have been raised in this chapter. The first consisted of defining the class of functions (signals, images) whose wavelet expansions are sparse, in one sense or another. These functions are adequately compressed with wavelets. Depending on the norm that was used to measure the appropriate approximation, several characterizations in terms of Besov spaces have been presented.

The second message of this chapter seems to be a success story for wavelet analysis: Whenever the a priori information on a given class of signals or images can be formulated as a bound on a Besov norm, then wavelet shrinkage provides an optimal denoising. On the other hand, if u is a smooth function inside finitely many domains with jump discontinuities across their boundaries, then one should shrink the ridgelet coefficients of $f = u + \sigma v$ to recover u (v is a standard Gaussian white noise).

These two statements seem to be contradictory, but they become consistent if one returns to the definition of the worst risk. This worst risk is the supremum of $\mathbb{E}[\|f - u\|_2^2]$ taken over the Besov ball $\|u\|_B \leq C$. Such a supremum can be attained for certain intricate functions u that do not correspond to our notion of a cartoon image. Besov balls are indeed very large sets. With the availability of ridgelets, new algorithms for optimal denoising should soon be available.

Another message is that there continues to be a need for new function spaces "adapted to edges," and this provides new goals for functional analysis.

CHAPTER 12

Wavelets and Astronomy

This final chapter is about the use of wavelets in astronomy and astrophysics. Wavelets are being applied in many fields of science and technology. We have selected astronomy as an example for several reasons: There are diverse applications within the field, and although they all involve some form of signal or image processing, the techniques vary from one application to another. Astronomy is driven by sophisticated technology for both ground-based and space-based observations, and this technology has led to problems that appear to be well suited to wavelet techniques. Finally, there is widespread popular interest in astronomy and cosmology, an interest that has been kindled by the richness of recent discoveries.

The chapter is based on our interpretation of the literature and on discussions with two groups of astronomers, the one directed by Albert Bijaoui (Observatoire de la Côte d'Azur, Nice) and the other led by André Lannes (Observatoire Midi-Pyrénées, Toulouse). In his review article on the uses of wavelets in astrophysics [36], Bijaoui discusses a number of problems where wavelet-based techniques are being applied; these include the analysis of solar time series; image compression, detection, and analysis of astronomical sources; and data fusion—as well as the study of the large-scale structure of the universe. We have selected three examples that illustrate different problems and techniques. In each case, wavelets are used in complex algorithms that are handcrafted by experts in astronomy to deal with specific problems. Roughly speaking, astronomical applications of wavelets differ from other applications because of the nature of astronomical images and signals.

12.1 The Hubble Space Telescope and deconvolving its images

Long in planning, greatly over budget, and fraught with management and scientific problems, the Hubble Space Telescope (HST) is today one of the scientific wonders of the world. It is not necessary to be an astronomer to be impressed with the images it produces. This was not always the case. Shortly after launch in April 1990, it came close to being the scientific laughingstock of the century. The first images were very disappointing, and the experts soon determined that the 2.4-meter primary mirror of the Ritchey–Crétien telescope had a serious spherical aberration. We will discuss this problem and its correction, but first we need to introduce the model and language astronomers use to describe the process of obtaining astronomical images.

12.1.1 The model

Suppose that f_i is a digital image received by an astronomer, say, by downloading it from the database at the Space Telescope Science Institute. (This is the agency that coordinates the use of the HST and the distribution of its data.) (Although we are focusing on the HST, f_i could be a digitized image from other sources; the model applies to many situations.) The astronomer's working assumption is that f_i is related to the "original object" f_o by the equation

$$f_i(x) = p * f_o(x) + n(x). \tag{12.1}$$

The function p is called the point-spread function. It is determined experimentally as the image of a "point source" star. For ground-based astronomy, p is determined, if possible, during each observing session; it includes the condition of the atmosphere and other parameters that can vary from observation to observation. The situation with the HST is different because there is no atmosphere, and in this case, p is quite stable. A "good" p closely approximates a delta function: Its support is concentrated around zero, and it decays rapidly to zero away from the origin. The width of the central spike is determined by the diffraction limitation of the optical system. A "bad" p will have serious side lobes, or wings, and it spreads the energy from a point source over a relatively large area. The function n denotes noise. In fact, n is a catch-all term that includes both random and systematic errors (errors in determining p, errors resulting from the linearity assumption, image sampling, etc.) and random noise not correlated with the signal (from the telescope, the detectors, the atmosphere, the pointing system, etc.). We write "original object" in quotes because trying to say exactly what it is leads to a philosophical debate. For our purposes, it is an element of the Hilbert space $L^2(\mathbb{R}^2)$. The mathematical problem is to recover f_o from the data f_i.

12.1.2 Discovering and fixing the problem

After the discovery of the aberration, the user community turned to deconvolution to restore the images. It soon became clear, however, that this approach was too costly and had limited success and that a hardware solution would be required. Nevertheless, these initial deconvolution efforts did produce useful data, and the analyses of the point-spread functions—which varied with position of the point source in the field—provided information that helped to uncover the original manufacturing mistake.

The mirror had been perfectly ground and polished but to the wrong function: The mirror was too flat. The problem was traced to an error in setting up the device, called a null corrector, used to test the shape of the mirror as it was being polished. By knowing the exact nature of the problem, it was possible to design optical systems to compensate for the aberrated mirror. The best known of these is COSTAR, which stands for Corrective Optics Space Telescope Axial Replacement. It is an optical device that intercepts the "aberrated" light just after it passes through the hole in the primary mirror and "corrects" it for use by the spectrographs and the Faint Object Camera. The original High Speed Photometer was removed to make space for COSTAR. Other corrective optics were built into a new Wide Field/Planetary Camera. These replacements, as well as other repairs, were done in December 1993 during the first servicing mission. The optical corrections proved to be wildly successful, and the overall performance was as good "as if the mirror were perfect."

One of the missions assigned to the HST is to explore the outer limits of the universe. We know, based on the time it takes the light to reach earth, that the most distant galaxies observed are relatively young. These distant galaxies are in the process of developing their geometric complexity, and the structure of these distant objects provides hints about the development of the universe. Unfortunately, these objects are extremely faint (low intensity), and the received images are particularly noisy. The signal-to-noise ratio is indeed poor. Noise is always a problem in observational astronomy; in fact, it is not an exaggeration to say that it is the central problem. Furthermore, a bad point-spread function leads to a poor signal-to-noise ratio.

In spite of the profound disappointment with the first images and the realization that the mirror was aberrated, the telescope provided some useful scientific information between 1990 and 1993. This was possible because the images could, to a certain degree, be deconvolved. Several algorithms have been used to reconstruct images from the HST—both before and after the installation of corrective optics. Two of these algorithms, the Richardson–Lucy method and the maximum entropy method, are probabilistic. We are going to describe how wavelets are being used to improve the performance of a deterministic approach called interactive deconvolution with error analysis (IDEA). This algorithm was developed in the late 1980s by Lannes and his colleagues [169]. As stressed by the astronomy community, the main advantage of IDEA over competing algorithms is the fact that it provides precise error bounds.

12.1.3 IDEA

The problem is to extract an image from the data f_i, which is modeled by (12.1). This happens to be an ill-posed inverse problem; it does not satisfy the three conditions of Hadamard, namely, the existence, uniqueness, and stability of the solution. To have a feeling for this situation, take the Fourier transform of both sides of (12.1). Then

$$\hat{f}_i(\xi) = \hat{p}(\xi)\hat{f}_o(\xi) + \hat{n}(\xi), \qquad (12.2)$$

and recovering $\hat{f}_o(\xi)$ means dividing both $\hat{f}_i(\xi)$ and $\hat{n}(\xi)$ by $\hat{p}(\xi)$. It is clear that problems arise where $\hat{p}(\xi)$ vanishes or where $|\hat{p}(\xi)| << |\hat{n}(\xi)|$. IDEA is a fairly complex algorithm designed to circumvent these problems. To apply IDEA, one must bring to the process information that does not reside in equation (12.1), so-called a priori information. This a priori information is used to force a solution of the ill-posed problem.

We will outline the main features IDEA, which existed as a stand-alone algorithm before wavelets entered the picture, and then we will show how wavelet techniques are being used to improve the performance of the original algorithm. We emphasize that IDEA is not a wavelet algorithm: Using IDEA means working with the Fourier transform and not the wavelet transform. (A detailed description of IDEA can be found in [169].)

Since IDEA is a regularization algorithm, we begin with a few words about Tikhonov's regularization of ill-posed problems (see [248]). As above, the problem to be solved is described by an equation of the form

$$Y = TX + \sigma Z, \qquad (12.3)$$

where T is a compact operator acting on some Hilbert space H, X is the object we wish to recover, Z is an additive noise, $\sigma > 0$ is a parameter, and Y is the observed data. Tikhonov's regularization can be described in the context of operator theory or in a more concrete form. In the abstract setting, the regularization depends on a positive number $\eta > 0$ and reads

$$\widehat{X}_\eta = (T^*T + \eta I)^{-1}T^*Y, \tag{12.4}$$

where T^* is the adjoint of T and I is the identity operator. Observe that $T^*T + \eta I$ has an inverse if $\eta > 0$. At a formal level, $\widehat{X}_\eta = T^{-1}Y$ if $\eta = 0$. But this inverse may not exist, and (12.4) provides us with an approximate inverse.

The second version of Tikhonov's regularization uses a singular-value decomposition. There exists an orthonormal basis $e_0, e_1, \ldots, e_n, \ldots$ for the Hilbert space H that consists of the eigenfunctions of the compact self-adjoint operator T^*T. Let λ_n^2 ($\lambda_n \geq 0$) be the corresponding eigenvalues. In both versions of Tikhonov's algorithm, T^*Y is decomposed as

$$T^*Y = \alpha_0 e_0 + \alpha_1 e_1 + \cdots + \alpha_n e_n + \cdots,$$

and in the first version we have

$$\widehat{X}_\eta = \alpha_0 w_0 \lambda_0^{-2} e_0 + \alpha_1 w_1 \lambda_1^{-2} e_1 + \cdots + \alpha_n w_n \lambda_n^{-2} e_n + \cdots, \tag{12.5}$$

where the weights $w_n = \frac{\lambda_n^2}{\eta^2 + \lambda_n^2}$ are in the interval $(0, 1)$. These weights serve to regularize the divergent series $\alpha_0 \lambda_0^{-2} e_0 + \alpha_1 \lambda_1^{-2} e_1 + \cdots + \alpha_n \lambda_n^{-2} e_n + \cdots$. We can go further and introduce other weights w_n, $w_n \in (0, 1)$ in (12.5). The data are the α_n, $n \geq 0$, and the weights indicate our trust in the data.

When T is a convolution operator, it is diagonalized by the Fourier transform. This transform plays the role of the eigenfunction expansion we have seen above. In this form, the weighting coefficients w_n are replaced by a weighting function g, which will appear in IDEA and plays a similar role. However, the IDEA algorithm is an improvement over pure Tikhonov regularization. Tikhonov's regularization is a linear algorithm, and it does not offer the possibility to use the specific (or a priori) information we may have about the object to be recovered.

Once the small parameter η or the weights w_n, $n \geq 0$, are fixed, they determine a smoothing operator W with the property that $W(e_n) = w_n e_n$. This smoothing or averaging serves to "kill" the noise and to compensate for the "bad" behavior of the unbounded operator T^{-1}. In image processing, this smoothing introduces a systematic blurring of the image and destroys the sharp localization of the edges.

The weighting function in the IDEA algorithm is defined in the Fourier domain, and it is determined by preprocessing the given image. Furthermore, IDEA uses geometric information about the image to be reconstructed that we introduce as a priori information. The IDEA algorithm acts in the Fourier domain, but at the same time, it keeps track of the a priori information, which is known in the space domain. We have described Tikhonov's regularization to provide general background about regularization algorithms, but we wish to stress again that the regularization used in IDEA has a different meaning. Here the regularization of the ill-posed problem involves imposing a priori constraints on the object we wish to reconstruct. With this background, we are ready to be more specific about the algorithm itself.

IDEA depends crucially on a function σ_i defined in the Fourier domain that provides a pointwise upper bound on the error function $\hat{n}$, that is,

$$|\hat{f}_i(\xi) - \hat{p}(\xi)\hat{f}_o(\xi)| \leq \sigma_i(\xi). \tag{12.6}$$

The quality of the performance of IDEA depends on the quality of this estimate, and it is here that wavelets enter the picture. More precisely, wavelets are used to determine σ_i. We will explain how this is done in a moment, after describing IDEA.

The first step in the IDEA algorithm is to "regularize" the support of the transfer function $\hat{p}$. If P is the essential support of $\hat{p}$, choose P_r to be a disc of radius r that contains P. (We are assuming two-dimensional optical images, although IDEA can be formulated more generally [169].) P_r will be the synthetic aperture of the system.

Because of the practical limitations of telescopes and other technology involved in modern astronomy, it is hopeless to expect that f_o can be reconstructed at "its highest level of resolution." The object to be reconstructed is thus defined to be a smoothed version of f_o, namely,

$$\hat{f}_s(\xi) = s(\xi)\hat{f}_o(\xi). \tag{12.7}$$

The main conditions on s are that most of its energy is concentrated in P_r and that $s(0) = 1$. One also wants the support of $\hat{s}$ to be as small as possible, concentrated around $x = 0$. It is shown in [169] that s can be taken to be a prolate spheroidal function. The support V of f_s, whose size and shape is determined interactively in a wavelet-assisted application of IDEA, plays an important role in the algorithm.

We stress that at this point neither $\hat{f}_s$ nor $\hat{f}_o$ is known. Our first approximation of $\hat{f}_s$ will be $\hat{f}_t$, which is defined below. This first "guess" mimics (12.7), but it also takes into account the fact that the data are noisy. The idea is that information buried in the noise should be discarded. This leads to the following procedure that relies on the computation of σ_i, which will be described in a moment.

The function

$$\text{SNR}(\xi) = \frac{|\hat{f}_i(\xi)|}{\sigma_i(\xi)} \tag{12.8}$$

defines a pointwise signal-to-noise ratio in the frequency space. This function is used to decide where the information given by $\hat{f}_i$ should be retained and where it should be discarded as being too noisy. To this end, one chooses a threshold value α_t that is greater than one, but of order one, and defines

$$\hat{f}_t(\xi) = \begin{cases} \dfrac{s(\xi)\hat{f}_i(\xi)}{\hat{p}(\xi)} & \text{if SNR}(\xi) \geq \alpha_t, \\ 0 & \text{otherwise.} \end{cases} \tag{12.9}$$

It is f_t that is now used to find the "reconstructed object" that we call f_r, and once again SNR enters the picture. This time SNR is used to define a weight function $g(\xi)$. Having defined g, f_r is defined to be the function that minimizes the functional

$$q(f) = \int g^2(\xi)|\hat{f}_t(\xi) - \hat{f}(\xi)|^2 \, d\xi. \tag{12.10}$$

The minimum is taken over all $f \in L^2(V)$ where V is the support of f_s. V is determined interactively and is part of the a priori information. The initial choice of V is described below.

One has some freedom in defining g. It should be a nondecreasing function of SNR such that $0 \leq g(\xi) \leq 1$. In addition, g must vanish on the parts of P_r where $\text{SNR}(\xi) < \alpha_t$ and be equal to one outside P_r. One way to define g is to select a threshold value α'_t with $\alpha_t < \alpha'_t \leq \sup_\xi \text{SNR}(\xi)$ and let

$$g(\xi) = \begin{cases} 1 & \text{if } \text{SNR}(\xi) \geq \alpha'_t, \\ \dfrac{(\text{SNR}(\xi) - \alpha_t)}{(\alpha'_t - \alpha_t)} & \text{if } \alpha_t \leq \text{SNR}(\xi) < \alpha'_t, \\ 0 & \text{if } \text{SNR}(\xi) < \alpha_t. \end{cases} \qquad (12.11)$$

It is clear that g measures the confidence that can be attributed to the spectral information furnished by the Fourier transform of the noisy image.

This is but a brief outline of the principal objects that are used in the IDEA algorithm. The algorithm itself is iterative, and we encourage interested readers to consult the cited papers for a detailed description. The purpose of this discussion has been to present just enough background so that one can show how wavelet techniques have been incorporated in IDEA, which, as mentioned above, existed as an effective algorithm before being wavelet assisted. In particular, we hope it is clear that the function SNR plays a key role in IDEA and that a good estimate for the function σ_i should contribute to the quality of the results.

In all versions of IDEA, it is necessary to estimate σ_i, and indeed there are prewavelet techniques for doing this. In the wavelet-assisted version of IDEA, σ_i is estimated using the denoising technique described in Chapter 11 called soft thresholding. This is how it is applied by Roques and her collaborators [232]:

Step 1: Compute the empirical wavelet coefficients z of the scaled noisy data $f_i/\sqrt{n}$, where n is the number of data points or pixels. This transform is computed using the two-dimensional version of the wavelets adapted to an interval introduced by Cohen, Daubechies, and Vial [58].

Step 2: Apply wavelet shrinkage (soft thresholding) to these empirical wavelet coefficients:

$$\eta_t(z) = \text{sign}(z)\big(\max\{0, |z| - t\}\big)$$

with the threshold

$$t = \sigma\sqrt{\dfrac{2 \log n}{n}}.$$

σ^2 is the variance of the noise; we address it below. Note that this operation "shrinks" the wavelet coefficients toward zero by t and sets the coefficients with modulus less than t equal to zero.

Step 3: Invert the wavelet transform to produce a denoised image f_d and define σ_i by

$$\sigma_i(\xi) = |\hat{f}_i(\xi) - \hat{f}_d(\xi)|.$$

The variance σ^2 used in the Donoho algorithm is estimated by analyzing a part of the field defined by f_i that contains no image. Recall that the denoising described in Chapter 11 is based on two assumptions: The noise is Gaussian and the image is geometric. The latter of these assumptions is clearly not satisfied for astronomical images, and the former is often violated. In particular, one of the components of noise in experimental astronomy may come from photon counters, where the noise is Poisson. In this case, astronomers transform the noisy data to make the noise "look" Gaussian and proceed to act as if it were Gaussian. They replace f_i by $2\sqrt{f_i + \frac{3}{8}}$; a more complicated transformation is used for mixed noise (see [4]).

There is another point, in addition to estimating σ^2, where wavelets "assist" IDEA: The denoised image f_d is used to choose the initial value of V, which is the support for the deconvolution and thus an important piece of a priori information. In the actual algorithm, the set V is improved dynamically. V also appears in the *interpolation parameter*

$$\nu = \Big(\int_V v(x)\,dx \Big)^{1/2} \Big(\int_{P_r} (1 - g^2(\xi))\,d\xi \Big)^{1/2},$$

where v is the characteristic function of V. The value of ν provides information about the stability of the reconstruction process.

An obvious question is, Why not just use the image f_d? As Roques and her colleagues show in [232], f_d is indeed a low-noise image, but the resolution has not been improved: The image has been denoised but not deblurred. The companion question is, Why not do the deconvolution using an estimate of the noise similar to the one used to apply shrinking? Again, it is shown in [232] that the combined processes produce better images, at least for the very faint images obtained with the HST. Of course, this brings up the question, What is a good image? Astronomers must judge the quality of the image based on experience. They also have more objective (mathematical) ways to measure the photometric and astrometric[13] quality of the restored image. One naturally wants to have as high a resolution as possible without introducing artifacts, but it is the astronomer who must differentiate artifact from image. As stressed before, IDEA has the advantage over competing algorithms of providing an estimate for the relative error

$$\frac{\|f_r - f_s\|}{\|f_r\|}.$$

We note that the people who invented IDEA have benefited from unforeseen good luck: They have been able to compare their deconvolved images with those obtained by the HST after the COSTAR correction was made. This comparison has led to these conclusions:

(1) The IDEA algorithm produces corrected images that are closer to the "true" images than the images obtained by denoising methods traditionally used in astronomy.

(2) The "true" images obtained after the COSTAR correction are better than those obtained by IDEA, which is not surprising.

(3) The IDEA algorithm allows one to improve further the images obtained by the corrected telescope.

Tests leading to these conclusions were made on images of the supernova SN1987A, which are particularly simple and spectacular. There is a bright core together with a well-delimited extended object, the ring (see [39]).

---

[13]Photometric refers to the local conservation of photons, and astrometric refers to the preservation of the geometry of the image.

12.2 Data compression

We are speaking about the problems of storing and transmitting the data acquired by the world's astronomical observatories. As in the last section, we are looking at a technologically driven problem: The overall quality of telescopes is much greater today than it was 50 years ago. Astronomers were able to capture ten million galaxies in 1950; today they can examine 100 million galaxies. The very nature of the images coming from these instruments had undergone a revolution. Charge coupled devices (CCDs) have replaced silver salts, and chemical photography is almost a thing of the past. We read in [38] that telescopes typically use 2048 × 2048 CCDs at their focus. With 16 bits per pixel this leads to an 8 megabyte image. As an example of the amount of data generated, the Canada–France–Hawaii Telescope generates about 100 images each night, which translates to as much as 800 megabytes per night [251]. Planned future telescopes will generate about 10 gigabytes per night. All of this data must be stored, preferably in a form that offers reasonably easy access. These problems are reminiscent of those posed by the storage of fingerprints. For comparison, the FBI database contains about 200 million fingerprint records, and they receive on the order of 30,000 new cards per day, which is about 300 gigabytes each day [41]. (A set of fingerprints amounts to around 10 megabytes.) The comparison does not end there. With the advent of computers and communication networks, both astronomical images and fingerprints are now transmitted around the world, and compression is an economic necessity for both storage and transmission.

Wavelets have been used to compress astronomical images since the late 1970s. G. M. Richter and others used Haar functions to compress astronomical data, which at that time came mainly from Schmidt plates (see [230] and [120]). These were scanned automatically and the data were compressed. This was before modern wavelet theory, in particular, before the introduction of multiresolution analysis, and Richter's transform differed from the two-dimensional Haar transform related to a multiresolution analysis. Since then the transform has been revised to be a "true" wavelet transform associated with a multiresolution analysis.

The Space Telescope Science Institute uses an algorithm called *hcompress* to compress the Digital Sky Survey, which is a database of images covering the whole sky. This algorithm consists in taking the two-dimensional Haar transform (called the H-transform by astronomers) of the digitized image and then quantizing the wavelet coefficients $w_{j,k}$ (called H-coefficients) using an arbitrary threshold. We will describe a more elaborate version of this algorithm called *ht_compress* that was developed by Yves Bobichon and Albert Bijaoui [38]. The *ht_compress* scheme differs from *hcompress* in the way the thresholds for the wavelet coefficients are determined. After describing the compression scheme, we will describe a regularized decompression algorithm also proposed by Bobichon and Bijaoui [38]. As in the case of IDEA, the regularization is not pure Tikhonov, since the Bobichon–Bijaoui scheme uses a priori information to provide a smooth restored image.

12.2.1 *ht_compress*

We illustrate the algorithm in one dimension. Thus, assume that f is a (noisy) signal defined on the integers $l = 0, 1, 2, \ldots, N-1$, where $N = 2^p$ for some positive integer p.

The first step is to estimate the standard deviation σ_0 of the noise in the original signal f. If the noise is not Gaussian, it is transformed as indicated in the last section so that it can be treated as Gaussian [4]. Knowing σ_0, and assuming uncorrelated Gaussian noise, one can deduce the standard deviation σ_j of the noise in the wavelet coefficients of the $w_{j,k}$ at scale j. With these assumptions, the standard deviation at scale $j+1$ is related to that at scale j by the relation

$$\sigma_{j+1} = \frac{1}{\sqrt{2}}\sigma_j.$$

The second step is to compute the Haar transform:

$$f_{j+1,k} = \sum_l f_{j,l} h(l - 2k), \tag{12.12}$$

$$w_{j+1,k} = \sum_l f_{j,l} g(l - 2k). \tag{12.13}$$

The two-term filters are defined by

$$h(n) = \begin{cases} \frac{1}{2} & \text{if } n = 0 \text{ or } 1, \\ 0 & \text{otherwise}, \end{cases} \qquad g(n) = \begin{cases} \frac{1}{2} & \text{if } n = 0, \\ -\frac{1}{2} & \text{if } n = 1, \\ 0 & \text{otherwise}. \end{cases}$$

Note that the original function f can be recovered using the equations

$$f_{j,k} = 2\sum_l [f_{j+1,l}\tilde{h}(k - 2l) + w_{j+1,l}\tilde{g}(k - 2l)], \tag{12.14}$$

where $\tilde{h} = h$ and $\tilde{g} = g$ for the Haar transform.

In the next step, the Haar coefficients $w_{j,k}$ are replaced with the $w'_{j,k}$ defined by

$$w'_{j,k} = \begin{cases} 0 & \text{if } |w_{j,k}| < \kappa\sigma_j, \\ w_{j,k} & \text{if } |w_{j,k}| \geq \kappa\sigma_j. \end{cases}$$

The positive parameter κ controls the compression ratio, once σ_0 is determined.

The coefficients $w'_{j,k}$ are quantized by forming the quotient

$$q_{j,k} = \frac{w'_{j,k}}{\kappa\sigma_j} \tag{12.15}$$

and defining $q'_{j,k}$ to be the integer nearest to $q_{j,k}$. (To avoid ambiguity, shrink the $q_{j,k}$ toward zero when it falls exactly between two integers.) Finally, the coefficients $q'_{j,k}$ are coded using a lossless 4-bit hierarchical coding scheme (see [146]).

The coded coefficients can now be stored or transmitted. For example, it is possible to buy the complete Digital Sky Survey on 102 CD-ROMs compressed by a factor of 10 or on 18 CD-ROMs (8 for the northern sky and 10 for the southern sky) compressed by a factor of 100. These are available from the Space Telescope Science Institute, and, as indicated above, the compression algorithm is *hcompress*. Astronomers can also download images from the Space Telescope Science Institute. Furthermore, because the compression is based on a multiresolution analysis, the images can be downloaded and restored scale by scale, beginning with the largest scale. This means that an astronomer can stop the process once it is determined that the image is good enough for the task at hand. Bijaoui points out that it is essential to have a correct idea of transmitted images as fast as possible for control during astronomical observations [36].

Unfortunately, the direct restoration of the transmitted (or stored) data using (12.14) can lead to some unpleasant images. To obtain reasonable compression ratios, many of the original wavelet coefficients are set equal to zero, and others are quantized as multiples of $\kappa\sigma_j$. The result is that the restored image contains relatively large fields of pixels having the same value with abrupt discontinuities between the fields. These blocking effects are the signature of Haar compression (see Figure 2.1). One might expect that the use of smooth wavelets would give better results; however, in this case, going to a longer filter does not seem to be the solution. As Bijaoui remarked [36, p. 85]:

> Press [227] has introduced the Daubechies filter of length 4. The compression and uncompression algorithms take more time than *hcompress* and the quality of the resulting measurements is generally less than those obtained with the simple Haar transform for astronomical images. This could be due to the specificities of these images, mainly compound of peaks due to the stars. The correlation length is very short, and it is not relevant to process the data with long filters.

This may be a victory for the Haar transform, but if a longer filter is not the answer, what is? Several solutions for producing a smoother image have been proposed; see, for example, [176] where Kalman filtering is applied and [259] where interpolation is used. We will outline a solution proposed by Bobichon and Bijaoui [38]; it is an inverse for their *ht_compress* algorithm.

12.2.2 Smooth restoration

Recall that the final coding was lossless, which means that we can recover the coefficients $q'_{j,k}$ exactly. We can also multiply the $q'_{j,k}$ by $\kappa\sigma_j$ to obtain a new set of wavelet coefficients $\hat{w}_{j,k}$:[14]

$$\hat{w}_{j,k} = \kappa\sigma_j q'_{j,k}.$$

If the inverse Haar transform (12.14) is applied directly to the truncated and quantized coefficients $\hat{w}_{j,k}$, the resulting image will certainly have unpleasant blocking effects. Bobichon and Bijaoui produce a smooth restored image scale by scale, beginning with the largest scale $j = p$. We speak of images, but for simplicity, we continue to illustrate the algorithm in the one-dimensional case.

The Bobichon–Bijaoui algorithm produces a smooth restored image $\hat{f}_j$ at each scale j by minimizing the energy of the gradient of $\hat{f}_j$ subject to certain constraints. To see how this works, we write (12.14) as the operator equation

$$f_j = \tilde{H} f_{j+1} + \tilde{G} w_{j+1} \tag{12.16}$$

and let D denote the first derivative (difference) operator. The restored image at scale j is defined to be the solution of the minimization problem

$$\inf_{v_{j+1}} [D(\tilde{H} f_{j+1} + \tilde{G} v_{j+1})]^2, \tag{12.17}$$

---

[14]In this section, "^" does not indicate the Fourier transform.

subject to the following constraints.

The first constraint is that $\hat{f}_{j,k} \geq 0$. This is the a priori information that the image (without noise) is given by a positive function that measures gray levels.

The second constraint limits the values $v_{j,k}$ can take in (12.17). If the coefficient $\hat{w}_{j,k} = 0$, we know that the original wavelet coefficient with index j, k satisfied the condition $|w_{j,k}| < \kappa\sigma_j$, and this condition is imposed on $v_{j,k}$ as it competes in the minimization (12.17). Similarly, if $\hat{w}_{j,k} = \kappa\sigma_j q'_{j,k} \neq 0$, we know that

$$\kappa\sigma_j\left(q'_{j,k} - \frac{1}{2}\right) < w_{j,k} \leq \kappa\sigma_j\left(q'_{j,k} + \frac{1}{2}\right),$$

and the same condition is imposed on $v_{j,k}$.

The algorithm used to solve this minimization problem is an iterative process that passes back and forth between physical space and wavelet space, using the constraints in the two spaces. It would take us too far afield to go further into the details of the algorithm; we encourage the interested reader to consult [38].

12.2.3 Comments

We emphasize that this algorithm proceeds scale by scale, and as pointed out above, this is important to the astronomer: It can save time and money. We mentioned in the last section that astronomers have ways to measure the astrometric and photometric qualities of a restored image. The restoration algorithm we have outlined scores well on both points.

The reader surely has noted the similarities between this restoration algorithm and IDEA. In both algorithms there were constraints imposed on the restored image (positivity in the Bobichon–Bijaoui algorithm and the support of the restored image in IDEA) and constraints imposed on the transform (P and s in IDEA and constraints on the $v_{j,k}$ in the Bobichon–Bijaoui algorithm).

We note that there are several other compression and decompression algorithms being proposed and used in astronomy. We mention, in particular, the pyramidal median transform developed by Jean-Luc Stark and his colleagues [241]. A study comparing compression algorithms using Schmidt plate data done at the Strasbourg Data Center has shown that a compression ratio of 260 to 1 can be obtained with acceptable quality using the pyramidal median transform algorithm. The same study showed that the limit for the JPEG standard with the same quality was only about 40 to 1. Readers interested in the details of these techniques can consult the recent book by Stark, Murtagh, and Bijaoui [240]; another source is the website www-dapnia.cea.fr.

12.3 The hierarchical organization of the universe

This last section concerns a much more ambitious program that demands considerable computing power as well as new ideas about how to deal with the information. The program, initiated and developed by Albert Bijaoui, seeks to determine the hierarchical structure in the universe. For example, our planetary system is part of the Milky Way, which itself is included in a much larger structure called the Local Group. According to Hubert Reeves [229, pp. 40, 41]:

> The Local Group consists of around twenty galaxies in the neighborhood of our own, within a radius of about five million light years. Andromeda and the two clouds of Magellan are part of this cluster.

The galactic clusters, are they themselves organized into larger units? It seems indeed to be the case. One then speaks of a supercluster. Our Local Group is part of the Virgo supercluster. A supercluster contains several thousand galaxies in a volume whose dimensions are measured in tens of millions of light years.

Ideas rarely have well-defined beginnings—consider wavelets and the historical account in Chapter 2. This is definitely the case for the idea of a hierarchically structured universe. Edward Harrison in his delightful book *Darkness at Night, a Riddle of the Universe* [139] cites several authors and sources where the notion of a hierarchical—or even fractal—structure is suggested more or less explicitly:

Emanuel Swedenborg, 1734, *Principia Rerum Naturalium.*

Immanuel Kant, 1755, *Universal Natural History and Theory of the Heavens.*

Johann Lambert, 1761, *Cosmological Letters.*

Edward Fournier d'Albe, 1907, *Two New Worlds.*

Charles Charlier, 1922, *How an Infinite World May Be Built Up.*

(The twentieth-century references are [113] and [48]. Detailed references to the eighteenth-century works can be found in [139].) Harrison referred to these sources in the context of his book, which is devoted to a historical and scientific account of the riddle: Why is the sky dark at night? We mention these sources to emphasize that the notion that the large-scale structure of the universe might be hierarchical goes back to at least the eighteenth century. By contrast, it is only as recently as 1924 that Edwin Hubble firmly established that ours is not the only galaxy. The history of cosmology is the history of competing views of the cosmos, and the idea that the Milky Way was the only galaxy was a popular model in the nineteenth century. Harrison points out that the famous astronomer William Herschel, who at one time supported the idea of many galaxies, later in life, "lost his confidence, renounced the many-island universe of Wright and Kant, and adopted a one-island universe. Following his lead, the one-island universe was widely adopted in astronomical circles in the nineteenth century."

The situation is vastly different today. The fact that the universe is expanding, as predicted by the Russian physicist Alexander Freidman in 1922, has been well established since the 1930s, and the controversy that thrived in the middle of the twentieth century between proponents of a steady-state universe and those who supported the notion of a big bang tilted definitively in favor of the latter with the discovery of the residual background radiation by Arno Penzias and Robert Wilson in 1965. This discovery led to the serious study of the implication of a big bang and the development of cosmological scenarios to explain how the universe got from $t = 0$ to what is observed today, which at certain scales is a rather lumpy universe. As noted by Slezak and others [237, p. 517]:

> The complexity of the distribution of galaxies and of clusters of galaxies is now clearly established up to scales of 50 h^{-1} Mpc ....
>
> The main feature of the galaxy distribution is the departure from homogeneity at all scales within reach. The topology of the distribution is characterized by a complex network of sharp structures, one-dimensional filaments (Giovanelli et al. 1986) or two-dimensional sheets (de Lapparent et al. 1986) suggesting a cell-like geometry .... The high-density structures appear to connect clusters of galaxies and delineate

large spherical regions which are devoid of bright galaxies (de Lapparent et al. 1986; Pellegrini, da Costa, & de Caravalho 1989).[15]

Qualitative observations like these lead to one of the outstanding problems in modern cosmology, which roughly stated is this: How, starting from a relatively homogeneous initial state, has the universe evolved into a structure that "departs from homogeneity at all scales within reach"? Particle physicists who speculate on the origins of the big bang tell us that the "initial conditions," or at least conditions at, say, $t = 10^{-30}$, were never homogeneous. At the top of the scientific hit parade for 1989 were the results provided by the Cosmic Background Explorer satellite, known as COBE, which showed that indeed there were very small variations (1 part in 100,000) in the residual radiation from the big bang. This evidence supports—does not contradict—the big bang theory, but it does not change the problem stated above. It does, however, provide limits within which the problem is to be resolved.

Given the problem, what experimental data exist with which one can start work? It is easier to say what does not exist: We do not have a nice three-dimensional map of the universe! The first data available were two-dimensional maps of galaxies. For example, in [236], Slezak, Bijaoui, and Mars identified about 7600 galaxies up to magnitude 19 from Schmidt plates in a $6° \times 6°$ field at the eastern end of the Coma supercluster. The data for [237] is a redshift survey that comes from the Center for Astrophysics. Each strip is $135°$ wide in right ascension and $6°$ thick in declination. This is again basically two-dimensional data, where the redshift measures the distance from earth.

Forget for the moment that the data are not ideal—that there are probably many low-surface-brightness galaxies that have been missed, and that the distance measurements are not perfect—and assume provisionally that we have a good three-dimensional map of the galaxies in a chunk of the universe. Ideally we would like to use this map to check various theories (scenarios) describing the evolution of the universe. One way to do this is to run numerical simulations of different scenarios, for example, the classical cold dark matter (CDM) model or the hot dark matter (HDM) model. This has been done, and one ends up with a simulated universe in a box 192 Mpc on a side. And indeed the results from the two scenarios *look* different (see [170] and [36]). But clearly it is not enough to look different; one wishes to have an objective measure, and this is one place where wavelet analysis can make a contribution.

Given real data, or even our ideal three-dimensional data, it is very difficult to use the data to define clusters, superclusters, and other perceived objects. To "see" these nested structures with objectivity is an extremely difficult problem. Slezak, Bijaoui, and Mars tell us some of the history of this research [236, p. 301]:

> After the visual identification of clusters on the POSS plates by Abell (1958) and Zwicky et al. (1961–1968), many computer algorithms were introduced to avoid a personal judgment, like cluster analysis (Materne, 1978; Huchra and Geller, 1982; Tago et al., 1984) or contrast analysis (Dodd and Mac Gillivray, 1986). In particular, with these tools or similar ones, the existence of substructures in clusters would be established for an important fraction of rich clusters (Geller and Beers, 1982; Baier, 1983; but see also West et al., 1988; Katgert et al., 1988). So, the distribution of galaxies cannot be reduced to the isolate groups identified

---

[15] References cited here are given in the original article.

by the cluster analysis, but to a fuzzy hierarchic structure for which the same galaxy can belong to many entities.[16]

We hope with this background on the astronomical problem and with the other applications presented in the book, particularly the work of Marie Farge, that the reader sees the introduction of wavelet analysis as a natural step. Bijaoui tells us that in the late 1980s, when he first heard about wavelets, their use was not so clear. It was, he says, after he heard a lecture that Alain Arneodo gave in Nice in 1987 that he decided to try wavelet analysis for studying the large-scale structure of the universe. Since then he and his group have pioneered the application of wavelet techniques in astronomy, including innovative mixes of wavelet and statistical techniques. In addition to showing that these techniques can be used to identify clusters and superclusters of galaxies, they have shown how to identify voids, which may ultimately prove to be more significant for differentiating cosmological scenarios [237]. Furthermore, they have introduced objective parameters to measure these voids. So far the results favor an intermediate scenario, somewhere between the CDM and the HDM models.

12.3.1 A fractal universe

We have talked about using wavelet techniques to determine hierarchical structures, but the complexity of the distribution of galaxies in the universe leads one naturally to think of a fractal structure. This is the path followed by Mandelbrot [192, 193], although, as we have seen, it had been suggested earlier by Fournier d'Albe and Charlier. In fact, they were quite specific in describing possible fractal arrangements that could lead to a dark sky at night. Although Mandelbrot suggested a distribution of matter leading to a fractal universe, it seems that a multifractal approach corresponds with reality [161]. We believe that wavelets are today the best tool for analyzing fractal and multifractal structures; in addition, there is some evidence that wavelet-based techniques have the potential for revealing the rules by which complex multifractal structures were constructed. This is a much more ambitious program than "simple analysis." We mentioned this kind of program in connection with turbulence in Chapter 9. We illustrate the idea with the following simple example.

The ideas come from Arneodo and his group at Bordeaux. They have been trying to elucidate the dynamical processes that generate complex fragmented structures like the Cantor triadic set. Here, briefly, is the proposed method, illustrated for the Cantor set [7]. One considers the canonical probability measure μ supported by the Cantor set. One then computes the wavelet transform of μ:

$$W(a,b) = \frac{1}{a} \int \psi\left(\frac{x-b}{a}\right) d\mu(x).$$

The set defined by $|W(a,b)| > \lambda$, where λ is a certain threshold, is represented in the half-plane, $a > 0$, $b \in \mathbb{R}$. One can also, for $a > 0$, determine the values b such that $b \to |W(a,b)|$ attains a maximum. When these maximal values are plotted, they are seen to be organized into more or less vertical lines with breaks and bifurcations.

In the two representations that we have just defined, the dynamics of the fragmentation appear in full force. Starting with the largest values of a, one moves

---
[16]References cited here are given in the original article.

toward the small values of a. One then observes a cascade of bifurcations that constitute a symbolic representation of the Cantor triadic set. Using the maximal lines, Arneodo has been able to reconstruct the process that leads to the measure μ; he also has been able to do this for more complex measures having support on more complex Cantor sets. We believe that in many cases all the necessary information to reconstruct the process is contained in these maximal skeletons. Can one in a similar way unravel the secrets of the fragmentation processes that have led to the structure of the galaxies? This surely seems to be an overly ambitious program. But is it today any more farfetched than were the ideas of Kant and others in the eighteenth century?

12.4 Conclusions

Wavelet-based techniques are being widely applied in astronomy and astrophysics. The review article [36] by Bijaoui cites 114 references. By now there must be well over 200 papers dealing with wavelets applied to astronomy. We believe that this work illustrates the flexibility of the ideas found in wavelet and multiresolution analysis. Astronomers have been particularly inventive in using wavelets to deal with the ubiquitous problem of noise. We have described the use of thresholding, but there are other techniques whereby the noise is dealt with in wavelet space rather than in the Fourier domain or in the original space. The number of different techniques invented are witness to the richness of the method.

While the wavelet transform itself plays an important role in the astronomer's algorithms, the general notion of multiscale processing seems to us to be more pervasive. We have noted the usefulness of progressive reconstruction in practical astronomy and the fact that the popular reconstruction algorithms, which are nonlinear regularization schemes, proceed scale by scale. Another divergence from classical wavelet theory is the use of redundant algorithms rather than, for example, algorithms with decimation. As Bijaoui notes [36, p. 78]:

> The wavelet transform is a tool widely used today by astrophysicists, but they do not apply only the discrete transform resulting from a multiresolution analysis but a large range of discrete transforms: Morlet's transform, for time-frequency analysis, the *à trous* algorithm and the pyramidal transform for image restoration and analysis, pyramidal with Fourier transform for synthesis aperture imaging. Physical constraints generally play an important part in applying the discrete transform.

And we emphasize again the unique character of astronomical images and the need to tune the algorithms to the image and the task at hand.

Our last remark is a prediction. We have seen in Chapter 11 the intimate relations that exist between wavelet theory and nonlinear approximation. We also have noted the use of nonlinear techniques being applied scale by scale in astronomy. Considering the strong interest and inventiveness astronomers have shown so far in using wavelets and multiresolution analysis, we expect to see the more innovative applications of nonlinear techniques to follow.

APPENDIX A

Filter Fundamentals

This appendix is written for readers who are not familiar with the basic concepts and language of filters. It also provides a larger context for parts of Chapter 3.

A.1 The $l^2(\mathbb{Z})$ theory and definitions

We begin with a general result about linear operators on $l^2(\mathbb{Z})$.

THEOREM A.1. *If $F : l^2(\mathbb{Z}) \to l^2(\mathbb{Z})$ is a continuous linear operator that commutes with translations, then there exists a sequence $(h_k)_{k \in \mathbb{Z}} \in l^2(\mathbb{Z})$ such that*

$$(Fx)_k = \sum_{n \in \mathbb{Z}} h_{k-n} x_n \tag{A.1}$$

for all $x = (x_k)_{k \in \mathbb{Z}} \in l^2(\mathbb{Z})$. Furthermore, the function $H(\omega) = \sum_{k \in \mathbb{Z}} h_k e^{ik\omega}$ is in $L^\infty(0, 2\pi)$, and $\|F\| = \|H\|_\infty$. Conversely, if $(h_k)_{k \in \mathbb{Z}} \in l^2(\mathbb{Z})$ is such that $H \in L^\infty(0, 2\pi)$, then (A.1) defines a continuous linear operator that commutes with translations, and $\|F\| = \|H\|_\infty$.

Proof. Assume that $F : l^2(\mathbb{Z}) \to l^2(\mathbb{Z})$ is a continuous linear operator that commutes with translations, and let $\{e_k\}_{k \in \mathbb{Z}}$ denote the canonical basis for $l^2(\mathbb{Z})$ defined by $e_k(n) = 0$ if $n \neq k$ and $e_k(k) = 1$. Then Fe_0 is an element of $l^2(\mathbb{Z})$, which we denote by $h = (h_k)$. Since F commutes with translations, we have

$$Fe_k = \sum_{n \in \mathbb{Z}} h_{n-k} e_n \tag{A.2}$$

for all $k \in \mathbb{Z}$. We go to the spectral domain and define the operator

$$\widetilde{F} : L^2(0, 2\pi) \to L^2(0, 2\pi)$$

in the obvious way: For $X(\omega) = \sum_{k \in \mathbb{Z}} x_k e^{ik\omega}$ in $L^2(0, 2\pi)$, define

$$\widetilde{F}X(\omega) = Y(\omega) = \sum_{k \in \mathbb{Z}} y_k e^{ik\omega},$$

where $(y_k) = F(x_k)$. Since the Fourier transform is an isometry, $\widetilde{F}$ is a bounded linear operator with the same norm as F, that is, $\|\widetilde{F}\| = \|F\|$. Taking the Fourier transform of both sides of (A.2) shows that

$$\widehat{Fe_k} = \sum_{n \in \mathbb{Z}} h_{n-k} e^{in\omega} = H(\omega) e^{ik\omega},$$

so by the definition of $\widetilde{F}$, $\widetilde{F}e^{ik\omega} = H(\omega)e^{ik\omega}$. By linearity, $\widetilde{F}X_N = H(\omega)X_N$ for any finite trigonometric sum X_N. We wish to show that this relation is true for all $X \in L^2(0, 2\pi)$. This follows directly from the continuity of $\widetilde{F}$: Assume that X_N is a finite trigonometric sum such that $X_N \to X$ in $L^2(0, 2\pi)$. Then by continuity, $\widetilde{F}X_N \to \widetilde{F}X$ in $L^2(0, 2\pi)$. Since $H \in L^2(0, 2\pi)$, $HX \in L^1(0, 2\pi)$, and we have the following inequality:

$$\|H(X_N - X)\|_1 \le \|H\|_2 \|X_N - X\|_2.$$

The right-hand side tends to zero as $N \to \infty$, so $HX_N \to HX$ in $L^1(0, 2\pi)$. This and the fact that $HX_N = \widetilde{F}X_N \to \widetilde{F}X$ in $L^2(0, 2\pi)$ imply that the two functions FX and HX are equal almost everywhere, that is,

$$(\widetilde{F}X)(\omega) = H(\omega)X(\omega)$$

for almost every $\omega \in (0, 2\pi)$.

At this point, it is purely a matter of measure, integration, and functional analysis to prove that $H \in L^\infty(0, 2\pi)$ and that $\|H\|_\infty = \|\widetilde{F}\|$. The general result is this: Let (X, μ) be a measure space and assume that $g \in L^2(X, \mu)$ is such that $gf \in L^2(X, \mu)$ for all $g \in L^2(X, \mu)$. Then

(i) the mapping $G : L^2(X, \mu) \to L^2(X, \mu)$ defined by $f \mapsto gf$ is a bounded linear transformation, and

(ii) the function g is in $L^\infty(X, \mu)$. Furthermore, $\|G\| = \|g\|_\infty$.

However, for the case at hand, one has the richness of the group structure of the integers and its dual group $\mathbb{T}$, and there is a more elegant way to proceed.

To prove that $H \in L^\infty(0, 2\pi)$, consider the special unit vectors

$$X_N(\omega - \xi) = \frac{1}{\sqrt{N}}\left[1 + e^{i(\omega-\xi)} + \cdots + e^{i(N-1)(\omega-\xi)}\right].$$

Since $\|X_N\|_2 = 1$ and $\widetilde{F}X_N = HX_N$, $\|\widetilde{F}X_N\|_2 \le \|\widetilde{F}\|$. When we compute the norm of $\widetilde{F}X_N$, we get

$$\|\widetilde{F}X_N(\omega)\|_2^2 = \frac{1}{2\pi}\int_0^{2\pi} K_N(\omega - \xi)|H(\xi)|^2\, d\xi,$$

where

$$K_N(\omega - \xi) = |X_N(\omega - \xi)|^2 = \frac{1}{N}\left(\frac{\sin N\frac{\omega-\xi}{2}}{\sin\frac{\omega-\xi}{2}}\right)^2$$

is the Fejér kernel. Hence, $K_N * |H|^2(\omega) = \|\widetilde{F}X_N(\omega)\|_2^2 \le \|\widetilde{F}\|^2$, so $K_N * |H|^2$ is in $L^\infty(0, 2\pi)$ uniformly in N. Since $|H|^2$ belongs to $L^1(0, 2\pi)$, $K_N * |H|^2$ tends to $|H|^2$ in $L^1(0, 2\pi)$, and we conclude that $|H|^2$ belongs to $L^\infty(0, 2\pi)$. To recapitulate, knowing that $\widetilde{F}X_N = HX_N$ for trigonometric polynomials, we have shown that H is in $L^\infty(0, 2\pi)$ and that $\|H\|_\infty \le \|\widetilde{F}\|$. Once we know that $H \in L^\infty(0, 2\pi)$, it follows directly that $\widetilde{F}X = HX$ for all $X \in L^2(0, 2\pi)$ and that $\|\widetilde{F}\| \le \|H\|_\infty$, which proves the result in one direction.

To prove the result in the other direction, we assume that the sequence (h_k) in $l^2(\mathbb{Z})$ is such that $H \in L^\infty(0, 2\pi)$. For $x \in l^2(\mathbb{Z})$, define the mapping F by

$$(Fx)_k = \sum_{n \in \mathbb{Z}} h_{k-n} x_n = h * x. \tag{A.3}$$

(Note that $\sum_{n \in \mathbb{Z}} h_{k-n} x_n$ is often called the discrete convolution of h and x.) We need to show that $Fx \in l^2(\mathbb{Z})$ and that F is bounded with $\|F\| = \|H\|_\infty$. (It is clearly linear and it commutes with translations.) But this follows directly from the fact that

$$\frac{1}{2\pi} \int_0^{2\pi} H(\omega) X(\omega) e^{ik\omega} \, d\omega = \sum_{n \in \mathbb{Z}} h_{k-n} x_n \tag{A.4}$$

and the arguments that were given for the proof in the other direction. □

This theorem provides the basis for the l^2 theory of discrete filters, and, in fact, we define a *filter* to be a continuous linear mapping $F : l^2(\mathbb{Z}) \to l^2(\mathbb{Z})$ that commutes with translations. There are other definitions for filters that involve different domains, ranges, and topologies, but whatever the setting, filters are always translation invariant and continuous. The l^2 context suits our objectives.

The *impulse response* of a filter F is defined to be $Fe_0 = h$, and the sequence $h = (h_k)_{k \in \mathbb{Z}}$ also is called the filter. If all but a finite number of the h_k are zero, we say that the filter has finite impulse response (FIR) and that it is an FIR filter. If not, it has infinite impulse response (IIR). In practice, filters are finite. This does not mean that IIR filters are of no interest; they are important theoretically, and they can often be approximated by finite filters for applications. There are, however, finite filters that are finite by design, such as the finite filters associated with compactly supported wavelets. For the moment we will stay with the general case and only assume that $(h_k)_{k \in \mathbb{Z}} \in l^2(\mathbb{Z})$.

We define the *transfer function* of F to be the 2π-periodic function

$$H(\omega) = \sum_{k \in \mathbb{Z}} h_k e^{ik\omega}.$$

For convenience, the transfer function of F is more often denoted by $F(\omega)$.

If T is a bounded linear operator on $l^2(\mathbb{Z})$, its adjoint T^* is defined in the usual way by

$$\langle Tx, y \rangle = \langle x, T^*y \rangle$$

for all $x, y \in l^2(\mathbb{Z})$. A simple computation shows that if F is the filter (h_k), then F^* is the filter $(\overline{h}_{-k})$. Thus,

$$F^*(\omega) = \overline{F}(\omega).$$

A.2 The general two-channel filter bank

The general two-channel filter bank is illustrated in Figure A.1.

We are not concerned here with the quantization and transmission, which are assumed to be perfect, although we wish to emphasize that these are serious problems in practice. We assume that the outputs of the analyzing filters F_0 and F_1, followed by *downsampling* (operator D in section 3.3), go directly to the *upsampling*

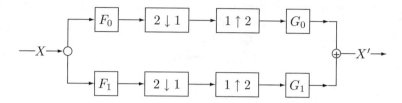

FIG. A.1. *The general two-channel filter bank.*

(operator D^* in section 3.3) and then to the synthesizing (or reconstruction) filters G_0 and G_1.

If $Y(\omega) = F_0(\omega)X(\omega)$ is the output of the filter F_0, then after downsampling, the signal is represented by $Y_0(\omega) = \sum_{k \in \mathbb{Z}} y_{2k} e^{i2k\omega}$. This can be written as

$$Y_0(\omega) = \frac{1}{2}\{F_0(\omega)X(\omega) + F_0(\omega + \pi)X(\omega + \pi)\},$$

and similarly

$$Y_1(\omega) = \frac{1}{2}\{F_1(\omega)X(\omega) + F_1(\omega + \pi)X(\omega + \pi)\}.$$

Thus, the output X' is given by

$$X'(\omega) = \frac{1}{2}\{(F_0(\omega)G_0(\omega) + F_1(\omega)G_1(\omega))X(\omega) \\ + (F_0(\omega + \pi)G_0(\omega) + F_1(\omega + \pi)G_1(\omega))X(\omega + \pi)\}. \tag{A.5}$$

Note that this output involves two forms of the input: the original signal $X(\omega)$ plus $X(\omega + \pi)$, which is called an *aliased* version of $X(\omega)$. Experts in signal processing tell us that this part of the output is undesirable, so the first step toward perfect reconstruction is to set the coefficient of $X(\omega + \pi)$ equal to zero:

$$F_0(\omega + \pi)G_0(\omega) + F_1(\omega + \pi)G_1(\omega) = 0. \tag{A.6}$$

Then to have the output exactly equal the input, we must have

$$F_0(\omega)G_0(\omega) + F_1(\omega)G_1(\omega) = 2. \tag{A.7}$$

This requirement is loosened in practice to be

$$F_0(\omega)G_0(\omega) + F_1(\omega)G_1(\omega) = 2e^{-in\omega}, \qquad n \in \mathbb{Z}, \tag{A.8}$$

which means that the original signal is allowed to be delayed.

The relations (A.6) and (A.8) are now classic, and they have been "solved" in various ways over the last two decades. We will describe several of these solutions. Esteban and Galand [99] took

$$\begin{aligned} F_1(\omega) &= F_0(\omega + \pi), \\ G_0(\omega) &= F_0(\omega), \\ G_1(\omega) &= -F_0(\omega + \pi). \end{aligned}$$

It is easy to see that these choices satisfy (A.6) and that condition (A.8) becomes

$$|F_0(\omega)|^2 - |F_0(\omega+\pi)|^2 = 2e^{-in\omega},$$

where n must now be odd, since $\omega \mapsto \omega + \pi$ changes the sign of the left-hand side. Esteban and Galand called these filters *quadrature mirror filters* (QMFs).

The name "mirror" comes about as follows: If we extend the function F_0 to a holomorphic function in an annulus about the unit circle by defining

$$\mathcal{F}_0(z) = \sum_{k \in \mathbb{Z}} h_k z^n,$$

then $\mathcal{F}_1(z) = \mathcal{F}_0(-z)$ and the filters are mirrored through the origin by the transform $z \to -z$. The idea behind "quadrature" is only slightly more complicated: Esteban and Galand were interested in real, symmetric FIR filters of the form

$$F_0(\omega) = \sum_{k=0}^{2N+1} h_k e^{ik\omega}$$

with $h_{N+k+1} = h_{N-k}$ for $0 \leq k \leq N$. These conditions imply that the phases of the filters F_0 and F_1 differ by $\pm\frac{\pi}{2}$; hence the phases are in quadrature.

Unfortunately, these conditions cannot be satisfied except for the Haar filter (see [253]). To fix this situation, Smith and Barnwell introduced the following conditions (for real filters) [238], [239]:

$$F_1(\omega) = -e^{-in\omega}\overline{F_0}(\omega+\pi), \quad n \text{ odd},$$
$$G_0(\omega) = F_1(\omega+\pi) = e^{-in\omega}\overline{F_0}(\omega),$$
$$G_1(\omega) = -F_0(\omega+\pi) = e^{-in\omega}\overline{F_1}(\omega).$$

These filters are often called *conjugate quadrature filters* (CQFs). Relation (A.6) is satisfied, and relation (A.8) becomes

$$|F_0(\omega)|^2 + |F_0(\omega+\pi)|^2 = 2.$$

The problem reduces to finding a filter F_0 that satisfies this relation. In practice, one would like F_0 to be finite (FIR) and "causal." Causal means that there is no output before there is an input or, formally, that $x_k = 0$ for $k < 0$ implies that $(Fx)_k = 0$ for $k < 0$. Then it is easy to see that a finite causal filter must be of the form

$$F_0(\omega) = \sum_{k=0}^{N} h_k e^{ik\omega}.$$

The development in Chapter 3 proceeds slightly differently. We begin by specifying that $G_0 = F_0^*$ and the $G_1 = F_1^*$. Then the problem is to find F_0 and F_1 that satisfy (A.6) and (A.8). But these are now

$$F_0(\omega+\pi)\overline{F_0}(\omega) + F_1(\omega+\pi)\overline{F_1}(\omega) = 0, \tag{A.6'}$$
$$|F_0(\omega)|^2 + |F_1(\omega)|^2 = 2, \tag{A.8'}$$

which are exactly the conditions for the matrix in Theorem 3.1 to be unitary. These conditions figure prominently in the proof of Theorem 3.1. A straightforward computation shows that (A.6′) and (A.8′) imply (3.1). The implication in the other direction is a bit more technical.

Finally, the novice is warned that there are many conventions in this business. Some authors (see [60], for example) save the odd coefficients for $Y_1(\omega)$ instead of the even ones, in which case $Y_1(\omega) = \frac{1}{2}\{F_1(\omega)X(\omega) - F_1(\omega + \pi)X(\omega + \pi)\}$. There are also conventions about the definition of the transfer function: Sometimes it is defined to be $F_0(\omega) = \sum h_k e^{-ik\omega}$. These differences can be confusing, but they do not alter the fundamental results.

APPENDIX B

Wavelet Transforms

The purpose of this appendix is to present several of the basic theorems about wavelet transforms that have been used in the text but that have not been proved. The techniques used to establish these results are typical of those used in wavelet theory, and the proofs will illustrate where the different assumptions about the analyzing wavelet are used.

B.1 The L^2 theory

To simplify the notation, we present the results in one dimension, although the results are true for $\mathbb{R}^n$. We assume throughout this section that the analyzing wavelet ψ is in $L^2(\mathbb{R})$ and that the wavelets are defined by

$$\psi_{(a,b)}(x) = \frac{1}{\sqrt{a}} \psi\left(\frac{x-b}{a}\right), \quad a > 0, \quad b \in \mathbb{R}. \tag{B.1}$$

For future reference, we note that the mapping $(a,b) \mapsto \psi_{(a,b)}$ is continuous from $\mathbb{R}^* \times \mathbb{R}$ to $L^2(\mathbb{R})$.

The wavelet transform is defined for $f \in L^2(\mathbb{R})$ by

$$W_f(a,b) = \int_{-\infty}^{\infty} f(x)\overline{\psi}_{(a,b)}(x)\,dx, \tag{B.2}$$

where, as elsewhere, $\overline{\psi}_{(a,b)}(x) = \overline{\psi_{(a,b)}(x)}$. Then $|W_f(a,b)| \le \|f\|\|\psi\|$, and thus by our remark about the continuity of $(a,b) \mapsto \psi_{(a,b)}$, it is clear that $W_f(a,b)$ is continuous on $\mathbb{R}^* \times \mathbb{R}$.

We wish to prove that the mapping $f \mapsto W_f(a,b)$ is a partial isometry from $L^2(\mathbb{R})$ into $L^2(\mathbb{R}^+ \times \mathbb{R}, db\frac{da}{a^2})$. This is not true in general, so additional assumptions must be made about ψ. The assumption we make, which is called an *admissibility condition*, is that

$$\int_0^\infty |\hat{\psi}(t\xi)|^2 \frac{dt}{t} = C_\psi = 1 \tag{B.3}$$

for almost all $\xi \in \mathbb{R}^n$. We have written the admissibility condition so that it is clear how the results generalize to $\mathbb{R}^n$. For our case, we expect (B.3) to hold for $\xi = \pm 1$, which means that it holds for all $\xi \ne 0$. One can normalize either C_ψ or the norm of ψ, but not necessarily both. We have chosen to normalize ψ so that $C_\psi = 1$. Note that the factor $a^{-1/2}$ in the definition of $\psi_{(a,b)}$ is chosen so that $\|\psi_{(a,b)}\|_2 = \|\psi\|_2$. One may prefer different normalizations in other settings, but these choices do not affect the substance of the results. In fact, in the following sections we replace the factor $a^{-1/2}$ by a and have $\|\psi_{(a,b)}\|_1 = \|\psi\|_1$.

The first result shows that if ψ satisfies (B.3), then the mapping $f \mapsto W_f(a,b)$ is a partial isometry (that is, $\|f\| = \|W_f\|$) from $L^2(\mathbb{R})$ into $L^2(\mathbb{R}^+ \times \mathbb{R}, db\frac{da}{a^2})$.

THEOREM B.1. *Assume that the analyzing wavelet $\psi \in L^2(\mathbb{R})$ satisfies the admissibility condition (B.3). If $f, g \in L^2(\mathbb{R})$, then*

$$\int_0^\infty \int_{-\infty}^\infty W_f(a,b)\overline{W}_g(a,b) \, db \, \frac{da}{a^2} = \langle f, g \rangle. \tag{B.4}$$

Proof. Since both f and ψ are in $L^2(\mathbb{R})$, we can use Parseval's identity to express $W_f(a,b)$ as

$$W_f(a,b) = \frac{1}{2\pi} \int_{-\infty}^\infty \hat{f}(\xi)\overline{\hat{\psi}_{(a,b)}(\xi)} \, d\xi = \frac{1}{2\pi} \int_{-\infty}^\infty \hat{f}(\xi)\sqrt{a}\,\overline{\hat{\psi}(a\xi)} e^{ib\xi} \, d\xi. \tag{B.5}$$

If we let $F_a(\xi) = \frac{1}{2\pi}\hat{f}(\xi)\sqrt{a}\,\overline{\hat{\psi}(a\xi)}$, then $F_a(\xi) \in L^1(\mathbb{R}, d\xi)$ and

$$W_f(a,b) = \hat{F}_a(-b) \qquad \text{for all } a > 0. \tag{B.6}$$

Consider the integral

$$I = (2\pi)^2 \int_0^\infty \int_{-\infty}^\infty |F_a(\xi)|^2 \, d\xi \, \frac{da}{a^2}. \tag{B.7}$$

By the definition of F_a and property (B.3), we have

$$I = \int_0^\infty \int_{-\infty}^\infty |\hat{f}(\xi)|^2 |\hat{\psi}(a\xi)|^2 \, d\xi \, \frac{da}{a} = \|\hat{f}\|^2 = 2\pi \|f\|^2 < +\infty.$$

Since I is finite, Fubini's theorem implies that $F_a(\xi)$ is an element of $L^2(\mathbb{R})$ for almost every (a.e.) $a > 0$. Thus, $W_f(a,b) = \hat{F}_a(-b)$ is an element of $L^2(\mathbb{R})$ for a.e. $a > 0$, and

$$2\pi \int_{-\infty}^\infty |F_a(\xi)|^2 \, d\xi = \int_{-\infty}^\infty |\hat{F}_a(b)|^2 \, db \tag{B.8}$$

for a.e. $a > 0$. It follows by integrating both sides of (B.8) that

$$\int_0^\infty \int_{-\infty}^\infty |W_f(a,b)|^2 \, db \, \frac{da}{a^2} = \|f\|^2. \tag{B.9}$$

Equation (B.9) means that the mapping $W : L^2(\mathbb{R}) \to L^2(\mathbb{R}^+ \times \mathbb{R}, db\frac{da}{a^2})$ defined by $f \mapsto W_f$ is a partial isometry. Since (B.4) is the inner product form of (B.9), this proves the result. $\square$

B.2 Inversion formulas

We are going to prove two inversion formulas. The first one is an L^2 formula, and it is a direct consequence of Theorem B.1. In fact, we will prove a more general L^2 result that has nothing to do with wavelets; the wavelet result follows from the general result. The second inversion formula is special because the analyzing wavelet is the Lusin wavelet

$$\psi(x) = \frac{1}{\pi(x+i)^2}.$$

This is the inversion formula that is needed in Chapter 10 for studying the wavelet transform of Riemann's function. To be specific, Theorem 10.1 is essential for our analysis of Riemann's function, and the proof of Theorem 10.1 uses (10.7). Lusin's wavelet satisfies the hypothesis of Theorem 10.1, and Theorem B.5 establishes (10.7) when f is Riemann's function. Thus, Theorem 10.1 applies when f is Riemann's function and ψ is the Lusin wavelet.

B.2.1 L^2 inversion

We begin with an abstract setting, not for the sake of abstraction, but because the setting reveals the essentials and simplicity of the result. Thus, let Ω be a locally compact Hausdorff space and let μ be a positive Radon measure on Ω. It is possible that $\mu(\Omega) = +\infty$, but we require that $\mu(K) < +\infty$ on compact subsets K of Ω. We will be dealing with the two Hilbert spaces $L^2(\mathbb{R})$ and $L^2(\Omega, d\mu)$. The next assumption is that there is a family of functions $\psi_\omega \in L^2(\mathbb{R})$, $\omega \in \Omega$, such that the mapping $\omega \mapsto \psi_\omega$ is continuous from Ω to $L^2(\mathbb{R})$. We will state and prove two theorems within this setting, and then we will interpret the results in the context of section B.1.

THEOREM B.2. *Define the operator* $T : L^2(\mathbb{R}) \to L^2(\Omega, d\mu)$ *by* $Tf = \langle f, \psi_\omega \rangle$. *If T is a partial isometry, that is, if*

$$\int_\Omega |\langle f, \psi_\omega \rangle|^2 \, d\mu(\omega) = \|f\|_2^2, \tag{B.10}$$

then

$$f(x) = \int_\Omega \langle f, \psi_\omega \rangle \psi_\omega(x) \, d\mu(\omega) = \lim_{j \to \infty} \int_{K_j} \langle f, \psi_\omega \rangle \psi_\omega(x) \, d\mu(\omega), \tag{B.11}$$

where $K_j \in \Omega$ is any sequence of increasing compact sets such that $\cup_j K_j = \Omega$. The right-hand limit is in the sense of $L^2(\mathbb{R})$.

The general theoretical basis for establishing this is the following: If $T : H_0 \to H_1$ is a partial isometry from one Hilbert space H_0 to another H_1, then the adjoint operator $T^* : H_1 \to H_0$ is such that $\|T^*\| \leq 1$ and $T^*T = I$, where I denotes the identity.

The bilinear form of (B.10) is

$$\langle f, g \rangle = \int_\Omega \langle f, \psi_\omega \rangle \overline{\langle g, \psi_\omega \rangle} \, d\mu(\omega), \tag{B.12}$$

and formally we have

$$\langle f, g \rangle = \int_\mathbb{R} \left(\int_\Omega \langle f, \psi_\omega \rangle \psi_\omega(x) \, d\mu(\omega) \right) \overline{g}(x) \, dx. \tag{B.13}$$

Equation (B.13) is the weak form of (B.11), but to arrive at (B.13) we had to interchange the order of integration. This and more will be justified by the following stronger result.

THEOREM B.3. *If F is a continuous function on Ω with compact support, then $f(x) = \int_\Omega F(\omega)\psi_\omega(x)\, d\mu(\omega)$ satisfies*

$$\|f\|_{L^2(\mathbb{R})} \leq \left[\int_\Omega |F(\omega)|^2\, d\mu(\omega) \right]^{1/2}. \tag{B.14}$$

Proof. The continuity of the mapping $\omega \mapsto \psi_\omega$ and the assumptions on F ensure that the function f is well defined as an element of $L^2(\mathbb{R})$. The inequality

$$\int_\mathbb{R} \int_\Omega |F(\omega)||\psi_\omega(x)||g(x)|\, d\mu(\omega)\, dx \leq \sup_{\omega \in K} \|F(\omega)\psi_\omega\|_{L^2(\mathbb{R})} \|g\|_{L^2(\mathbb{R})}, \quad g \in L^2(\mathbb{R}),$$

allows us to invoke Fubini's theorem and interchange the order of integration in the following formula:

$$\langle F, \langle g, \psi_\omega \rangle \rangle = \langle \langle F, \overline{\psi}_\omega \rangle, g \rangle.$$

Since, $\langle T^*F, g \rangle = \langle F, Tg \rangle = \langle F, \langle g, \psi_\omega \rangle \rangle$, it follows that

$$f(x) = T^*F(x) = \int_\Omega F(\omega)\psi_\omega(x)\, d\mu(\omega).$$

With this representation, (B.14) is a restatement of $\|T^*\| \leq 1$. □

With the establishment of (B.14), the proof of Theorem B.2 is straightforward. The estimate (B.14) implies that the representation

$$T^*F(x) = \int_\Omega F(\omega)\psi_\omega(x)\, d\mu(\omega) \tag{B.15}$$

is true for all $F \in L^2(\Omega, d\mu)$. Indeed, since the continuous functions with compact support are dense in $L^2(\Omega, d\mu)$, (B.14) implies that the representation (B.15) is true for all $F \in L^2(\Omega, d\mu)$. By taking $F = Tf$, $f \in L^2(\mathbb{R})$, we see that equation (B.11) is just a restatement of $T^*T = I$.

An algorithm for computing the integral in (B.15) is equally easily established. Let Ω be the union of an increasing sequence of compact sets K_j, $j \in \mathbb{N}$. Then $\int_{K_j} F(\omega)\psi_\omega(x)\, d\mu(\omega) = \int_\Omega F_j(\omega)\psi_\omega(x)\, d\mu(\omega)$, where $F_j(\omega) = F(\omega)$ if $\omega \in K_j$ and $F_j(\omega) = 0$ if $\omega \notin K_j$. Clearly, the sequence $f_j(x) = \int_{K_j} F(\omega)\psi_\omega(x)\, d\mu(\omega)$ tends to $f(x) = \int_\Omega F(\omega)\psi_\omega(x)\, d\mu(\omega)$ in $L^2(\mathbb{R})$.

Finally, we interpret Theorem B.2 in the context of Theorem B.1. Thus, let $\Omega = \mathbb{R}^+ \times \mathbb{R}$, $\omega = (a, b)$, $\psi_\omega = \psi_{a,b}$, and $d\mu(\omega) = db\frac{da}{a^2}$. It is not difficult to show that the mapping $(a, b) \mapsto \psi_{a,b}$ from $\mathbb{R}^+ \times \mathbb{R}$ to $L^2(\mathbb{R})$ is continuous. Thus, in view of Theorem B.1, Theorem B.2 applies, and we have the following result for the continuous wavelet transform.

THEOREM B.4. *With the same hypotheses of Theorem B.1,*

$$f(x) = \int_0^\infty \int_{-\infty}^\infty W_f(a,b)\psi_{(a,b)}(x)\, db\, \frac{da}{a^2}. \tag{B.16}$$

The integral converges strongly in $L^2(\mathbb{R})$ in the sense indicated in Theorem B.2.

We can state a slightly different form of Theorem B.4. A consequence of Theorem B.1 (and Fubini's theorem) is that $W_f(a,b)$ is in $L^2(\mathbb{R}, db)$ for a.e. $a > 0$. Thus, for a.e. $a > 0$, the function

$$f_a(x) = \int_{-\infty}^{\infty} W_f(a,b)\psi_{(a,b)}(x)\,db \tag{B.17}$$

is well defined. This function f_a can be interpreted as the component of f at scale a for the decomposition given by the wavelet ψ. Another application of Theorem B.3 shows that

$$f(x) = \int_0^{\infty} f_a(x)\,\frac{da}{a^2}, \tag{B.18}$$

which says that f is the weighted sum of its components at scale a.

B.2.2 Inversion with the Lusin wavelet

The next result is a reconstruction theorem when the analyzing wavelet is the Lusin wavelet

$$\psi(x) = \frac{1}{\pi(x+i)^2}.$$

Lusin's wavelet is the restriction to the real line of the function $\Psi(z) = \frac{1}{\pi}(z+i)^{-2}$, which is holomorphic in the open half-plane $\Omega = \{z = x + iy \mid y > 0\}$. The same remark applies to the functions $\psi_{(a,b)} = \frac{1}{a}\psi(\frac{x-b}{a})$, where $a > 0$ and $b \in \mathbb{R}$. (Note the normalization $\|\psi_{(a,b)}\|_1 = \|\psi\|_1$.) Thus, we cannot expect a function f to belong to the linear span of the functions $\psi_{(a,b)}$, $a > 0$, $b \in \mathbb{R}$, unless f has a holomorphic extension in Ω. The reconstruction theorem is the converse of this statement.

THEOREM B.5. *Let f be a bounded continuous function defined on $\mathbb{R}$, and assume that there exists a bounded holomorphic function F defined in Ω with the following properties:*

(1) $F(x + iy) \to f(x)$ *uniformly for $x \in \mathbb{R}$ as $y \to 0$.*
(2) $F(x + iy) \to 0$ *uniformly for $x \in \mathbb{R}$ as $y \to +\infty$.*

Then for $x \in \mathbb{R}$,

$$f(x) = \lim_{\substack{\rho \to 0 \\ R \to +\infty}} \int_{\rho}^{R} \int_{-\infty}^{\infty} W_f(a,b)\psi_{(a,b)}(x)\,db\,\frac{da}{a}, \tag{B.19}$$

where $W_f(a,b)$ is the wavelet transform

$$W_f(a,b) = \langle f, \psi_{(a,b)} \rangle = \int_{-\infty}^{\infty} f(x)\overline{\psi}_{(a,b)}(x)\,dx. \tag{B.20}$$

Furthermore, the convergence is uniform on compact subsets of $\mathbb{R}$.

The proof is an exercise in classical complex analysis, and it follows directly from the following two lemmas.

LEMMA B.1. $yF'(x+iy) \to 0$ *uniformly for $x \in \mathbb{R}$ as $y \to +\infty$.*

Let Γ denote the circle $\{\zeta = z + \frac{y}{2}e^{i\theta} \mid 0 \leq \theta \leq 2\pi,\, z = x + iy,\, y > 0\}$. We use Cauchy's formula to write F' as

$$F'(x+iy) = \frac{1}{2\pi i}\int_\Gamma \frac{F(\zeta)}{(\zeta - x - iy)^2}\, d\zeta,$$

which implies that $y|F'(x+iy)| \leq 2\sup_{\zeta \in \Gamma}|F(\zeta)|$. The result follows from this estimate and property (2) of Theorem B.5.

LEMMA B.2. $yF'(x+iy) \to 0$ *uniformly on compact subsets of* $\mathbb{R}$ *as* $y \to 0$.

Cauchy's formula and a simple limiting argument show that

$$F'(x+iy) = \frac{1}{2\pi i}\int_{-\infty}^\infty \frac{f(t)}{(t - x - iy)^2}\, dt, \qquad y > 0.$$

Since $\int_{-\infty}^\infty \frac{1}{(t-x-iy)^2}\, dt = 0$, we can write

$$\int_{-\infty}^\infty \frac{f(t)}{(t-x-iy)^2}\, dt = \int_{-\infty}^\infty \frac{f(t) - f(x)}{(t-x-iy)^2}\, dt$$

$$= \frac{1}{y}\int_{-\infty}^\infty \frac{f(x+yu) - f(x)}{(u-i)^2}\, du.$$

Write the last integral as

$$\int_{|u|\leq R} \frac{f(x+yu) - f(x)}{(u-i)^2}\, du + \int_{|u|>R} \frac{f(x+yu) - f(x)}{(u-i)^2}\, du,$$

and fix R large enough so that

$$\left|\int_{|u|>R} \frac{f(x+yu) - f(x)}{(u-i)^2}\, du\right| \leq 2\|f\|_\infty \int_{|u|>R} \frac{du}{1+u^2} \leq \frac{\varepsilon}{2}.$$

With R fixed, $f(x+yu) - f(x) \to 0$ uniformly for $x \in K \subset \mathbb{R}$ and $|u| \leq R$ as $y \to 0$, where K is any compact subset of $\mathbb{R}$. This shows that

$$\left|\int_{|u|\leq R} \frac{f(x+yu) - f(x)}{(u-i)^2}\, du\right| \leq \frac{\varepsilon}{2}$$

uniformly for $x \in K$ whenever y is sufficiently small. Combining these estimates proves Lemma B.2.

We now return to the proof of the theorem. The first step is to observe that $2iaF'(b+ia) = \langle f, \psi_{(a,b)}\rangle$. Then Cauchy's formula yields

$$\int_{-\infty}^\infty W_f(a,b)\psi_{(a,b)}(x)\, db = \frac{2ia^2}{\pi}\int_{-\infty}^\infty \frac{F'(b+ia)}{(x-b+ia)^2}\, db = -4a^2 F''(x+2ia),$$

and our task is to show that

$$f(x) = -\lim_{\substack{\rho \to 0 \\ R \to +\infty}} \int_\rho^R F''(x+ia)a\, da.$$

Integration by parts yields

$$\int_\rho^R F''(x+ia)a\, da = \big[-iF'(x+ia)a\big]_\rho^R + i\int_\rho^R F'(x+ia)\, da.$$

The first term on the right-hand side tends to zero uniformly on compact subsets of $\mathbb{R}$ by the lemmas. The second term is

$$F(x + iR) - F(x + i\rho),$$

which converges by hypothesis to $-f(x)$ uniformly on $\mathbb{R}$ as $\rho \to 0$ and $R \to +\infty$.

B.3 Generalizations

The wavelet analysis of functions that are not square integrable is an important issue. It is not an academic problem, since it concerns many stochastic processes such as white noise and fractional Brownian motion. Ordinary Brownian motion is also an example.

We assume that the analyzing wavelet ψ belongs to the Schwartz class $\mathcal{S}(\mathbb{R})$. If we wish to analyze a tempered distribution $f \in \mathcal{S}'(\mathbb{R})$, we first compute its wavelet coefficients

$$W_f(a,b) = \langle f, \psi_{(a,b)} \rangle, \quad a > 0, \quad b \in \mathbb{R}, \tag{B.21}$$

where $\psi_{(a,b)}(x) = \frac{1}{a}\psi(\frac{x-b}{a})$. We then hope to recover f through the inversion formula

$$f(x) = \int_0^\infty \int_{-\infty}^\infty W_f(a,b)\psi_{(a,b)}(x)\, db\, \frac{da}{a} \tag{B.22}$$

whenever ψ satisfies the admissibility condition (B.3), which in this case is

$$\int_0^\infty |\hat{\psi}(\pm u)|^2 \frac{du}{u} = 1. \tag{B.23}$$

However, equation (B.22) is not true in general if f is not square integrable. More precisely, the validity of (B.22) is related to the behavior of f at infinity. This is rather surprising. One might have thought that (B.22) cannot be true because f was too irregular. This is not the case, and, in fact, complicated distributions can be represented locally thanks to the oscillating nature of the wavelet. A counter-example to (B.22) is simply the function $f(x) = 1$, or, more generally, $f(x) = P(x)$, where P is any polynomial. On the other hand, the Dirac mass at $x = x_0$ or any compactly supported distribution can be recovered from its wavelet coefficients using the inversion formula (B.22).

We are going to discuss these facts in a slightly more general setting. The analysis of f, which is the computation of the wavelet coefficients, will be done using an analyzing wavelet ψ, but the synthesis will be achieved with a second wavelet θ. (The usefulness of this generalization was stressed by Matthias Holschneider. This generalization also paves the way to discrete biorthogonal wavelet expansions [see Chapter 4].) The first step is to rewrite the admissibility condition as

$$\int_0^\infty \hat{\theta}(\varepsilon u)\overline{\hat{\psi}(\varepsilon u)} \frac{du}{u} = 1, \quad \varepsilon \pm 1. \tag{B.24}$$

If we write $\eta = \theta * \tilde{\psi}$, where $\tilde{\psi}(x) = \overline{\psi}(-x)$, then (B.24) is equivalent to the two conditions

$$\int_{-\infty}^0 \hat{\eta}(u) \frac{du}{u} = -1 \quad \text{and} \quad \int_0^\infty \hat{\eta}(u) \frac{du}{u} = 1. \tag{B.25}$$

We will assume that f is a tempered distribution and that both θ and ψ belong to the Schwartz class $\mathcal{S}(\mathbb{R})$; however, in certain cases it is sufficient to assume that $\eta \in \mathcal{S}(\mathbb{R})$. With these assumptions, we have the following result.

THEOREM B.6. *There exists a function φ in $\mathcal{S}(\mathbb{R})$ such that $\int_{-\infty}^{\infty} \varphi(x)\,dx = 1$ and*

$$f(x) = \int_0^1 \int_{-\infty}^{\infty} W_f(a,b)\theta_{(a,b)}(x)\,db\,\frac{da}{a} + f * \varphi(x). \tag{B.26}$$

We will prove this result, but first a few comments are in order. If, for instance, $f(x) = 1$ identically, we certainly have $W_f(a,b) = 0$, and (B.22) is not true. However, (B.26) is true; it reads $1 = 1$. Identity (B.26) appears several times in the book in various disguises. In the context of a multiresolution analysis, (B.26) corresponds to writing

$$f = g + \sum_{j=0}^{\infty} f_j, \tag{B.27}$$

where g belongs to V_0 and f_j belongs to W_j. More precisely,

$$g(x) = \sum_{k=-\infty}^{\infty} \langle f, \varphi_k \rangle \varphi(x - k), \tag{B.28}$$

which mimics the convolution product $f * \varphi$. In other words, (B.26) amounts to writing f as the sum of a trend, given by $f * \varphi$, and small scale details, given by the integral.

Proof of Theorem B.6. We will be considering the function

$$F_\varepsilon(x) = \int_\varepsilon^1 \int_{-\infty}^{\infty} W_f(a,b)\theta_{(a,b)}(x)\,db\,\frac{da}{a}. \tag{B.29}$$

Since $W_f(a,b)$ is infinitely differentiable on $(0, +\infty) \times (-\infty, +\infty)$ and since it grows no faster than a polynomial as $|b| \to \infty$, the integral in (B.29) raises no convergence issues and is well defined. As in the L^2 theory, we use Plancherel's identity (which is in fact used to define the Fourier transform of a tempered distribution) to write $W_f(a,b)$ as

$$W_f(a,b) = \frac{1}{2\pi} \int_{-\infty}^{\infty} \hat{f}(\xi)\overline{\hat{\psi}}(a\xi)e^{ib\xi}\,d\xi. \tag{B.30}$$

Now fix x and integrate with respect to b to obtain

$$\int_{-\infty}^{\infty} W_f(a,b)\theta_{(a,b)}(x)\,db = \frac{1}{2\pi} \int_{-\infty}^{\infty} \int_{-\infty}^{\infty} \hat{f}(\xi)\overline{\hat{\psi}}(a\xi)\theta_{(a,b)}(x)e^{ib\xi}\,d\xi\,db. \tag{B.31}$$

Since $\hat{f}$ can be viewed as a tempered distribution on $\mathbb{R}^2$, that is, $\hat{f} \in \mathcal{S}'(\mathbb{R}^2)$, and since for each fixed x, $\overline{\hat{\psi}}(a\xi)\theta_{(a,b)}(x)e^{ib\xi}$ is in $\mathcal{S}(\mathbb{R}^2)$, we can interchange the order of "integration" in (B.31). Thus, (B.31) can be written as

$$\begin{aligned}\int_{-\infty}^{\infty} W_f(a,b)\theta_{(a,b)}(x)\,dx &= \frac{1}{2\pi} \int_{-\infty}^{\infty} \int_{-\infty}^{\infty} \hat{f}(\xi)\overline{\hat{\psi}}(a\xi)\hat{\theta}(a\xi)e^{ix\xi}\,d\xi \\ &= \frac{1}{2\pi} \int_{-\infty}^{\infty} \hat{f}(\xi)\hat{\eta}(a\xi)e^{ix\xi}\,d\xi.\end{aligned} \tag{B.32}$$

We end the proof by doing the integration with respect to a. This becomes simpler once we introduce the function H defined by

$$H(\xi) = -\int_{-\infty}^{\xi} \hat{\eta}(u) \frac{du}{u}.$$

Observe that $\hat{\eta}(0) = \overline{\hat{\psi}(0)}\hat{\theta}(0) = 0$ because ψ has at least one vanishing moment. Thus $\frac{\hat{\eta}(u)}{u}$ is in the Schwartz class. Note that (B.25) implies that $\int_{-\infty}^{\infty} \hat{\eta}(u) \frac{du}{u} = 0$. It follows from these facts that H is in the Schwartz class. Furthermore, (B.25) implies that $H(0) = 1$.

Then we have

$$\int_{\varepsilon}^{1} \hat{\eta}(a\xi) \frac{da}{a} = H(\varepsilon\xi) - H(\xi), \tag{B.33}$$

and

$$F_\varepsilon(x) = \frac{1}{2\pi} \int_{-\infty}^{\infty} \hat{f}(\xi)[H(\varepsilon\xi) - H(\xi)] e^{ix\xi} \, d\xi. \tag{B.34}$$

The function H is the Fourier transform of a function φ that also belongs to the Schwartz class. Thus we can write (B.34) as

$$F_\varepsilon = f * \varphi_\varepsilon - f * \varphi, \tag{B.35}$$

where $\varphi_\varepsilon(x) = \frac{1}{\varepsilon}\varphi(\frac{x}{\varepsilon})$. The inversion formula follows directly from (B.35): Since $H(0) = \int_{-\infty}^{\infty} \varphi(x) \, dx = 1$, $f * \varphi_\varepsilon$ converges to f in the sense of distributions as ε tends to zero. This completes the proof of the theorem. As a final remark, note that it is possible to define φ directly as

$$\varphi(x) = \frac{1}{x}\int_{-\infty}^{x} \eta(s) \, ds = -\frac{1}{x}\int_{x}^{\infty} \eta(s) \, ds.$$

□

One might suspect that

$$F_{\varepsilon,R}(x) = \int_{\varepsilon}^{R} \int_{-\infty}^{\infty} W_f(a,b) \theta_{(a,b)}(x) \, db \, \frac{da}{a} \tag{B.36}$$

would converge to f as ε tends to 0 and R tends to ∞. A calculation that is almost identical to the one above yields

$$F_{\varepsilon,R} = f * \varphi_\varepsilon - f * \varphi_R. \tag{B.37}$$

As we have already seen, $f * \varphi_\varepsilon \to f$ in the sense of distributions as $\varepsilon \to 0$. If f has compact support, then $f * \varphi_R \to 0$ as $R \to \infty$, but this is not true in general if f does not have compact support. Thus we see that it is the behavior of f at infinity that accounts for the failure of (B.22).

We have stated and proved Theorem B.6 in the context of tempered distributions and wavelets in the Schwartz class. The strategy of the proof can be used to establish similar results under different assumptions about the analyzed object f and the wavelets ψ and θ.

In many problems concerning pointwise regularity, it is not necessary to have an inversion formula like (B.22) that includes all of the wavelet coefficients. Formula (B.26) is often sufficient. For example, the first term contains all of the information necessary to compute Hölder exponents at a given point, while the term $f * \varphi$ is the "smooth" part of the function.

Matthias Holschneider made the interesting observation that the flexibility offered by the choice of the second wavelet allows one to "cheat." For example, assume that the analyzing wavelet satisfies only $\int_{-\infty}^{\infty} \psi(x)\, dx = 0$ and that the wavelet θ used for the synthesis belongs to $\mathcal{S}(\mathbb{R})$ and is such that η satisfies the admissibility condition. Now assume that the wavelet coefficients $W_f(a, b)$ of the function f that we wish to analyze satisfy

$$|W_f(a,b)| \leq C a^\alpha \left(1 + \frac{|b - x_0|}{a}\right)^{\alpha'} \tag{B.38}$$

for all $0 < a \leq 1$ and $|b - x_0| \leq 1$, where $0 < \alpha' < \alpha$ and $\alpha > 1$. Then it is a straightforward application of Theorem B.6 to show that f belongs to $C^\alpha(x_0)$. The point is that since $\alpha > 1$ and ψ has only one vanishing moment—$\int_{-\infty}^{\infty} x\psi(x)\, dx$ may not exist, or, if it exists, it may not vanish—ψ cannot be used alone to conclude that $f \in C^\alpha(x_0)$.

An application of this strategy is provided by the Riemann function

$$\mathcal{R}(x) = \sum_{n=1}^{\infty} \frac{1}{n^2} \sin(\pi n^2 x).$$

For analyzing $\mathcal{R}$, it is convenient to use the Lusin wavelet $\psi(x) = \frac{1}{\pi(x+i)^2}$ because the wavelet coefficients are the values of the Jacobi theta function in the upper half-plane. In principle, the Lusin wavelet cannot be used to prove that $\mathcal{R}$ belongs to $C^{3/2}(x_0)$ when $x_0 = 1$: Although $\int_{-\infty}^{\infty} \psi(x)\, dx = 0$, $\int_{-\infty}^{\infty} x\psi(x)\, dx$ is not even finite. One can navigate around this obstacle by choosing a synthesizing wavelet $\theta \in \mathcal{S}(\mathbb{R})$ such that

$$\int_0^\infty \hat{\theta}(u) \overline{\hat{\psi}(u)}\, \frac{du}{u} = 1.$$

(The original paper is [144].) Another example of the flexibility provided by the "biorthogonal" continuous wavelet analysis was provided by Holschneider to invert the Radon transform [143].

APPENDIX C

A Counterexample

C.1 Introduction

A counterexample to Mallat's conjecture about zero-crossings (section 8.4) was found by Yves Meyer in the early 1990s. It appeared in conference notes, but it has never been published. Since there is continuing interest in analyzing signals using zero-crossings, we have elected to present a complete discussion rather than the outline given in the first edition of this book. The following counterexample is based on the one announced by Meyer. The development given here is more "constructive" than that presented by Meyer, but the price paid is that the proof requires considerable computation.

The counterexample for two dimensions follows rather easily from the one-dimensional case, where the real work must be done. The construction given here is reminiscent of the one given for the counterexample to Marr's conjecture in section 8.3. However, in the case of Mallat's conjecture, there are other conditions to be satisfied, since both the zero-crossings and the first derivatives of the functions must agree. Fortunately, these conditions must be met only for dyadic values of the scaling parameter ρ. This makes it possible to construct a smooth, compactly supported counterexample. The other difference between the two conjectures is that in Marr's case the kernel is the Gaussian and in Mallat's case the kernel is the basic cubic spline.

We begin with the function f_0 defined by

$$f_0(t) = \begin{cases} 1 + \cos t & \text{if } |t| \leq \pi, \\ 0 & \text{if } |t| \geq \pi. \end{cases} \tag{C.1}$$

We will show that there are infinitely many functions of the form

$$f(t) = f_0(t) + R(t) \tag{C.2}$$

such that $(f_0 * \theta_\rho)''$ and $(f * \theta_\rho)''$ have the same zeros when $\rho = 2\pi 2^{-j}$, $j \in \mathbb{Z}$, and such that at these zeros, $(f_0 * \theta_\rho)'(t) = (f * \theta_\rho)'(t)$. The function R will be a "small" C^∞ function whose support is $\frac{\pi}{8} \leq |t| \leq \frac{\pi}{4}$. To keep things symmetric, we will define R so that it is even. The function $\theta_\rho(t) = \rho^{-1}\theta(\rho^{-1}t)$, where θ is the basic cubic spline (Figure C.1).

The analysis centers on locating the zeros of $(f_0 * \theta_\rho)''$. We will show that there are only two simple zeros in the interval $(-\pi - 2\rho, \pi + 2\rho)$ for each value of $\rho = 2\pi 2^{-j}$. We will also show that, at these zeros, both $(R * \theta_\rho)''$ and $(R * \theta_\rho)'$ vanish. This proves that at these points the derivatives of $f_0 * \theta_\rho$ and $f * \theta_\rho$ agree. Observe that both $f_0 * \theta_\rho$ and $f * \theta_\rho$ vanish identically outside $(-\pi - 2\rho, \pi + 2\rho)$.

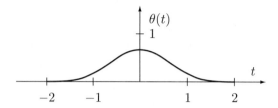

FIG. C.1. *The cubic spline $\theta = T * T$.*

The last step will be to show that the functions $(f_0 * \theta_\rho)''$ and $(f * \theta_\rho)''$ have the same zeros. For this we will argue that there is a constant M such that

$$|(R * \theta_\rho)''(t)| \leq M |(f_0 * \theta_\rho)''(t)| \tag{C.3}$$

for all $t \in \mathbb{R}$, uniformly in $\rho = 2\pi 2^{-j}$. Then for some $\lambda > 0$, we can replace R by λR and have

$$|(\lambda R * \theta_\rho)''(t)| < r |(f_0 * \theta_\rho)''(t)|, \tag{C.4}$$

where $r < 1$. The conclusion that $(f_0 * \theta_\rho)''(t)$ and $(f * \theta_\rho)''(t)$ have the same zeros follows from the following lemma.

LEMMA C.1. *If u and v are two continuous functions on $\mathbb{R}$ such that*

$$|v(t)| \leq r |u(t)|$$

for some $0 < r < 1$, then $u + v$ and u have the same zeros.

We begin by establishing several representations for the convolutions and their derivatives of the functions f_0, θ, and R. Once we have established these representations, the proof follows rather easily. Our approach is to develop explicit representations for the various functions. The objective is to reveal the geometry of the situation, which we hope leads to an understanding of how the example works. We begin with observations about the kernel θ.

C.2 The function θ

As before, θ is the cubic spline $\theta = T * T$, where T is the triangular function that is equal to $1 - |t|$ if $|t| < 1$ and is equal to zero if $|t| > 1$. Recall that $\theta_\rho(t) = \rho^{-1}\theta(\rho^{-1}t)$. The support of θ, and of its derivatives, is $[-2, 2]$, and thus the support of $\theta_\rho(t) = \rho^{-1}\theta(\rho^{-1}t)$ is $[-2\rho, 2\rho]$. In what follows, we will change scale and let $\rho = 2\pi 2^{-j}$, $j \in \mathbb{Z}$, rather than having $\rho = 2^{-j}$. This makes the supports of the functions θ_ρ commensurate with the support of f_0.

Note that θ and θ'' are even functions. Since it is useful to visualize θ'', which is the analyzing wavelet, it is shown explicitly in Figure C.2.

The fourth derivative of θ_ρ, which is the distribution

$$\theta_\rho^{(4)} = \rho^{-4}[\delta_{-2\rho} - 4\delta_{-\rho} + 6\delta - 4\delta_\rho + \delta_{2\rho}], \tag{C.5}$$

plays a featured role in our analysis. Here, δ is the usual "delta function," and we write δ_a to indicate that the "Dirac mass" is at the point a. We use the notation

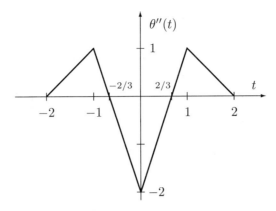

FIG. C.2. *The second derivative of θ.*

$\tau_t \theta_\rho^{(4)}$ to denote the distribution $\theta_\rho^{(4)}$ shifted to the right by t. Thus by definition, $\varphi * \theta_\rho^{(4)}(t) = \langle \tau_t \theta_\rho^{(4)}, \varphi \rangle$, and for any continuous function φ,

$$\varphi * \theta_\rho^{(4)}(t) = \rho^{-4}[\varphi(t - 2\rho) - 4\varphi(t - \rho) + 6\varphi(t) - 4\varphi(t + \rho) + \varphi(t + 2\rho)].$$

Note that the "filter" $\theta_\rho^{(4)}$ has the following important property:

$$P * \theta_\rho^{(4)}(t) = 0 \qquad (C.6)$$

for all t whenever P is a polynomial of degree ≤ 3. Also note that $\int \theta_\rho(t)\,dt = 1$.

C.3 Representations of $f_0 * \theta_\rho$ and its derivatives

Although the counterexample depends on the discrete values $\rho = 2\pi 2^{-j}$, the functions $f_0 * \theta_\rho$ and $(f_0 * \theta_\rho)''$ can be analyzed for all $\rho > 0$. Thus, in this and the following section we will consider ranges of ρ rather than ranges of j. More will be said about this distinction as we progress through the demonstration.

We note once and for all that f_0, θ_ρ, and $f_0 * \theta_\rho$ are even functions, as are their even derivatives. We use this fact freely without further comment.

Here are several expressions for $f_0 * \theta_\rho$ and $(f_0 * \theta_\rho)''$ that will be used in what follows:

$$f_0 * \theta_\rho(t) = \rho^{-1} \int f_0(t - s)\theta(\rho^{-1}s)\,ds, \qquad (C.7)$$

$$(f_0 * \theta_\rho)''(\rho t) = \rho^{-2} \int f_0(\rho(s - t))\theta''(s)\,ds, \qquad (C.8)$$

$$(f_0 * \theta_\rho)''(t) = F_0 * \theta_\rho^{(4)}(t), \qquad (C.9)$$

where F_0 is any C^2 function such that $F_0''(t) = f_0(t)$ for all $t \in \mathbb{R}$. In particular, an obvious definition for F_0 is this:

$$F_0(t) = \begin{cases} 0 & \text{if } t \leq -\pi, \\ \frac{1}{2}(t + \pi)^2 - (1 + \cos t) & \text{if } |t| \leq \pi, \\ 2\pi t & \text{if } t \geq \pi. \end{cases} \qquad (C.10)$$

However, another useful function is $F_0(-t)$; the choice of using $F_0(t)$ or $F_0(-t)$ in (C.9) depends on the computation one is doing, which in turn depends on the value of ρ.

We emphasize that relations (C.9) and (C.10) can be used to give an explicit representation of $(f_0 * \theta_\rho)''$ for all ρ and all t, and hence that the values of the function and its zeros can be computed to any degree of accuracy for a given ρ.

When the support of $\tau_t \theta_\rho^{(4)}$ is in $[-\pi, \pi]$, that is, when the five points $t-2\rho$, $t-\rho$, t, $t+\rho$, and $t+2\rho$ lie in $[-\pi, \pi]$, there is another useful representation for $(f_0 * \theta_\rho)''$. In this case, $F_0 * \theta_\rho^{(4)}(t) = -\cos t * \theta_\rho^{(4)}(t)$, since the filter $\theta_\rho^{(4)}$ "kills" the quadratic part of F_0, as (C.6) shows. Thus from (C.9),

$$\rho^4 (f_0 * \theta_\rho)''(t) = -\cos(t-2\rho) + 4\cos(t-\rho) - 6\cos t + 4\cos(t+\rho) - \cos(t+2\rho)$$
$$= -2^4 \left(\sin \tfrac{\rho}{2}\right)^4 \cos t,$$

and we have

$$(f_0 * \theta_\rho)''(t) = -\left(\frac{\sin \tfrac{\rho}{2}}{\tfrac{\rho}{2}}\right)^4 \cos t. \tag{C.11}$$

This representation holds for all $\rho \leq \frac{\pi}{4}$ and $|t| \leq \pi - 2\rho$.

Since the theme of our program is to understand the behavior of the functions involved, we list for future reference the explicit representation of $(f_0 * \theta_\rho)''(t)$ for $\rho \geq 2\pi$. This expansion is based on (C.9) and (C.10). Since $(f_0 * \theta_\rho)''$ is an even function, we consider only positive values of t.

For $0 \leq t \leq \pi$,

$$\rho^4 (f_0 * \theta_\rho)''(t) = 3t^2 + 3\pi^2 - 4\pi\rho - 6(1 + \cos t). \tag{C.12}$$

For $\pi \leq t \leq \rho - \pi$,

$$\rho^4 (f_0 * \theta_\rho)''(t) = 6\pi t - 4\pi\rho. \tag{C.13}$$

For $\rho - \pi \leq t \leq \rho + \pi$,

$$\rho^4 (f_0 * \theta_\rho)''(t) = -2(t - \rho + \pi)^2 + 6\pi t - 4\pi\rho + 4(1 + \cos(t - \rho)). \tag{C.14}$$

For $\rho + \pi \leq t \leq 2\rho - \pi$,

$$\rho^4 (f_0 * \theta_\rho)''(t) = -2\pi t + 4\pi\rho. \tag{C.15}$$

For $2\rho - \pi \leq t \leq 2\rho + \pi$,

$$\rho^4 (f_0 * \theta_\rho)''(t) = \frac{1}{2}(t - 2\rho - \pi)^2 - (1 + \cos(t - 2\rho)). \tag{C.16}$$

Using this representation of $\rho^4(f_0 * \theta_\rho)''$ (and the representation of its derivative) it is an easy piece of analysis to show that $\rho^4(f_0 * \theta_\rho)''$ has the following properties (see Figure C.3):

- $\rho^4(f_0 * \theta_\rho)''$ has a minimum value of $3\pi^2 - 12 - 4\pi\rho$ at $t = 0$.
- $\rho^4(f_0 * \theta_\rho)''$ has a zero at $t = \frac{2}{3}\rho$ when $\rho \geq 3\pi$.
- $\rho^4(f_0 * \theta_\rho)''$ is monotonic increasing from $t = 0$ to $t = t_m$, where $t_m = \mu + \rho$ and where μ is the unique solution of $t + \sin t = \frac{\pi}{2}$ in the interval $(-\pi, \pi)$.
- $\rho^4(f_0 * \theta_\rho)''$ is monotonic decreasing from t_m to $t = 2\rho + \pi$, after which it vanishes identically.

The point $t_m = \mu + \rho$ will appear again in section C.7.

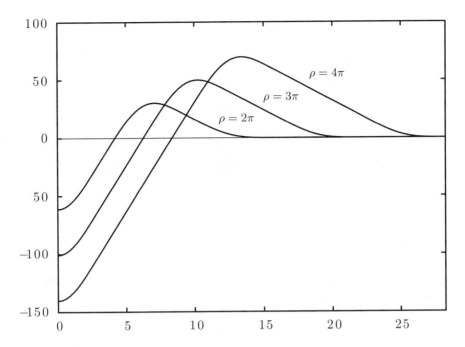

FIG. C.3. *Plots of $\rho^4(f_0 * \theta_\rho)''$ for three values of ρ.*

C.4 Hunting the zeros of $(f_0 * \theta_\rho)''$

There are two problems: to find the zeros and to show that they are simple. The method used depends on the size of ρ. It is relatively easy to do the analysis for $\rho \leq \frac{\pi}{4}$ and for $\rho \geq 3\pi$. The intermediate cases require more computation. We will not present all of the computations involved in this analysis, but we will indicate at least one way to proceed in each case. As before, we will consider only $t \geq 0$. In each case, we will prove that there is one simple zero between 0 and $\pi + 2\rho$. All of the functions vanish identically for $t \geq \pi + 2\rho$. We will consider three cases.

Case 1: $\rho \leq \frac{\pi}{4}$

The representation (C.11) shows that $\frac{\pi}{2}$ is a simple zero if $\rho < \frac{\pi}{4}$. It also shows that $\frac{\pi}{2}$ is a zero when $\rho = \frac{\pi}{4}$; that it is a simple zero follows from the continuity of the function $(f_0 * \theta_\rho)'''$. Moreover, there are no other zeros in $[0, \pi + 2\rho)$.

The fact that there are no other zeros between $\frac{\pi}{2}$ and $\pi + 2\rho$ is easily established for $\rho \leq \frac{\pi}{8}$ if one writes

$$(f_0 * \theta_\rho)''(t) = \int f_0''(s)\theta_\rho(s-t)\,ds = -\int_{-\pi}^{\pi} \cos(s)\theta_\rho(s-t)\,ds. \qquad \text{(C.17)}$$

When $\rho \leq \frac{\pi}{8}$, the last integral equals $-\int_{\frac{\pi}{2}}^{\pi} \cos(s)\theta_\rho(s-t)\,ds$, which is strictly positive for $\frac{\pi}{2} < t < \pi + 2\rho$.

The case $\rho = \frac{\pi}{4}$ also follows from (C.17), but the argument is less obvious. There is no problem when $t \geq \pi$, since for these values the integrand is positive. Thus, we consider $\frac{\pi}{2} \leq t \leq \pi$. In this case, the integral is

$$\int_{t-\frac{\pi}{2}}^{\pi} -\cos(s)\theta_{\frac{\pi}{4}}(s-t)\,ds = \int_{t-\frac{\pi}{2}}^{\frac{\pi}{2}} + \int_{\frac{\pi}{2}}^{t} + \int_{t}^{\pi} -\cos(s)\theta_{\frac{\pi}{4}}(s-t)\,ds.$$

The second and third integrals are positive, but the first is negative. We compare the values of the first and third integrals to show that the first is smaller in absolute value than the third. For this we write

$$A(t) = \int_{t-\frac{\pi}{2}}^{\frac{\pi}{2}} -\cos(s)\theta_{\frac{\pi}{4}}(s-t)\,ds = \int_{t}^{\pi} -\cos\left(s - \frac{\pi}{2}\right)\theta_{\frac{\pi}{4}}\left(s - \frac{\pi}{2} - t\right)ds$$

and

$$B(t) = \int_{t}^{\pi} -\cos(s)\theta_{\frac{\pi}{4}}(s-t)\,ds = \int_{t}^{\pi} -\cos(s-\pi-t)\theta_{\frac{\pi}{4}}(s-\pi)\,ds.$$

Since $\frac{\pi}{2} \leq t \leq s \leq \pi$, it is not difficult to see that $|-\cos(s-\frac{\pi}{2})| \leq -\cos(s-\pi-t)$ and that $\theta_{\frac{\pi}{4}}(s - \frac{\pi}{2} - t) \leq \theta_{\frac{\pi}{4}}(s-\pi)$. Hence $|A(t)| \leq B(t)$, and since the second integral is positive, the sum is positive.

It is not necessary for the counterexample (and thus they are not included), but arguments similar to the last one can be used to show that $(f_0 * \theta_\rho)''$ is strictly positive for $\frac{\pi}{2} < t < \pi + 2\rho$ when $\frac{\pi}{8} < \rho < \frac{\pi}{4}$. This establishes that for $\rho \leq \frac{\pi}{4}$, $(f_0 * \theta_\rho)''$ has only one simple zero when $0 < t < \pi + 2\rho$.

Case 2: $\rho \geq 3\pi$

We have essentially dealt with this case in section C.3. It is clear from the analysis of the representation (C.12)–(C.16) that $\rho^4(f_0 * \theta_\rho)''$ has only one zero, $t = \frac{2}{3}\rho$, in $[0, \pi + 2\rho]$ and that it is simple.

Case 3: $\frac{\pi}{4} < \rho < 3\pi$

This is the no-man's-land case where the supports of f_0 and θ are about the same size, and consequently the computations become more difficult. We have done specific computations for $\rho = \frac{\pi}{2}, \pi$, and 2π using the representation (C.9) and developing explicit formulas similar to (C.12)–(C.16). The result, as expected, is that there is exactly one simple zero in the interval $[0, \pi + 2\rho)$ in each case. We also have located these zeros well enough for the task at hand, which is to show that these zeros are also zeros of $(R * \theta_\rho)''$. These computations, while perhaps tedious, are completely elementary. The results on the zeros of $(f_0 * \theta_\rho)''$ are summarized as a lemma.

LEMMA C.2. *For each $\rho = 2\pi 2^{-j}$, $j \in \mathbb{Z}$, $(f_0 * \theta_\rho)''$ has only two (symmetric) zeros in the interval $(-\pi - 2\rho, \pi + 2\rho)$ and these zeros are simple.*

- *For $j \leq -1$, the zeros are $\pm\frac{2}{3}\rho$.*
- *For $j = 0$, the zeros are located in the intervals $\frac{4}{3}\pi < |t| < \frac{3}{2}\pi$.*
- *For $j = 1$, the zeros are located in the intervals $\frac{1}{2}\pi < |t| < \frac{3}{4}\pi$.*
- *For $j = 2$, the zeros are located in the intervals $\frac{1}{2}\pi < |t| < \frac{5}{8}\pi$.*
- *For $j \geq 3$, the zeros are $\pm\frac{1}{2}\pi$.*

C.5   The functions R, $R * \theta_\rho$, $(R * \theta_\rho)'$, and $(R * \theta_\rho)''$

The function R is defined in terms of a small C^∞ function whose support is contained in $\frac{\pi}{8} \leq |t| \leq \frac{\pi}{4}$. Thus let h be an arbitrary function in C^∞ with support in $[\frac{\pi}{8}, \frac{\pi}{4}]$. Define

$$g(t) = \begin{cases} 0 & \text{if } 0 \leq t \leq \frac{\pi}{8}, \\ h(t) & \text{if } \frac{\pi}{8} \leq t \leq \frac{\pi}{4}, \\ 0 & \text{if } t \geq \frac{\pi}{4}, \\ -h(-t) & \text{if } t \leq 0. \end{cases} \tag{C.18}$$

Define $R(t) = g'''(t)$. Then R is an even C^∞ function defined on the whole real line. The function R has been defined as a third derivative so that the representations

$$(R * \theta_\rho)''(t) = g' * \theta_\rho^{(4)}(t), \tag{C.19}$$

$$(R * \theta_\rho)'(t) = -g * \theta_\rho^{(4)}(t) \tag{C.20}$$

hold for all ρ and all $t \in \mathbb{R}$. The utility of these representations, which play a central role in our arguments, is based on the following two facts: First, the supports of both g' and $-g$ are contained in the set $K = [-\frac{\pi}{4}, -\frac{\pi}{8}] \cup [\frac{\pi}{8}, \frac{\pi}{4}]$. Second (which will be proved later), the support of the filter $\tau_{t_0} \theta_\rho^{(4)}$ when t_0 is an isolated zero of $(f_0 * \theta_\rho)''$ does not intersect the interior of K. This means that $g' * \theta_\rho^{(4)}(t_0) = 0$ and $-g * \theta_\rho^{(4)}(t_0) = 0$, and thus, by (C.19) and (C.20), that $(R * \theta_\rho)''$ and $(R * \theta_\rho)'$ vanish at these zeros. This shows that the fact that $(R * \theta_\rho)''$ and $(R * \theta_\rho)'$ vanish at the zeros of $(f_0 * \theta_\rho)''$ depends only on the support of h.

C.6 $(R * \theta_\rho)''$ and $(R * \theta_\rho)'$ vanish at the zeros of $(f_0 * \theta_\rho)''$

The idea of the proof is to locate the zeros of $(f_0 * \theta_\rho)''$ and then to use the representations (C.19) and (C.20) to show that these zeros are also zeros of $(R * \theta_\rho)''$ and $(R * \theta_\rho)'$. The motivation for this approach is to see explicitly the conditions that must be imposed on h to make the counterexample work. In this part of the proof, it is only the support of h that counts.

Having located the zeros of $(f_0 * \theta_\rho)''$, it is a simple exercise to show that they are zeros of $(R * \theta_\rho)''$ and $(R * \theta_\rho)'$. In fact, if $t_0 > 0$ is an isolated zero of $(f_0 * \theta_\rho)''$, then the points $t_0 - 2\rho$, $t_0 - \rho$, t_0, $t_0 + \rho$, and $t_0 + 2\rho$ do not intersect the interior of the support of g or g', which is K in both cases. In only two cases, $j = 3$ and $j = 4$, does one of these points even intersect the boundary of K (at $\frac{\pi}{4}$), and both g and g' vanish (along with all of their derivatives) at this point. For $j \geq 5$, the five points completely miss K. If $t_0 = \frac{2}{3}\rho$ for $j \leq -1$, then the five points in question are $-\frac{4}{3}\rho$, $-\frac{1}{3}\rho$, $\frac{2}{3}\rho$, $\frac{5}{3}\rho$, and $\frac{8}{3}\rho$, and none of these points come even close to K. The cases $j = 0, 1, 2$ are easily checked, although here one must check that the five points of the filter miss K for the range of values indicated in the "zero summary." In short, the support of $\tau_{t_0} \theta_\rho^{(4)}$ misses K whenever t_0 is a zero of $(f_0 * \theta_\rho)''$. We note that the isolated zeros of $(f_0 * \theta_\rho)''$ are all "infinite" zeros of $(R * \theta_\rho)''$. Since the supports of g and g' are contained in K, the zeros of $(f_0 * \theta_\rho)''$ are also zeros of $(R * \theta_\rho)'$.

This proves that having $(R * \theta_\rho)''$ and $(R * \theta_\rho)'$ vanish at the zeros of $(f_0 * \theta_\rho)''$ depends only on the support of h. Proving that the zeros of $(f_0 * \theta_\rho)''$ and $(f * \theta_\rho)''$ are the same for the discrete values $\rho = 2\pi 2^{-j}$, $j \in \mathbb{Z}$, is another matter.

C.7 The behavior of $(R * \theta_\rho)''/(f_0 * \theta_\rho)''$

As indicated in the introduction, we wish to show that $\frac{(R*\theta_\rho)''}{(f_0*\theta_\rho)''}$ is bounded uniformly in $\rho = 2\pi 2^{-j}$, $j \in \mathbb{Z}$. But before getting into the details, we make some preliminary observations. The range for t will always be $[-\pi - 2\rho, \pi + 2\rho]$, but because of the symmetry we only look at $0 \leq t \leq \pi + 2\rho$. For each ρ, $(R * \theta_\rho)''(t) = 0$ for all $t \geq \frac{\pi}{4} + 2\rho$ and $(f_0 * \theta_\rho)''(t) = 0$ for all $t \geq \pi + 2\rho$. For each fixed $\rho = 2\pi 2^{-j}$, the function $\frac{(R*\theta_\rho)''}{(f_0*\theta_\rho)''}$ is continuous on the interval $0 \leq t \leq \pi + 2\rho$. This follows from the fact that the isolated, simple zero of $(f_0 * \theta_\rho)''$ is a zero of $(R * \theta_\rho)''$. The "other" zero of $(f_0 * \theta_\rho)''$, that is, $t_0 = \pi + 2\rho$, offers no problem, since $(R * \theta_\rho)''$ is identically zero in a neighborhood of t_0. Thus, for each j, there is a constant $M_j > 0$ such that

$$\left|\frac{(R*\theta_\rho)''(t)}{(f_0*\theta_\rho)''(t)}\right| \leq M_j$$

for $0 \leq t \leq \pi + 2\rho$. Our immediate goal is to show that there is an M such that $M_j \leq M$ for all j. Again, we consider cases.

For $j \geq 5$, we use the representations (C.11) and (C.19). We know from (C.19) that $(R * \theta_\rho)''(t) = 0$ for $t \geq \frac{\pi}{4} + 2\rho$ for all ρ. Thus for all $j \geq 5$, $(R * \theta_\rho)''(t) = 0$ when $t \geq \frac{3}{8}\pi$. For $0 \leq t \leq \frac{3}{8}\pi$, it is clear from (C.11) that

$$|(f_0 * \theta_\rho)''(t)| \geq \left(\frac{\sin\frac{\rho}{2}}{\frac{\rho}{2}}\right)^4 \cos\left(\frac{3}{8}\pi\right) \geq \left(\frac{2}{\pi}\right)^4 \cos\left(\frac{3}{8}\right) = C > 0.$$

Thus, whenever $(R * \theta_\rho)''(t) \neq 0$,

$$\left|\frac{(R*\theta_\rho)''(t)}{(f_0*\theta_\rho)''(t)}\right| \leq C^{-1}|(R*\theta_\rho)''(t)|.$$

Since $(R * \theta_\rho)''(t) = R'' * \theta_\rho(t)$, and since $|R'' * \theta_\rho(t)| \leq \max|R''(t)| = \max|g^{(5)}(t)|$, we have

$$\left|\frac{(R*\theta_\rho)''(t)}{(f_0*\theta_\rho)''(t)}\right| \leq C^{-1} \max|g^{(5)}(t)|. \tag{C.21}$$

Hence, there is an $M_{j\geq 5} > 0$ such that

$$\left|\frac{(R*\theta_\rho)''(t)}{(f_0*\theta_\rho)''(t)}\right| \leq M_{j\geq 5}$$

for all $j \geq 5$.

When $j \leq -1$, we continue with our program of "explicit representation" and use the representation (C.12)–(C.16) to show that $|\rho^4(f_0 * \theta_\rho)''(t)|$ is bounded away from zero, uniformly in ρ, for any t where $(R * \theta_\rho)''(t) \neq 0$.

First, use (C.19) to deduce that $(R * \theta_\rho)''(t) = 0$ for $\pi \leq t \leq \rho - \pi$ and use (C.13) to see that

$$\rho^4(f_0 * \theta_\rho)''(\pi) = 6\pi^2 - 4\pi\rho \leq -10\pi^2,$$

and

$$\rho^4(f_0 * \theta_\rho)''(\rho - \pi) = 2\pi\rho - 6\pi^2 \geq 2\pi^2.$$

This takes care of any possibility that $\rho^4(f_0 * \theta_\rho)''(t)$ comes close to zero on the support of $(R * \theta_\rho)''$ between $t = 0$ and $t = t_m$, where $\rho^4(f_0 * \theta_\rho)''(t)$ is maximum.

Next, we must see what happens for $t > t_m$. Recall that $(R * \theta_\rho)''(t) = 0$ for all $t \geq \frac{\pi}{4} + 2\rho$. Thus, we wish to investigate the value of $\rho^4(f_0 * \theta_\rho)''(t)$ at $t = \frac{\pi}{4} + 2\rho$. For this, we use (C.16) and discover that

$$\rho^4(f_0 * \theta_\rho)''\left(\frac{\pi}{4} + 2\rho\right) = \frac{9}{32}\pi^2 - \frac{2 + \sqrt{2}}{2} > 1.$$

This means that $|\rho^4(f_0 * \theta_\rho)''(t)|$ is bounded away from zero uniformly on the support of $(R * \theta_\rho)''$. The representation (C.19) and the fact that only one point of the filter can intersect K when $\rho \geq 4\pi$ ($j \leq -1$) imply that

$$|\rho^4(R * \theta_\rho)''(t)| \leq 6 \max |g'(t)|.$$

Hence,

$$\left|\frac{(R * \theta_\rho)''(t)}{(f_0 * \theta_\rho)''(t)}\right| \leq M_{j \leq -1}, \qquad (C.22)$$

where $M_{j \leq -1} \leq 6 \max |g'(t)|$.

For each $j = 0, 1, 2, 3, 4$ there is an M_j such that

$$\left|\frac{(R * \theta_\rho)''(t)}{(f_0 * \theta_\rho)''(t)}\right| \leq M_j.$$

By taking $M = \max\{M_{j \leq -1}, M_0, M_1, M_2, M_3, M_4, M_{j \geq 5}\}$, we see that (C.3) is satisfied uniformly in ρ. The result follows as indicated in the introduction by choosing λ so that $\lambda M < 1$ and using the Lemma C.1.

C.8 Remarks

We have analyzed the zeros of the function $(f_0 * \theta_\rho)''$ in considerable detail. To summarize, let $t_0(\rho)$ denote the unique zero of $(f_0 * \theta_\rho)''$ in the interval $[0, \pi + 2\rho)$. Then

$$t_0(\rho) = \begin{cases} \frac{\pi}{2} & \text{if } 0 < \rho \leq \frac{\pi}{4}, \\ \frac{\pi}{2} + \varepsilon(\rho) & \text{if } \frac{\pi}{4}\rho \leq \frac{\pi}{2}, \\ \frac{2}{3}\rho + \varepsilon(\rho) & \text{if } \frac{\pi}{2}\rho \leq 3\pi, \\ \frac{2}{3}\rho & \text{if } 3\pi \leq \rho < \infty. \end{cases} \qquad (C.23)$$

We have not analyzed the function $\varepsilon(\rho)$, but we know that it is relatively small and positive. For example, $\varepsilon(2\pi) < (0.005)\pi$.

This behavior of $t_0(\rho)$ is qualitatively typical in the following sense: If θ is replaced by any symmetric kernel η in C^3 having compact support $[-T, T]$ and having the property that η'' has exactly one simple zero, say, at $t = r$, $0 < r < T$, then $\frac{\pi}{2}$ is the unique simple zero of $(f_0 * \eta_\rho)''(t)$ for all sufficiently small ρ and $t_0(\rho)$ is asymptotic to the linear function $r\rho$ as $\rho \to +\infty$. This is the case, for example, if one takes $\eta = f_0$. That the zeros behave as claimed can be seen by using the representation

$$(f_0 * \eta_\rho)''(\rho t) = \rho^{-2} \int f_0(\rho(t-s))\eta''(s)\,ds$$

for large ρ and the representation

$$(f_0 * \eta_\rho)''(t) = \rho^{-1} \int f_0''(t-s)\eta(\rho^{-1}s)\,ds$$

for small ρ. The point here is that this part of the counterexample is not particularly sensitive to the kernel θ. However, the fact that $\theta_\rho^{(4)}$ is a linear combination of delta functions makes it easy to evaluate the zeros for the finite number of intermediate values that we needed to locate for our particular construction.

On the other hand, the nature of $\theta_\rho^{(4)}$ was critical for other parts of the counterexample. We constructed the perturbation R so that $\tau_{t_0}\theta_\rho^{(4)}$ would not intersect the (interior) of the support of R for $t_0 = t_0(2\pi 2^{-j})$, $j \in \mathbb{Z}$. This depended on the fact that $\tau_{t_0}\theta_\rho^{(4)}$ is concentrated at five points, and it is was easy to avoid these points for the discrete values $\rho = 2\pi 2^{-j}$, $j \in \mathbb{Z}$. Note that this cannot happen if we consider all values of ρ, which means that we do not have a counterexample for the continuous case. In fact, it is clear from (C.23) that the five points $t_0(\rho) - 2\rho$, $t_0(\rho) - \rho$, $t_0(\rho)$, $t_0(\rho) + \rho$, $t_0(\rho) + 2\rho$ sweep out the whole real line $\mathbb{R}$ as ρ traverses the real axis, and thus there is no place to "hide" the support of a perturbation R.

Assuming that we have a kernel η satisfying the conditions indicated above, there is no problem in having the support of R miss the support of η_ρ for small ρ. The problem arises when ρ is large. For example, if the support $\tau_{t_0}\eta_\rho^{(4)}$ includes all of $[t_0 - \rho T, t_0 + \rho T]$, then this support will ultimately cover all of R, and again there is no place to hide a perturbation. This is the case, for example, if $\eta = f_0$, the Tukey kernel. Having said this, it is conceivable that there are other combinations of perturbations R and kernels η such that $R * \eta_{\rho_j}^{(4)}(t_0) = 0$ for some sequence $\rho_j \to +\infty$. In our example, we have attributed this to the fact that the supports of R and $\tau_t \theta_\rho^{(4)}$ do not intersect. This, however, is just a reflection of the fact that R was constructed as the third derivative of a function g with compact support, and hence that $\int t^n R(t)\,dt = 0$ for $0 \leq n \leq 3$.

By now it should be clear that the assumption $\rho = 2\pi 2^{-j}$ is not necessary for our construction. We could have used any sequence ρ_j such that $\rho_j \to +\infty$ as $j \to -\infty$ and $\rho_j \to 0$ as $j \to +\infty$. The essential point is that there are only a finite number of ρ_j, $j \in J$, that must be checked "by hand." In fact, it is not necessary to do the computations, as we have done. One can argue as follows: For each $j \in J$, the function $(f_0 * \theta_{\rho_j})''(t)$ has only a finite number of isolated zeros. This follows from the fact that $F_0 * \theta_{\rho_j}^{(4)}$ is an entire function on each subinterval on which it has an analytic form, and an entire function has only finitely many zeros on a compact interval. Ensuring that $(R * \theta_{\rho_j})''$ and $(R * \theta_{\rho_j})'$ vanish at these zeros amounts to writing finitely many linear equations of the form $l_1(R) = 0, \ldots, l_N(R) = 0$. Since the vector space of our perturbations R is infinite dimensional, there are infinitely many nontrivial R that satisfy the conditions.

Ensuring that the zeros of $(f_0 * \theta_\rho)''$ are zeros of $(R * \theta_\rho)''$ and $(R * \theta_\rho)'$ is only part of the problem. The other part is to guarantee that the perturbation R does not introduce new zeros. In our example, we were able to bound

$$\frac{(R * \theta_\rho)''(t)}{(f_0 * \theta_\rho)''(t)}$$

uniformly for large ρ and for small ρ, and we were faced with only a finite number of intermediate values. In this part of the argument, the size of R'', or the fifth derivative of h, enters the picture. In fact, it is not difficult to see that a large value of $g^{(5)}(t)$ can introduce new zeros in the cases where ρ is small. On the other hand, we see from (C.22) that controlling g' is sufficient when ρ is large. But since g has compact support, we have $\sup |g'(t)| \leq (\frac{\pi}{8})^4 \sup |g^{(5)}(t)|$, and it is fairly clear that the counterexample depends on having $|g^{(5)}(t)|$ sufficiently small. We have not, however, carried the analysis to the point where we can say exactly how small $|g^{(5)}(t)|$ must be.

C.9 A case of perfect reconstruction

We mentioned in Chapter 8 that perfect reconstruction is possible if the analyzed function f has compact support and if the kernel θ is the Tukey window. Here is the precise statement and proof of that case of perfect reconstruction.

THEOREM C.1. *Assume that f is a real-valued function in $L^1(\mathbb{R})$ with compact support. If θ is the kernel*

$$\theta(t) = \begin{cases} 1 + \cos t & \text{for } |t| \leq \pi, \\ 0 & \text{for } |t| > \pi, \end{cases}$$

then f is uniquely determined by knowing the location of the zeros of $(f * \theta_{\rho_j})''$ and the values of $(f * \theta_{\rho_j})'$ at these zeros for any sequence ρ_j such that $\rho_j \to +\infty$ as $j \to +\infty$.

Proof. The proof depends on the fact that the Fourier transform $\hat{f}$ is the restriction to the real line of the entire function $\hat{f}(z) = \int f(x) e^{-zx}\, dx$, and thus that $\hat{f}$ is uniquely determined by the values of $\hat{f}$ at a sequence of points $\frac{1}{\rho_j}$ that tends to zero as $j \to +\infty$. We are going to assume that we know a value of $R > 0$ such that $f(t) = 0$ for $|t| \geq R$.

The first step is to compute $(f * \theta_\rho)''(t)$ and $(f * \theta_\rho)'(t)$, and for this we assume that $|t| \leq \pi\rho - R$. For these values of t, we have

$$(f * \theta_\rho)''(t) = -\rho^{-3} \int_{-\pi\rho}^{\pi\rho} f(t-s) \cos(\rho^{-1} s)\, ds$$

$$= -\rho^{-3} \int_{-\infty}^{\infty} f(t-s) \cos(\rho^{-1} s)\, ds$$

$$= -\frac{\rho^{-3}}{2} \left[e^{it\rho^{-1}} \hat{f}(\rho^{-1}) + e^{-it\rho^{-1}} \hat{f}(-\rho^{-1}) \right]. \tag{C.24}$$

Write $\hat{f}(x) = A(x) e^{i\alpha(x)}$ with the conventions that

$$A(x) \geq 0, \tag{C.25}$$

$$-\pi < \alpha(x) \leq \pi. \tag{C.26}$$

Since f is real-valued, $\hat{f}(-x) = \overline{\hat{f}(x)} = A(x) e^{-i\alpha(x)}$, and thus from (C.24) we have

$$(f * \theta_\rho)''(t) = -\frac{1}{\rho^3} A(\rho^{-1}) \cos\left[t\rho^{-1} + \alpha(\rho^{-1}) \right]. \tag{C.27}$$

A similar computation shows that

$$(f * \theta_\rho)'(t) = -\frac{1}{\rho^2} A(\rho^{-1}) \sin\left[t\rho^{-1} + \alpha(\rho^{-1}) \right]. \tag{C.28}$$

These expressions hold for $t \in [-\pi\rho + R, \pi\rho - R]$. In particular, they hold in the interval $I_\rho = [-\frac{\pi}{2}\rho, \frac{\pi}{2}\rho)$ whenever $\rho \geq \frac{2}{\pi}R$. The next point is central to the argument, and we state it as a lemma.

LEMMA C.3. *Either $(f * \theta_\rho)''$ vanishes identically on I_ρ or $(f * \theta_\rho)''$ has exactly one zero, $t = t_\rho$, in I_ρ. In the latter case, $t_\rho \rho^{-1} + \alpha(\rho^{-1}) = \pm\frac{\pi}{2}$.*

It is clear from (C.27) that $(f * \theta_\rho)''$ vanishes identically on I_ρ if and only if $A(\rho^{-1}) = 0$. Thus, assume that $A(\rho^{-1}) \neq 0$. We consider the linear function $l(t) = t\rho^{-1} + \alpha(\rho^{-1})$. l maps I_ρ into the half-open interval

$$l(I_\rho) = \left[-\frac{\pi}{2} + \alpha(\rho^{-1}), \frac{\pi}{2} + \alpha(\rho^{-1})\right). \tag{C.29}$$

$l(I_\rho)$ contains one, and only one, number of the form $\frac{\pi}{2} + k\pi$, $k \in \mathbb{Z}$—irrespective of the value of $\alpha(\rho^{-1})$. Furthermore, since we require that $\alpha(x)$ be contained in $(-\pi, \pi]$, the only points of this form that appear in $l(I_\rho)$ are $\pm\frac{\pi}{2}$. This proves the lemma.

One part of the hypothesis is that we "know" the zeros of $(f * \theta_\rho)''$. Thus, given ρ (sufficiently large) we know if $(f * \theta_\rho)''$ vanishes identically on I_ρ or not. If it vanishes, then $A(\rho^{-1}) = 0$, and we know that $\hat{f}(\rho^{-1}) = 0$. If $(f * \theta_\rho)''$ does not vanish identically on I_ρ, then we know from the lemma that there is exactly one $t_\rho \in I_\rho$ where $(f * \theta_\rho)''(t_\rho) = 0$ and that

$$t_\rho \rho^{-1} + \alpha(\rho^{-1}) = \pm\frac{\pi}{2}. \tag{C.30}$$

The other part of the hypothesis is that we know the value of $(f * \theta_\rho)'(t_\rho)$. In particular, we know if $(f * \theta_\rho)'(t_\rho) > 0$ or if $(f * \theta_\rho)'(t_\rho) < 0$.

If $(f * \theta_\rho)'(t_\rho) > 0$, then we know from (C.28) that

$$t_\rho \rho^{-1} + \alpha(\rho^{-1}) = -\frac{\pi}{2}.$$

Similarly, if $(f * \theta_\rho)'(t_\rho) < 0$, then we know that

$$t_\rho \rho^{-1} + \alpha(\rho^{-1}) = +\frac{\pi}{2}.$$

Thus we know unambiguously that

$$\alpha(\rho^{-1}) = -\left[\operatorname{sign}(f * \theta_\rho)'(t_\rho)\right]\frac{\pi}{2} - t_\rho \rho^{-1} \tag{C.31}$$

and that

$$A(\rho^{-1}) = \rho^2 |(f * \theta_\rho)'(t_\rho)|. \tag{C.32}$$

This completes the proof, but it is perhaps useful to state what we have done as an algorithm.

Assume that we are given a sequence of positive numbers $\{\rho_j\}_{j \in \mathbb{N}}$ such that $\rho_{j+1} > \rho_j$ and $\rho_j \to +\infty$ as $j \to +\infty$. The algorithm reads as follows:

Step 1: For all sufficiently large ρ_j ($\rho_j \geq \frac{2}{\pi}R$ if we know R), examine the zeros of $(f * \theta_{\rho_j})''$ in the interval $[-\frac{\pi}{2}\rho_j, \frac{\pi}{2}\rho_j)$. If $(f * \theta_{\rho_j})''$ vanishes identically on this interval, then $\hat{f}(\rho_j^{-1}) = 0$, and we know the value of $\hat{f}(\rho_j^{-1})$. In this case, go to the next value of j and repeat Step 1. If $(f * \theta_{\rho_j})''$ does not vanish identically on $[-\frac{\pi}{2}\rho_j, \frac{\pi}{2}\rho_j)$, go to Step 2.

Step 2: Denote the unique zero of $(f * \theta_{\rho_j})''$ in the interval $[-\frac{\pi}{2}\rho_j, \frac{\pi}{2}\rho_j)$ by t_j. If $(f * \theta_{\rho_j})'(t_j) > 0$, then $\alpha(\rho_j^{-1}) = -\frac{\pi}{2} - t_j\rho_j^{-1}$ and $A(\rho_j^{-1}) = \rho_j^2 (f * \theta_{\rho_j})'(t_j)$, in which case we know $\hat{f}(\rho_j^{-1})$. If $(f * \theta_{\rho_j})'(t_j) < 0$, then $\alpha(\rho_j^{-1}) = \frac{\pi}{2} - t_j\rho_j^{-1}$ and $A(\rho_j^{-1}) = \rho_j^2 |(f * \theta_{\rho_j})'(t_j)|$, and again we know the value of $\hat{f}(\rho_j^{-1})$. Go to the next value of j and return to Step 1.

This algorithm produces the sequence $\{\hat{f}(\rho_j^{-1})\}_{j \in \mathbb{N}}$, and this sequence determines the entire function $\hat{f}(z) = \int f(x) e^{-zx} dx$ in the following sense: If g_1 and g_2 are entire functions and if $g_1(\rho_j^{-1}) = g_2(\rho_j^{-1})$ for infinitely many $j \in \mathbb{N}$, then $g_1 = g_2$. There is another way in which the $\hat{f}(\rho_j^{-1})$ determine $\hat{f}$: If $\hat{f}(z) = \sum_{n=0}^{\infty} a_n z^n$, then the coefficients a_n can be computed inductively by the relation

$$a_N = \lim_{j \to \infty} \frac{\hat{f}(\rho_j^{-1}) - \sum_{n=0}^{N-1} a_n \rho_j^{-n}}{\rho_j^{-N}}.$$

Finally, since the Fourier transform $f \mapsto \hat{f}$ is one-to-one, f is uniquely determined by $\hat{f}$. □

Strict constructionists may find these "determinations" or "reconstructions" less than satisfactory, and, indeed, as it is stated, all we have is a uniqueness theorem.

Mallat's conjecture is proved for this specific window. We note, however, that large values of ρ played a key role, since we needed to have information for an infinite number of points ρ_j that tend to infinity. It would be interesting to know if having the same information for a sequence ρ_j that tends to zero would also guarantee uniqueness.

APPENDIX D

Hölder Spaces and Besov Spaces

This appendix contains the definitions and some fundamental results about Besov spaces and related spaces that are mentioned in Chapter 9 and again prominently in Chapter 11.

D.1 Hölder spaces

We begin by defining the homogeneous Hölder spaces because they are simple and they lead naturally to the Besov spaces.

For a given α, $0 < \alpha < 1$, $\dot{C}^\alpha(\mathbb{R}^n)$ is defined to be the set of all continuous functions f such that

$$|f(y) - f(x)| \leq C|y - x|^\alpha \quad \text{for all } x, y \in \mathbb{R}^n.$$

If we let

$$\|f\|_{\dot{C}^\alpha} = \sup_{x,y \in \mathbb{R}} \frac{|f(y) - f(x)|}{|y - x|^\alpha},$$

then $\|\cdot\|_{\dot{C}^\alpha}$ is a norm and $\dot{C}^\alpha(\mathbb{R}^n)$ is a Banach space in this norm, modulo the constant functions.

This definition can be reformulated using the modulus of continuity $\omega_\infty(f, h)$, which is defined as follows:

$$\omega_\infty(f, h) = \sup_{\substack{x \in \mathbb{R}^n \\ |y| \leq h}} |f(x + y) - f(x)|.$$

Then f belongs to $\dot{C}^\alpha(\mathbb{R}^n)$ if and only if $\omega_\infty(f, h) \leq Ch^\alpha$. It is easy to see that

$$\sup_{h > 0} \frac{\omega_\infty(f, h)}{h^\alpha} = \|f\|_{\dot{C}^\alpha}.$$

If $1 \leq \alpha < 2$, the definition is similar, but $[\Delta_y f](x) = f(x+y) - f(x)$ is replaced by $[\Delta_y^2 f](x) = f(x + 2y) - 2f(x + y) + f(x)$. The space $\dot{C}^\alpha(\mathbb{R}^n)$ is again defined by the condition $\omega_\infty(f, h) \leq Ch^\alpha$. It is a Banach space, but now the elements are modulo the affine functions. For $N \leq \alpha < N + 1$, $[\Delta_y f](x)$ is replaced by the iterated difference $[\Delta_y^{N+1} f](x)$. $\dot{C}^\alpha(\mathbb{R}^n)$ is then a Banach space of functions modulo polynomials P_N, where $\deg P_N \leq N$.

The spaces $\dot{C}^\alpha(\mathbb{R}^n)$ (with the dot) are said to be homogeneous for the following reason: If $0 < \lambda < \infty$ and $f_\lambda(x) = f(\lambda x)$, then $\|f_\lambda\|_{\dot{C}^\alpha} = \lambda^\alpha \|f\|_{\dot{C}^\alpha}$. The non-homogeneous Hölder spaces $C^\alpha(\mathbb{R}^n)$ (without the dot) are defined by the relation

$C^\alpha(\mathbb{R}^n) = \dot{C}^\alpha(\mathbb{R}^n) \cap L^\infty(\mathbb{R}^n)$. To be more precise, we should say that a function f is in $C^\alpha(\mathbb{R}^n)$ if and only if it is a bounded representative of an element of $\dot{C}^\alpha(\mathbb{R}^n)$. The norm of $f \in C^\alpha(\mathbb{R}^n)$ is defined by $\|f\|_{C^\alpha} = \|f\|_\infty + \|f\|_{\dot{C}^\alpha}$, and $C^\alpha(\mathbb{R}^n)$ is a Banach space in this norm, but the norm does not satisfy the homogeneous property. If the spaces are defined on a compact subset $K \subset \mathbb{R}^n$, then $C^\alpha(K) = \dot{C}^\alpha(K)$, and there is no distinction. In this case, f_λ does not make sense, and the homogeneous property is lost.

Both $C^\alpha(\mathbb{R}^n)$ and $\dot{C}^\alpha(\mathbb{R}^n)$ have advantages in analysis. The advantage of $C^\alpha(\mathbb{R}^n)$ is that it is an algebra: If $f, g \in C^\alpha(\mathbb{R}^n)$, then $fg \in C^\alpha(\mathbb{R}^n)$. This is not true for $\dot{C}^\alpha(\mathbb{R}^n)$. On the other hand, there are examples where the large-scale behavior (or complete self-similarity) is important, and where it is necessary to admit functions that are unbounded at infinity. Examples include fractional Brownian motion or, more generally, $\frac{1}{f}$ processes. In other situations, it is only necessary to focus on small scales.

D.2 Besov spaces

We will move from the homogeneous Hölder spaces $\dot{C}^\alpha(\mathbb{R}^n)$ to the homogeneous Besov spaces $\dot{B}^{\alpha,q}(L^p)(\mathbb{R}^n)$ (which are often denoted by $\dot{B}_p^{\alpha,q}(\mathbb{R}^n)$) in two steps: The modulus of continuity $\omega_\infty(f, h)$ is replaced by

$$\omega_p(f, h) = \sup_{|y| \leq h} \|f(x + y) - f(x)\|_{L^p(\mathbb{R}^n)},$$

and the condition $\omega_\infty(f, h) \leq Ch^\alpha$ is replaced by $\omega_p(f, h) \leq \varepsilon(h) h^\alpha$, where $\varepsilon(h)$ must satisfy the condition

$$\left(\int_0^\infty \varepsilon^q(h) \frac{dh}{h} \right)^{1/q} < \infty.$$

The norm of $f \in \dot{B}^{\alpha,q}(L^p)$ is naturally defined as

$$\|f\|_{\dot{B}_p^{\alpha,q}} = \left(\int_0^\infty \frac{\omega_p^q(f, h)}{h^{\alpha q}} \frac{dh}{h} \right)^{1/q},$$

and this norm is homogeneous with $\|f_\lambda\|_{\dot{B}_p^{\alpha,q}} = \lambda^{\alpha - \frac{n}{p}} \|f\|_{\dot{B}_p^{\alpha,q}}$.

This new definition is for the case $1 \leq p \leq \infty$, $1 \leq q \leq \infty$, and $0 < \alpha < 1$. For $N \leq \alpha < N + 1$, $n \geq 1$, it is necessary to replace $[\Delta_y f](x)$ by $[\Delta_y^{N+1} f](x)$. If $1 \leq p \leq \infty$, $1 \leq q \leq \infty$, and $0 < \alpha < \infty$, then $\dot{B}_p^{\alpha,q}(\mathbb{R}^n)$ is a Banach space of functions modulo polynomials of degree N, where N is the largest integer in $\alpha - \frac{n}{p}$. (There are no polynomials when $\alpha < \frac{n}{p}$.)

In parallel with what was done for Hölder spaces, the nonhomogeneous Besov space $B_p^{\alpha,q}(\mathbb{R}^n)$ is by definition the intersection $\dot{B}_p^{\alpha,q}(\mathbb{R}^n) \cap L^p(\mathbb{R}^n)$, and

$$\|f\|_{B_p^{\alpha,q}} = \|f\|_{L^p} + \|f\|_{\dot{B}_p^{\alpha,q}}.$$

Besov spaces are easily characterized by size estimates on wavelet coefficients. Let $\psi_{(i,j,k)}(x) = 2^{nj/2} \psi_i(2^j x - k)$, $j \in \mathbb{Z}$, $k \in \mathbb{Z}^n$, $i = 1, \ldots, 2^n - 1$, be an orthonormal wavelet basis for $L^2(\mathbb{R}^n)$, where each ψ_i belongs to the Schwartz class $\mathcal{S}(\mathbb{R}^n)$. Then all of the moments of the wavelets ψ_i vanish, and hence $\int P(x) \psi(x) \, dx = 0$ for all polynomials P. As is customary, we simplify the notation by dropping the index i. We also change the normalization of the wavelet coefficients of f and write

$$c(j,k) = 2^{nj} \int_{\mathbb{R}^n} f(x)\overline{\psi}(2^j x - k)\,dx. \tag{D.1}$$

The integral in (D.1) is unambiguously defined for elements of the Besov space $\dot{B}^{\alpha,q}(L^p)$, since any two representatives of f and g of the same element differ by a polynomial of degree less than or equal to $\alpha - \frac{n}{p}$. With these conventions, we have the following result.

THEOREM D.1. *If f belongs to $\dot{B}^{\alpha,q}(L^p)$, then the sequence ε_j defined by*

$$\left(\sum_{k\in\mathbb{Z}^n} |c(j,k)|^p\right)^{1/p} = 2^{-j(\alpha-\frac{n}{p})}\varepsilon_j \tag{D.2}$$

belongs to $l^q(\mathbb{Z})$. Conversely, if the wavelet coefficients of a function f satisfy this condition, then $f = g + P$, where $g \in \dot{B}^{\alpha,q}(L^p)$ and P is polynomial. (There is no restriction on the degree of P.)

A simplification occurs when $\alpha = \frac{n}{p}$ and $p = q$. In this case, condition (D.2) becomes

$$\sum_{j\in\mathbb{Z}}\sum_{k\in\mathbb{Z}^n} |c(j,k)|^p < \infty, \tag{D.3}$$

and the Besov space $\dot{B}_p^{n/p,p}(\mathbb{R}^n)$ is isomorphic to $l^p(\mathbb{Z})$.

When $0 < p < 1$, $p = q$, and $\alpha = \frac{n}{p}$, the corresponding Besov space can be defined either by (D.3) or by the following growth property of the dyadic blocks $\Delta_j(f)$ that occur in the Littlewood–Paley expansion of f. This growth property reads

$$\|\Delta_j(f)\|_p \leq \varepsilon_j 2^{-\alpha j}, \qquad \varepsilon_j \in l^q(\mathbb{Z}), \tag{D.4}$$

and it characterizes the Besov space $\dot{B}_p^{\alpha,q}(\mathbb{R}^n)$. In the particular case $p = q$ and $\alpha = \frac{n}{p}$, condition (D.4) becomes

$$\sum_{j=-\infty}^{\infty} 2^{nj}\|\Delta_j(f)\|_p^p < \infty. \tag{D.5}$$

The wavelet coefficients $c(j,k)$ of f can be interpreted as a sample of $\Delta_j(f)$ on the grid $2^{-j}\mathbb{Z}^n$, and this heuristic leads to replacing $\|\Delta_j(f)\|_p^p$ by the Riemann sum $\sum_{k\in\mathbb{Z}^n} |c(j,k)|^p 2^{-nj}$. By carrying out this program, one can show that (D.5) is equivalent to $\sum_{j\in\mathbb{Z}}\sum_{k\in\mathbb{Z}^n} |c(j,k)|^p < \infty$. The details can be found in [203].

D.3 Examples

We are going to illustrate the use of Besov spaces for modeling and denoising with a textbook example. The signal we wish to denoise is written as the sum of two terms $s(x) = \theta(x) + \lambda g(x)\cos(\omega x)$, where the signal θ is the characteristic function of an interval (a,b), the noise is a modulated Gaussian $g_\omega(x) = g(x)\cos(\omega x)$ (ω large), and the coefficient λ is a small parameter. Clearly, this noise is an academic simplification, but the following discussion also applies to more realistic situations.

We will try to extract the signal by using a regularity criterion. From the usual point of view, the signal θ is not a regular function, whereas the noise g_ω is infinitely differentiable. In trying to extract the signal θ from the sum $\theta(x) + \lambda g(x)\cos(\omega x)$, which represents the noisy signal, we have a problem where the "good" function is irregular and the "bad" function is regular. We will see that the judicious use of Besov spaces lets us reverse the order of things.

From the point of view of the Sobolev spaces H^s, the *relative* regularity of θ, compared with that of the product $g(x)\cos(\omega x)$, increases with ω. If, for example, the exponent s of the Sobolev space is less than $\frac{1}{2}$, the Sobolev norm of θ is bounded while that of $g(x)\cos(\omega x)$ tends to infinity like ω^s as $\omega \to \infty$. In this sense, the Sobolev norm knows how to distinguish edges from textures. This contrast is even greater if one uses the Besov spaces $B^{s,q}(L^p)$. In fact, if $0 < p \le 1$, $s = \frac{1}{p}$, and $q = \infty$, then θ belongs to the corresponding Besov space while the norm of $g(x)\cos(\omega x)$ is of the order $\omega^{1/p}$. (With Sobolev spaces, the best one can do is $\omega^{1/2-\varepsilon}$.) This means that if λ is small enough and ω is large enough, one can extract the signal θ from the noisy signal $\theta(x) + \lambda g(x)\cos(\omega x)$ using a criterion based on the optimization of a Besov norm. In this case, one is using the Besov space $B_p^{1/p,\infty}$, and the closer p is to zero, the sharper is the discrimination between the main term $\theta(x)$ and the error term $\lambda g(x)\cos(\omega x)$. We note that the jump discontinuities of $\theta(x)$ do not prevent it from belonging to the Besov space $B_p^{1/p,\infty}$. We see then how this approach is preferable to low-pass filtering, which would indeed eliminate $\lambda g(x)\cos(\omega x)$ but at the same time would blur the edges of $\theta(x)$.

A similar example in two dimensions is given by $f = u + v$, where u is the characteristic function of the unit disc and v is a Gaussian white noise. Measured in the Besov norm $B_p^{s,q}$, where $s < \frac{1}{p}$, the function u has a relatively small norm while the norm of v is infinite whenever $s > -1$. The discrepancy between u and v becomes more apparent as p tends to zero.

This leads us to a program that discriminates the edges from the textures (and from the noise) by using a criterion given by the Besov norm and that requires the final result to have a "small" Besov norm. This is indeed the viewpoint adopted by Donoho. For Donoho, the a priori knowledge is modeled by membership in certain Besov spaces.

Bibliography

Numbers in brackets following an entry indicate the pages where the reference is cited. Several books and articles, which are not cited in the text, have also been listed.

[1] P. ABRY, *Ondelettes et turbulences: Multiresolutions, algorithms de décomposition, invariance d'échelle et signaux de pression*, Diderot Editeur, Arts et Sciences, Paris, 1997. [140]

[2] P. ABRY AND F. SELLAN, *The wavelet-based synthesis for the fractional Brownian motion proposed by F. Sellan and Y. Meyer: Remarks and fast implementation*, Appl. Comput. Harmon. Anal., 3 (1996), pp. 377–383. [21, 65, 181]

[3] E. H. ADELSON, E. SIMONCELLI, AND R. HINGORANI, *Orthogonal pyramid transforms for image coding*, in Proc. SPIE Conf. Visual Comm. Image Process. II, vol. 845, 1987, pp. 50–58. Reprinted in *Selected Papers in Image Coding and Compression*, M. Rabbani, ed., SPIE Milestone Series, SPIE Press, Bellingham, WA, pp. 331–339, 1992. [36, 49]

[4] F. ANSCOMBE, *The transformation of Poisson, binomial and negative-binomial data*, Biometrika, 35 (1948), pp. 246–254. [193, 195]

[5] F. ANSELMET, Y. GAGNE, E. J. HOPFINGER, AND R. A. ANTONIA, *High-order velocity structure functions in turbulent shear flow*, J. Fluid Mech., 140 (1984), pp. 63–89. [130]

[6] A. ANTONIADIS AND G. OPPENHEIM, EDS., *Wavelets and Statistics*, Lecture Notes in Statistics 103, Springer-Verlag, New York, 1995.

[7] A. ARNEODO, F. ARGOUL, E. BACRY, J. ELEZGARAY, AND J.-F. MUZY, *Ondelettes, multifractals et turbulences: De l'ADN aux croissances cristallines*, Diderot Editor, Arts et Sciences, Paris, 1995. [55, 135, 200]

[8] A. ARNEODO, B. AUDIT, E. BACRY, S. MANNEVILLE, J.-F. MUZY, AND S. G. ROUX, *Thermodynamics of fractal signals based on wavelet analysis; applications to fully developed turbulence data and DNA sequences*, Phys. A, 254 (1998), pp. 24–45. [137]

[9] A. ARNEODO, E. BACRY, S. JAFFARD, AND J.-F. MUZY, *Oscillating singularities on Cantor sets. A grand-canonical multifractal formalism*, J. Statist. Phys., 87 (1997), pp. 179–209. [142]

[10] A. ARNEODO, E. BACRY, AND J.-F. MUZY, *The thermodynamics of fractals revisited with wavelets*, Phys. A, 213 (1995), pp. 232–275. [140]

[11] ———, *Random cascades on wavelet dyadic trees*, J. Math. Phys., 39 (1998), pp. 4142–4164. [136]

[12] A. ARNEODO, Y. D'AUBENTON CARAFA, B. AUDIT, E. BACRY, J.-F. MUZY, AND C. THERMES, *What can we learn with wavelets about DNA sequences?*, Phys. A, 249 (1998), pp. 439–448. [137]

[13] A. ARNEODO, Y. D'AUBENTON CARAFA, E. BACRY, P. V. GRAVES, J.-F. MUZY, AND C. THERMES, *Wavelet based fractal analysis of DNA sequences*, Phys. D, 96 (1996), pp. 291–320. [137]

[14] J.-M. AUBRY, *Traces of oscillating functions*, J. Fourier Anal. Appl., 5 (1999), pp. 331–345. [143]

[15] E. BACRY, A. ARNEODO, U. FRISCH, Y. GAGNE, AND E. HOPFINGER, *Wavelet analysis of fully developed turbulence data and measurement of scaling exponents*, in Turbulence and Coherent Structures, O. Métais and M. Lesieur, eds., Kluwer Academic Press, Norwell, MA, 1991, pp. 203–215. [130, 139]

[16] E. BACRY, J.-F. MUZY, AND A. ARNEODO, *Singularity spectrum of fractal signals from wavelet analysis: Exact results*, J. Statist. Phys., 70 (1993), pp. 635–674. [134, 135, 140]

[17] R. BALIAN, *Un principe d'incertitude fort en théorie du signal ou en mécanique quantique*, C. R. Acad. Sci. Paris Sér. II, 292 (1981), pp. 1357–1361. [9, 90, 67]

[18] R. G. BARANIUK AND D. L. JONES, *New dimensions in wavelet analysis*, in Proc. IEEE Internat. Conf. Acoust. Speech Signal Process., IEEE Press, Piscataway, NJ, 1992. [100]

[19] ———, *New signal-space orthonormal bases via the metaplectic transform*, in IEEE-SP Internat. Symposium on Time-Frequency and Time-Scale Analysis, IEEE Press, Piscataway, NJ, 1992. [100]

[20] ———, *Unitary equivalence: A new twist on signal processing*, IEEE Trans. Signal Process., 43 (1995), pp. 2269–2282. [100]

[21] ———, *Wigner-based formulation of the chirplet transform*, IEEE Trans. Signal Process., 44 (1996), pp. 3129–3135. [100]

[22] J. BARRAL, *Moments, continuité et analyse multifractale des martingales de Mandelbrot*, Probab. Theory Related Fields, 113 (1999), pp. 535–570. [131]

[23] M. BASSEVILLE, A. BENVENISTE, K. CHOU, S. GOLDEN, R. NIKOUKHAH, AND A. WILLSKY, *Modeling and estimation of multiresolution stochastic processes*, Special issue of IEEE Trans. Inform. Theory on Wavelet Transforms and Multiresolution Signal Analysis, 38 (1992), pp. 766–784. [133]

[24] M. BASSEVILLE, A. BENVENISTE, AND A. WILLSKY, *Multiscale autoregressive processes, part I: Schur–Levinson parametrizations*, IEEE Trans. Signal Process., 40 (1992), pp. 1915–1934. [133]

[25] ———, *Multiscale autoregressive processes, part II: Lattice structures for whitening and modeling*, IEEE Trans. Signal Process., 40 (1992), pp. 1935–1954. [133]

[26] G. BATCHELOR AND A. A. TOWNSEND, *The nature of turbulent motion at large wave numbers*, Proc. Roy. Soc. London Ser. A, 199 (1949), pp. 238–255. [130]

[27] G. BATTLE, *A block spin construction of ondelettes, part II: The QFT connection*, Comm. Math. Phys., 114 (1988), pp. 93–102. [32]

[28] ———, *Wavelet refinement of the Wilson recursion formula*, in Recent Advances in Wavelet Analysis, L. Schumaker and G. Webb, eds., Academic Press, Norwell, MA, 1994, pp. 87–118. [32]

[29] G. BATTLE AND P. FEDERBUSH, *Ondelettes and phase cluster expansions, a vindication*, Comm. Math. Phys., 109 (1987), pp. 417–419. [32]

[30] ———, *Divergence-free vector wavelets*, Michigan Math. J., 40 (1993), pp. 181–195. [145]

[31] A. BENASSI, S. JAFFARD, AND D. ROUX, *Analyse multi-échelle des champs gaussiens markoviens d'ordre p indexés par [0, 1]*, C. R. Acad. Sci. Paris Sér. I, (1991), pp. 403–406. [21]

[32] P. BENDJOYA, E. SLEZAK, AND C. FROESCHLÈ, *The wavelet transform, a new tool for asteroid family determination*, Astronom. Astrophys., 251 (1991), pp. 312–330.

[33] J. J. BENEDETTO AND M. W. FRAZIER, EDS., *Wavelets: Mathematics and Applications*, CRC Press, Boca Raton, FL, 1993.

[34] J. BERGER, R. R. COIFMAN, AND M. J. GOLDBERG, *Removing noise from music using local trigonometric bases and wavelet packets*, J. Audio Eng. Soc., 42 (1994), pp. 808–818. [72, 104, 105]

[35] J. BERTOIN, *The inviscid Burgers equation with Brownian initial velocity*, Comm. Math. Phys., 193 (1998), pp. 397–406. [165]

[36] A. BIJAOUI, *Wavelets and astrophysical applications*, in Wavelets in Physics, H. C. van den Berg, ed., Cambridge Univ. Press, Cambridge, U. K., 1997, pp. 77–115. [187, 196, 199, 201]

[37] R. E. BLAHUT, W. MILLER JR, AND C. H. WILCOX, EDS., *Radar and Sonar, Part I*, Springer-Verlag, New York, 1991. [74]

[38] Y. BOBICHON AND A. BIJAOUI, *A regularized image restoration algorithm for lossy compression in astronomy*, Experiment. Astronom., 7 (1997), pp. 239–255. [194, 196, 197]

[39] K. BOUYOUCEF, D. FRAIX-BURNET, AND S. ROQUES, *Interactive deconvolution with error analysis (IDEA) in astronomical imaging: Application to aberrated HST images on SN1987A, M87 and 3C66B*, Astronom. Astrophys. Suppl. Ser., 121 (1997), pp. 575–585. [193]

[40] L. BRILLOUIN, *Science and Information Theory*, Academic Press, New York, 1956. [9]

[41] C. M. BRISLAWN, *Fingerprints go digital*, Notices Amer. Math. Soc., 42 (1995), pp. 1278–1283. [6, 70, 97, 194]

[42] P. J. BURT AND E. H. ADELSON, *The Laplacian pyramid as a compact image code*, IEEE Trans. Comm., 31 (1983), pp. 532–540. [50]

[43] P. L. BUTZER AND E. L. STARK, *"Riemann's example" of a continuous nondifferentiable function in the light of two letters (1865) of Christoffel to Prym*, Bull. Soc. Math. Belg., 38 (1986), pp. 45–73. [150]

[44] J. S. BYRNES, *Quadrature mirror filters, low crest factor arrays, functions achieving optimal uncertainty principle bounds, and complete orthonormal sequences—A unified approach*, Appl. Comput. Harmon. Anal., 1 (1994), pp. 261–266. [111]

[45] M. CANNONE, *Ondelettes, paraproduits et Navier–Stokes*, Diderot Editeur, Arts et Sciences, Paris, 1995. [147]

[46] R. CARMONA, W.-L. HWANG, AND B. TORRÉSANI, *Practical Time-Frequency Analysis*, vol. 9 of Wavelet Analysis and Its Applications, Academic Press, San Diego, CA, 1998.

[47] B. CASTAING AND B. DUBRULLE, *Fully developed turbulence: A unifying point of view*, J. Physique II (Paris), 5 (1995), p. 895. [131]

[48] C. V. L. CHARLIER, *How an infinite world may be built up*, Ark. Mat. Astron. Fys., 16 (1922), pp. 1–34. [198]

[49] E. CHASSANDE-MOTTIN AND P. FLANDRIN, *On the time-frequency detection of chirps*, Appl. Comput. Harmon. Anal., 6 (1999), pp. 252–281. [80, 103]

[50] J.-Y. CHEMIN, *Calcul paradifférentiel précisé et application à des équations aux dérivées partielles non semi-linéaires*, Duke Math. J., 56 (1988), pp. 431–469. [147]

[51] ———, *Persistance de structures géométriques dans les fluides incompressibles bidimensionnels*, Ann. Ecole Normale Supérieure, 26 (1993), pp. 1–26. [147]

[52] A. J. CHORIN AND J. E. MARSDEN, *A Mathematical Introduction to Fluid Mechanics*, Springer-Verlag, New York, 1979. [139]

[53] Z. CIESIELSKI, *Hölder conditions for realizations of Gaussian processes*, Trans. Amer. Math. Soc., 99 (1961), pp. 403–413. [21]

[54] ———, *Properties of the orthonormal Franklin system*, Studia Math., 23 (1963), pp. 141–157. [24]

[55] ———, *Properties of the orthonormal Franklin system*, II, Studia Math., 27 (1966), pp. 289–323. [24]

[56] A. COHEN, W. DAHMEN, AND R. DEVORE, *Adaptive wavelet methods for elliptic operator equations: Convergence rates*, Math. Comput., to appear. [147]

[57] A. COHEN, I. DAUBECHIES, AND J.-C. FEAUVEAU, *Biorthogonal bases of compactly supported wavelets*, Comm. Pure Appl. Math., 44 (1992), pp. 485–560. [63]

[58] A. COHEN, I. DAUBECHIES, AND P. VIAL, *Wavelets and fast wavelet transform on an interval*, Appl. Comput. Harmon. Anal., 1 (1993), pp. 54–81. [180, 192]

[59] A. COHEN, R. DEVORE, P. PETRUSHEV, AND H. XU, *Nonlinear approximation and the space $BV(\mathbb{R}^2)$*, Amer. J. Math., 121 (1999), pp. 587–628. [177, 183]

[60] A. COHEN AND R. D. RYAN, *Wavelets and Multiscale Signal Processing*, Chapman & Hall, London, 1995. [45, 47, 55, 62, 208]

[61] R. R. COIFMAN, *Adapted multiresolution analysis, computation, signal processing and operator theory*, in Proc. Internat. Congr. Math., Kyoto, Japan, 1990, vol. II, Springer-Verlag, New York, 1991, pp. 879–887.

[62] R. R. COIFMAN AND D. DONOHO, *Translation-invariant de-noising*, in Wavelets in Statistics, A. Antoniadis and G. Oppenheim, eds., Springer-Verlag, New York, 1995, pp. 125–150. [99]

[63] R. R. COIFMAN, G. MATVIYENKO, AND Y. MEYER, *Modulated Malvar–Wilson bases*, Appl. Comput. Harmon. Anal., 4 (1997), pp. 58–61. [100]

[64] R. R. COIFMAN AND Y. MEYER, *Remarques sur l'analyse de Fourier à fenêtre*, C. R. Acad. Sci. Paris Sér. I Math., 312 (1991), pp. 259–261. [92]

[65] R. R. COIFMAN, Y. MEYER, AND V. WICKERHAUSER, *Size properties of wavelet packets*, in Wavelets and Their Applications, M. B. Ruskai et al., eds., Jones and Bartlett, Boston, MA, 1992, pp. 453–470. [108, 109, 111]

[66] R. R. COIFMANN, Y. MEYER, S. QUAKE, AND V. WICKERHAUSER, *Signal processing and compression with wavelet packets*, in Progress in Wavelet Analysis and Applications, Y. Meyer and S. Roques, eds., Editions Frontières, Gif-sur-Yvette, France, 1993, pp. 77–93.

[67] A. CORDOBA AND C. FEFFERMAN, *Wave packets and Fourier integral operators*, Comm. Partial Differential Equations, 3 (1978), pp. 979–1005. [87]

[68] A. CROISIER, D. ESTEBAN, AND C. GALAND, *Perfect channel splitting by use of interpolation/decimation/tree decomposition techniques*, in Internat. Conf. Inform. Sci. Systems, Patras, Greece, 1976, pp. 443–446. [35]

[69] M. DÆHLEN AND T. LYCHE, *Decomposition of splines*, in Mathematical Methods in CAGD and Image Processing, T. Lyche and L. Schumaker, eds., Academic Press, Boston, MA, 1992, pp. 135–160.

[70] K. DAOUDI, A. FRAKT, AND A. WILLSKY, *Multiscale autoregressive models and wavelets*, Special issue of IEEE Trans. Inform. Theory on Multiscale Statistical Signal Analysis and its Applications, 45 (1999), pp. 828–845. [133]

[71] I. DAUBECHIES, *Orthonormal bases of compactly supported wavelets*, Comm. Pure Appl. Math., 41 (1988), pp. 909–996. [9]

[72] ———, *The wavelet transform, time-frequency localization and signal analysis*, IEEE Trans. Inform. Theory, 36 (1990), pp. 961–1005. [106]

[73] ———, *Ten Lectures on Wavelets*, SIAM, Philadelphia, 1992. [9, 46]

[74] I. DAUBECHIES, S. JAFFARD, AND J.-L. JOURNÉ, *A simple Wilson orthonormal basis with exponential decay*, SIAM J. Math. Anal., 22 (1991), pp. 554–573. [91]

[75] G. DAVIS, S. G. MALLAT, AND M. AVELLANEDA, *Adaptive greedy approximations*, Constr. Approx., 13 (1997), pp. 57–98. [71]

[76] N. G. DE BRUIJN, *Uncertainty principals in Fourier analysis*, in Inequalities, O. Shisha, ed., Academic Press, New York, 1967, pp. 57–71. [87]

[77] J.-M. DELORT, *FBI-Transformation, Second Microlocalization and Semilinear Caustics*, Lecture Notes in Math. 1522, Springer-Verlag, New York, 1992. [87]

[78] R. DEVORE, *Adaptive wavelet bases for image compression*, in Curves and Surfaces in Geometric Design, P.-J. Laurent, A. Le Méhauté, and L. Schumaker, eds., A K Peters, Natick, MA, 1994, pp. 1–16.

[79] ———, *Nonlinear approximation*, Acta Numer., 7 (1998), pp. 51–150. [71]

[80] R. DEVORE, B. JAWERTH, AND B. LUCIER, *Surface compression*, Comput. Aided Geom. Design, 9 (1992), pp. 219–239. [175]

[81] R. DEVORE, B. JAWERTH, AND V. POPOV, *Compression of wavelet decompositions*, Amer. J. Math., 114 (1992), pp. 737–785. [169, 170, 174]

[82] R. DEVORE AND G. G. LORENTZ, *Constructive Approximation*, Springer-Verlag, New York, 1993.

[83] R. DEVORE AND B. LUCIER, *Fast wavelet techniques for near-optimal image processing*, in Proc. 1992 IEEE Military Comm. Conf., IEEE Press, Piscataway, NJ, 1992, pp. 1129–1135. [183]

[84] R. DEVORE, B. LUCIER, M. KALLERGI, W. QUIN, R. CLARK, E. SAFF, AND L. P. CLARKE, *Wavelet compression and segmentation of mammographic images*, J. Digital Imag., 7 (1994), pp. 27–38. [175]

[85] R. DEVORE, B. LUCIER, AND Z. YANG, *Feature extraction in digital mammography*, in Wavelets in Biology and Medicine, A. Aldroubi and M. Unser, eds., CRC Press, Boca Raton, FL, 1996, pp. 145–156. [175]

[86] R. DEVORE AND V. POPOV, *Interpolation spaces and non-linear approximation*, in Function Spaces and Applications, 1986, M. Cwikel et al., eds., Lecture Notes in Math. 1302, Springer-Verlag, New York, 1988. [172]

[87] R. DEVORE, Z. YANG, M. KALLERGI, B. LUCIER, W. QIAN, R. CLARK, AND L. P. CLARKE, *The effect of wavelet bases on the compression of digital mammograms*, IEEE Engrg. Med. Biol., 15 (1995), pp. 570–577. [175]

[88] D. DONOHO, *Wavelet shrinkage and W.V.D.: A ten-minute tour*, in Progress in Wavelet Analysis and Applications, Y. Meyer and S. Roques, eds., Editions Frontières, Gif-sur-Yvette, France, 1993. [168]

[89] ———, *Denoising by soft thresholding*, IEEE Trans. Inform. Theory, 41 (1995), pp. 613–627. [102]

[90] ———, *Nonlinear solution of linear inverse problems by wavelet-vaguelette decomposition*, Appl. Comput. Harmon. Anal., 2 (1995), pp. 101–126. [168]

[91] ———, *Tight frames of k-plane ridgelets and the problem of representing objects that are smooth away from d-dimensional singularities in $\mathbb{R}^n$*, Proc. Nat. Acad. Sci., U.S.A., 96 (1999), pp. 1828–1833. [185]

[92] D. DONOHO AND I. JOHNSTONE, *Ideal denoising in an orthonormal basis chosen from a library of bases*, C. R. Acad. Sci. Paris Sér. A, 319 (1994), pp. 1317–1322. [168]

[93] ———, *Ideal spatial adaption via wavelet shrinkage*, Biometrika, 81 (1994), pp. 425–455. [168]

[94] D. DONOHO, I. M. JOHNSTONE, G. KERKYACHARIAN, AND D. PICARD, *Wavelet shrinkage: Asymptopia?*, J. Roy. Statist. Soc. Ser. B, 57 (1995), pp. 301–369. [168]

[95] D. DONOHO, M. VETTERLI, R. DEVORE, AND I. DAUBECHIES, *Data compression and harmonic analysis*, IEEE Trans. Inform. Theory, 44 (1998), pp. 2435–2476. [36]

[96] P. DU BOIS-REYMOND, *Versuch einer Classification der willkürlichen Functionen reeller Argumente nach ihren Aenderungen in den kleinsten Intervallen*, J. Reine Angew. Math., 79 (1875), pp. 21–37. [150]

[97] J. J. DUISTERMAAT, *Selfsimilarity of 'Riemann's nondifferentiable function'*, Nieuw Arch. Wisk., 9 (1991), pp. 303–337. [150, 158, 159]

[98] E. ESCALERA, E. SLEZAK, AND A. MAZURE, *New evidence for subclustering in the Coma cluster using the wavelet analysis*, Astronom. Astrophys., 269 (1992), pp. 379–384.

[99] D. ESTEBAN AND C. GALAND, *Application of quadrature mirror filters to split band voice coding systems*, in Proc. IEEE Internat. Conf. Acoust. Speech Signal Process., IEEE Press, Piscataway, NJ, 1977, pp. 191–195. [206]

[100] G. FABER, *Über die orthogonal Funktionen des Herrn Haar*, Jahresber. Deutsch. Math.-Verein., 19 (1910), pp. 104–112. [18]

[101] K. FALCONER, *Fractal Geometry: Mathematical Foundations and Applications*, John Wiley & Sons, West Sussex, U. K., 1993. [147, 162]

[102] A. FAN, *Moyene de localization fréquentielle des paquets d'ondelettes*, Rev. Mat. Iberoamericana, 14 (1998), pp. 63–70. [107]

[103] M. FARGE, *The continuous wavelet transform of two-dimensional turbulent flows*, in Wavelets and Their Applications, M. B. Ruskai et al., eds., Jones and Bartlett, Boston, MA, 1992, pp. 275–302. [140]

[104] M. FARGE, E. GOIRAND, Y. MEYER, F. PASCAL, AND V. WICKERHAUSER, *Improved predictability of two-dimensional turbulent flows using wavelet packet compression*, Fluid Dynam. Res., 10 (1992), pp. 229–250. [141]

[105] M. FARGE, N. KEVLAHAN, V. PERRIER, AND E. GOIRAND, *Wavelets and Turbulence*, Proc. IEEE, 84 (1996), pp. 639–669. [137]

[106] S. FAUVE, C. LAROCHE, AND B. CASTAING, *Pressure fluctuations in swirling turbulent flows*, J. Physique II (Paris), 3 (1993), pp. 271–278. [140]

[107] J.-C. FEAUVEAU, *Analyse multirésolution par ondelettes non orthogonales et bases de filtres numérique*, Ph.D. thesis, Univ. of Paris-South, Orsay, France, 1990. [63]

[108] P. FEDERBUSH, *Quantum theory in ninety minutes*, Bull. Amer. Math. Soc., 17 (1987), pp. 93–103. [32, 33]

[109] C. FEFFERMAN, *The multiplier problem for the ball*, Ann. of Math., 94 (1971), pp. 330–336.

[110] P. FLANDRIN, *Some aspects of non-stationary signal processing with emphasis on time-frequency and time-scale methods*, in Wavelets: Time-Frequency Methods and Phase Space, J.-M. Combes, A. Grossman, and P. Tchamitchian, eds., Springer-Verlag, Berlin, 1989, pp. 68–98. [167]

[111] ———, *Wavelet analysis and synthesis of fractional Brownian motion*, IEEE Trans. Inform. Theory, 38 (1992), pp. 910–917. [21]

[112] ———, *Time-Frequency/Time-Scale Analysis*, Academic Press, San Diego, CA, 1998. [73, 76, 80, 85, 86, 103]

[113] E. FOURNIER D'ALBE, *Two New Worlds*, Longmans, Green, London, 1907. [198]

[114] M. FRAZER, B. JAWERTH, AND G. WEISS, *Littlewood–Paley Theory and the Study of Function Spaces*, AMS, Providence, RI, 1991. [23]

[115] G. FREUD, *Über trigonometrische approximation und fouriersche reihen*, Math. Z., 78 (1962), pp. 252–262. [150]

[116] P. FRICK AND V. ZIMIN, *Hierarchical models of turbulence*, in Wavelets, Fractals, and Fourier Transforms, M. Farge, J. C. R. Hunt, and J. C. Vassilicos, eds., vol. 43 of Inst. Math. Appl. Conf. Ser. New Ser., The Clarendon Press, Oxford, U. K., 1993, pp. 265–283. [145]

[117] J. FRIEDMAN AND W. STUETZLE, *Projection pursuit regression*, J. Amer. Statist. Assoc., 76 (1981), pp. 817–823. [71]

[118] U. FRISCH, *Turbulence: The legacy of A. N. Kolmogorov*, Cambridge Univ. Press, Cambridge, U. K., 1995. [127]

[119] U. FRISCH, P. L. SULEM, AND M. NELKIN, *A simple dynamical model of intermittent fully developed turbulence*, J. Fluid Mech., 87 (1978), pp. 719–736.

[120] K. FRITZE, M. LANGE, H. OLEAK, AND G. M. RICHTER, *A scanning microphotometer with an on-line data reduction for large field Schmidt plates*, Astron. Nach., 298 (1977), pp. 189–196. [194]

[121] J. FROMENT, *Traitement d'images et applications de la transformée en ondelettes*, Ph.D. thesis, Univ. of Paris-Dauphine, Paris, 1990.

[122] ———, *A functional analysis model for natural images permitting structured compression*, ESAIM: Control, Optimization and Calculus of Variations, 4 (1999), pp. 473–495. Available on-line at http://www.emath.fr. [115]

[123] J. FROMENT AND J.-M. MOREL, *Analyse multiéchelle, vision stéréo et ondelettes*, in Les ondelettes en 1989, P. G. Lemarié, ed., Lecture Notes in Math. 1438, Springer-Verlag, Berlin, 1990, pp. 51–80.

[124] D. GABOR, *Theory of communication*, J. IEE, 93 (1946), pp. 429–457. [9, 67, 90]

[125] C. GALAND, *Codage en sous-bandes: théorie et applications à la compression numérique du signal de parole*, Ph.D. thesis, Univ. of Nice, Nice, France, 1983. [35]

[126] C. GASQUET AND P. WITOMSKI, *Fourier Analysis and Applications: Filtering, Numerical Computation, Wavelets*, Springer-Verlag, New York, 1998.

[127] J. GERVER, *The differentiability of the Riemann function at certain rational multiples of π*, Amer. J. Math., 92 (1970), pp. 33–55. [20, 158]

[128] ———, *More on the differentiability of the Riemann function*, Amer. J. Math., 93 (1970), pp. 33–41. [20, 158]

[129] ———, *On Cubic Lacunary Fourier Series*. Rutgers Univ., Camden, NJ, preprint, 1999. [164]

[130] J. GLIMM AND A. JAFFE, *Quantum Physics: A Functional Integral Point of View*, 2nd ed., Springer-Verlag, New York, 1987. [33, 78]

[131] H. H. GOLDSTINE AND J. VON NEUMANN, *On the principles of large scale computing machines*, in John von Neumann: Collected Works, vol. 5, A. Taub, ed., Pergamon Press, Oxford, U. K., 1963, pp. 1–32. [This paper was never published elsewhere. It contains material presented by von Neumann in a number of lectures, in particular, one at a meeting on 15 May 1946 of the Mathematical Computing Advisory Board, Office of Research and Inventions, Navy Department, which in 1947 became the Office of Naval Research.] [138]

[132] R. GRIBONVAL, *Approximations non-linéaires pour l'analyse des signaux sonores*, Ph.D. thesis, Univ. of Paris-Dauphine, Paris, 1999. [71]

[133] A. GROSSMANN AND J. MORLET, *Decomposition of Hardy functions into square integrable wavelets of constant shape*, SIAM J. Math. Anal., 15 (1984), pp. 723–736. [8, 27]

[134] A. HAAR, *Zur Theorie der orthogonalen Functionensysteme*, Math. Ann., 69 (1910), pp. 331–371. [18]

[135] W. HÄRDLE, G. KERKYACHARIAN, D. PICARD, AND A. TSYBAKOV, EDS., *Wavelets, Approximation, and Statistical Applications*, Lecture Notes in Statistics 129, Springer-Verlag, New York, 1998.

[136] G. H. HARDY, *Weierstrass's non-differentiable function*, Trans. Amer. Math. Soc., 17 (1916), pp. 301–325. [157]

[137] G. H. HARDY AND J. E. LITTLEWOOD, *Some problems in Diaphantine approximation II*, Acta Math., 37 (1914), pp. 194–238. [157]

[138] G. H. HARDY AND E. M. WRIGHT, *An Introduction to the Theory of Numbers*, 4th ed., Oxford Univ. Press, London, 1962. [161]

[139] E. HARRISON, *Darkness at Night, A Riddle of the Universe*, Harvard Univ. Press, Cambridge, MA, 1987. [198]

[140] F. HAUSDORFF, *Dimension und äusseres Mass*, Math. Ann., 79 (1919), pp. 157–179. [148]

[141] W. HEISENBERG, *Zur statistischen Theorie der Turbulenz*, Z. Phys., 124 (1948), pp. 628–657. [128]

[142] E. HERNANDES AND G. WEISS, *A First Course on Wavelets*, CRC Press, Boca Raton, FL, 1996.

[143] M. HOLSCHNEIDER, *Inverse Radon transforms through inverse wavelet transforms*, Inverse Problems, 7 (1991), pp. 853–861. [218]

[144] M. HOLSCHNEIDER AND P. TCHAMITCHIAN, *Pointwise analysis of Riemann's "non differentiable" function*, Invent. Math., 105 (1991), pp. 157–176. [20, 218]

[145] C. HOUDRE AND R. AVERKAMP, *Wavelet Thresholding for Non (necessarily) Gaussian Noise: Idealism*. Georgia Institute of Technology, Atlanta, GA, preprint, 1999. [179]

[146] L. HUANG AND A. BIJAOUI, *Astronomical image data compression by morphological skeleton transformations*, Experiment. Astronom., 1 (1991), pp. 311–327. [195]

[147] J. C. R. HUNT, N. K.-R. KEVLAHAN, J. C. VASSILICOS, AND M. FARGE, *Wavelets, fractals and fourier transforms: Detection and analysis of structures*, in Wavelets, Fractals, and Fourier Transforms, M. Farge, J. C. R. Hunt, and J. C. Vassilicos, eds., vol. 43 of Inst. Math. Appl. Conf. Ser. New Ser., The Clarendon Press, Oxford, U. K., 1993, pp. 1–38. [104]

[148] J.-M. INNOCENT AND B. TORRÉSANI, *Wavelets and binary coalescences detection*, Appl. Comput. Harmon. Anal., 4 (1997), pp. 113–116. [103, 143]

[149] S. ITATSU, *The differentiability of the Riemann function*, Proc. Japan Acad., Ser. A Math. Sci., 57 (1981), pp. 492–495. [158, 164]

[150] S. JAFFARD, *Propriétés des matrices "bien localisées" près de leur diagonale et quelques applications*, Ann. Inst. H. Poincaré, Anal. Non Linéaire, 7 (1990), pp. 461–476. [24]

[151] ———, *Pointwise smoothness, two microlocalization and wavelet coefficients*, Publ. Mat., 35 (1991), pp. 155–168. [153, 158]

[152] ———, *Local behavior of Riemann's function*, Contemp. Math., 189 (1995), pp. 278–307. [162]

[153] ———, *The spectrum of singularities of Riemann's function*, Rev. Mat. Iberoamericana, 12 (1996), pp. 441–460. [20, 132, 163, 165]

[154] ———, *Multifractal formalism for functions, Part 1: Results valid for all functions, Part 2: Self-similar functions*, SIAM J. Math. Anal., 28 (1997), pp. 944–998. [133, 134, 136, 149, 176]

[155] ———, *Old friends revisited: The multifractal nature of some classical functions*, J. Fourier Anal. Appl., 3 (1997), pp. 1–22. [132]

[156] ———, *Oscillation spaces: Properties and applications to fractal and multifractal functions*, J. Math. Phys., 39 (1998), pp. 4129–4141. [143, 145]

[157] ———, *Beyond Besov Spaces*. Univ. of Paris XII, Créteil, France, preprint, 1999. [136]

[158] ———, *The multifractal nature of Lévy processes*, Probab. Theory Related Fields, 114 (1999), pp. 207–227. [165]

[159] S. JAFFARD AND B. MANDELBROT, *Peano–Poyla motion, when time is intrinsic or binomial (uniform or multifractal)*, Math. Intelligencer, 19 (1997), pp. 21–26. [132]

[160] S. JAFFARD AND Y. MEYER, *Wavelet methods for pointwise regularity and local oscillations of functions*, Mem. Amer. Math. Soc. 123, No. 587, AMS, Providence, RI, 1996. [142, 152]

[161] B. J. T. JONES, V. J. MARTINEZ, E. SAAR, AND J. EINASTO, *Multifractal description of the large-scale structure of the universe*, Astrophys. J., 332 (1988), pp. 1–5. [200]

[162] L. JONES, *On a conjecture of Huber concerning the convergence of projection pursuit regression*, Ann. Statist., 15 (1987), pp. 880–882. [71]

[163] ———, *A simple lemma on greedy approximation in Hilbert space and convergence results for projection pursuit regression and neural network training*, Ann. Statist., 20 (1992), pp. 608–613. [71]

[164] J.-P. KAHANE AND P. G. LEMARIÉ-RIEUSSET, *Fourier Series and Wavelets*, vol. 3 of Stud. Devel. Modern Math., Gordon and Breach, London, 1995. [16]

[165] J.-P. KAHANE AND J. PEYRIÈRE, *Sur certaines martingales de Benoît Mandelbrot*, Adv. Math., 2 (1976), pp. 131–145. [131]

[166] C. J. KICEY AND C. J. LENNARD, *Unique reconstruction of band-limited signals by a Mallat-Zhong wavelet transform algorithm*, J. Fourier Anal. Appl., 3 (1997), pp. 63–82. [125]

[167] A. N. KOLMOGOROV, *The local structure of turbulence in incompressible viscous fluid for very large Reynolds numbers*, Dokl. Akad. Nauk SSSR, 30 (1941). Proc. Roy. Soc. London Ser. A, 434 (1991), pp. 9–13, reprint. [128]

[168] ———, *A refinement of previous hypotheses concerning the local structure of turbulence in viscous incompressible fluid at a high Reynolds number*, J. Fluid Mech., 13 (1962), pp. 82–85. [128]

[169] A. LANNES, S. ROQUES, AND M. J. CASANOVE, *Resolution and robustness in image processing: A new regularization principle*, J. Opt. Soc. Amer., 4 (1987), pp. 189–199. [189, 191]

[170] E. LEGA, H. SCHOLL, J. M. ALIMI, A. BIJAOUI, AND P. BURY, *A parallel algorithm for structure detection based on wavelet and segmentation algorithm*, Parallel Comput., 21 (1995), pp. 265–285. [199]

[171] P. G. LEMARIÉ-RIEUSSET, *Analysis multi-résolution non orthogonal, commutations entre projecteurs et derivation et ondelettes vecteurs à divergence nulle*, Rev. Mat. Iberoamericana, 8 (1992), pp. 221–237. [65, 145]

[172] J. LERAY, *Etudes de diverses équations intégrales non-linéaires et de quelques problèmes que pose l'hydrodynamique*, J. Math. Pures Appl., 9 (1933), pp. 1–82. [128]

[173] J.-S. LIÉNARD, *Speech analysis and reconstruction using short-time, elementary waveforms*, in Proc. IEEE Internat. Conf. Acoust. Speech Signal Process., IEEE Press, Piscataway, NJ, 1987, pp. 948–951. [68]

[174] J.-L. LIONS, *El Planeta Tierra, El papel de les matematicas y de los superordenadores*, Espasa Calpe, Madrid, 1990. [Lectures given at the Instituto de España.] [5]

[175] G. G. LORENTZ, *Approximation of Functions*, 2nd ed., Chelsea Publishing Co., New York, 1986. [171]

[176] H. LORENZ, G. M. RICHTER, M. CAPPACCIOLI, AND G. LONGO, *Adaptive filtering in astronomical image processing*, Astronom. Astrophys., 277 (1993), pp. 321–330. [196]

[177] F. LOW, *Complete sets of wave packets*, in A Passion for Physics—Essays in Honor of Geoffrey Chew, World Scientific, Singapore, 1985, pp. 17–22. [9]

[178] T. LYCHE AND K. MØRKEN, *Knot removal for parametric B-spline curves and surfaces*, Comput. Aided Geom. Design, 4 (1987), pp. 217–230. [175]

[179] S. G. MALLAT, *Multifrequency channel decompositions of images and wavelet models*, IEEE Trans. Acoust. Speech Signal Process., 37 (1989), pp. 2091–2110.

[180] ———, *A theory for multiresolution signal decomposition: The wavelet representation*, IEEE Trans. Patt. Anal. Mach. Intell., 11 (1989), pp. 674–693.

[181] ———, *A Wavelet Tour of Image Processing*, Academic Press, New York, 1998. [71, 135]

[182] S. G. MALLAT AND W.-L. HWANG, *Singularity detection and processing with wavelets*, IEEE Trans. Inform. Theory, 38 (1992), pp. 617–643. [135]

[183] S. G. MALLAT AND Z. ZHANG, *Matching pursuits with time-frequency dictionaries*, IEEE Trans. Signal Process., 41 (1993), pp. 3397–3415. [71]

[184] S. G. MALLAT AND S. ZHONG, *Characterization of signals from multiscale edges*, IEEE Trans. Patt. Anal. Mach. Intell., 14 (1992), pp. 710–732. [135]

[185] H. S. MALVAR, *Lapped transforms for efficient transform/subband coding*, IEEE Trans. Acoust. Speech Signal Process., 38 (1990), pp. 969–978. [90]

[186] ———, *Fast algorithm for modulated lapped transform*, Electron. Lett., 27 (1991), pp. 775–776. [90]

[187] ———, *Signal Processing with Lapped Transforms*, Artech House, Norwood, MA, 1991. [90]

[188] H. S. MALVAR AND D. H. STAELIN, *Reduction of blocking effects in image coding with a lapped orthogonal transform*, in Proc. IEEE Internat. Conf. Acoust. Speech Signal Process., IEEE Press, Piscataway, NJ, 1988, pp. 781–784. [90]

[189] ———, *The lot: transform coding without blocking effects*, IEEE Trans. Acoust. Speech Signal Process., 37 (1989), pp. 553–559. [90]

[190] B. MANDELBROT, *Possible refinement of the lognormal hypothesis concerning the distribution of energy dissipation in intermittent turbulence*, in Statistical Models and Turbulence, M. Rosenblatt and C. W. Van Atta, eds., Lecture Notes in Physics 12, Springer-Verlag, Berlin, 1972, pp. 333–351. [130]

[191] ———, *Intermittent turbulence in self-similar cascades: Divergence of high moments and dimension of carrier*, J. Fluid Mech., 62 (1974), pp. 331–358. [130]

[192] ———, *The Fractal Geometry of Nature*, Freeman, San Francisco, 1982. [200]

[193] ———, *Les objects fractals*, Flammarion, Paris, 1995. [200]

[194] S. MANN AND S. HAYKIN, *The chirplet transform—A generalization of Gabor's logon transform*, in Vision Interface '91, Canadian Inform. Process. Society, Toronto, Canada, 1991. [100]

[195] ———, *Adaptive chirplet transform: An adaptive generalization of the wavelet transform*, Optical Engineering, 31 (1992), pp. 1243–1256. [100]

[196] ———, *Time-frequency perspectives: The chirplet transform*, in Proc. IEEE Internat. Conf. Acoust. Speech Signal Process., IEEE Press, Piscataway, NJ, 1992. [100]

[197] M. W. MARCELLIN, *private communication*, October 1999. [Prof. Marcellin is a member of the JPEG-2000 committee.] [65]

[198] D. MARR, *Vision: A Computational Investigation into the Human Representation and Processing of Visual Information*, W. H. Freeman and Co., New York, 1982. [6, 12, 49, 117, 120, 121, 123]

[199] D. MARR AND E. HILDRETH, *Theory of edge detection*, Proc. Roy. Soc. London Ser. B, 207 (1980), pp. 187–217. [120]

[200] F. G. MEYER, *Image compression in libraries of bases*, 1998. [Lecture notes for a course given at the Institut Henri Poincaré, Paris.] [114]

[201] F. G. MEYER, A. Z. AVERBUCH, J.-O. STRÖMBERG, AND R. R. COIFMAN, *Multi-layered image representation: Application to image compression*, in Internat. Conf. Image Process., ICIP'98, Chicago, IL, IEEE Press, Piscataway, NJ, 1998. [115]

[202] F. G. MEYER AND R. R. COIFMAN, *Brushlets: A tool for directional image analysis and image compression*, Appl. Comput. Harmon. Anal., 4 (1997), pp. 147–187. [115]

[203] Y. MEYER, *Ondelettes et Opérateurs* I: *Ondelettes*, Hermann, Paris, 1990 (in French). *Wavelets and Operators*, Cambridge Univ. Press, Cambridge, U. K., 1992 (in English). [29, 167, 235]

[204] ———, *Ondelettes et Opérateurs* II: *Opérateurs de Calderón-Zygmund*, Hermann, Paris, 1990 (in French). *Wavelets*, Cambridge Univ. Press, Cambridge, U. K., 1997 (in English).

[205] ———, *L'analyse par ondelettes d'un objet multifractal: La function $\sum_1^\infty \frac{1}{n^2} \sin n^2 t$ de Riemann*, Math. Colloquium of the Univ. of Rennes, Rennes, France, 1991.

[206] ———, *Ondelettes et algorithmes concurrents*, Hermann, Paris, 1992.

[207] ———, *Wavelets, paraproducts, and Navier–Stokes equations*, in Current Developments in Mathematics 1996, International Press, Cambridge, MA, 1997. [145, 147]

[208] ———, *Wavelets, Vibrations and Scalings*, CRM Monogr. Ser. 9, AMS, Providence, RI, 1998. [132, 152]

[209] Y. MEYER AND R. R. COIFMAN, *Ondelettes et Opérateurs* III: *Opérateurs multilinéaires*, Hermann, Paris, 1991 (in French). *Wavelets*, Cambridge Univ. Press, Cambridge, U. K., 1997 (in English).

[210] Y. MEYER AND F. PAIVA, *Convergence de l'algorithme de Mallat*, J. Anal. Math., 60 (1993), pp. 227–240. [44]

[211] Y. MEYER, F. SELLAN, AND M. TAQQU, *Wavelets, generalized white noise and fractional integration: The synthesis of fractional Brownian motion*, J. Fourier Anal. Appl., 5 (1999), pp. 465–494. [181]

[212] G. M. MOLCHAN, *Scaling exponents and multifractal dimensions for independent random cascades*, Comm. Math. Phys., 179 (1996), pp. 681–702. [131]

[213] J.-M. MOREL AND S. SOLIMINI, *Variational Methods in Image Segmentation*, Birkhäuser, Boston, MA, 1995. [182]

[214] J. E. MOYAL, *Quantum mechanics as a statistical theory*, Proc. Cambridge Philos. Soc., 45 (1949), pp. 99–124. [87]

[215] D. MUMFORD AND A. DESOLNEUX, *Pattern Theory through Examples*. Forthcoming. [5]

[216] D. MUMFORD AND B. GIDAS, *Stochastic models for generic images*, Quart. Appl. Math. to appear. [5]

[217] J.-F. MUZY, E. BACRY, AND A. ARNEODO, *The multifractal formalism revisited with wavelets*, Internat. J. Bifur. Chaos Appl. Sci. Engrg., 4 (1994), pp. 245–302.

[218] D. J. NEWMAN, *Rational approximation of $|x|$*, Michigan Math. J., 11 (1964), pp. 11–14. [169]

[219] F. NICOLLEAU AND C. VASSILICOS, *The Topology of Intermittency*, tech. report, Department of Applied Mathematics and Theoretical Physics, Cambridge Univ., Cambridge, U. K., 1999. [137]

[220] A. M. OBUKHOV, *On the distribution of energy in the spectrum of turbulent flow*, Dokl. Akad. Nauk SSSR, 32 (1941), pp. 22–24. [128]

[221] L. ONSAGER, *The distribution of energy in turbulence*, Phys. Rev., 68 (1945), p. 286. [128]

[222] A. PAPOULIS, *Signal Analysis*, 4th ed., McGraw-Hill, New York, 1988. [25]

[223] G. PARISI AND U. FRISCH, *On the singularity structure of fully developed turbulence*, in Turbulence and Predictability in Geophysical Fluid Dynamics, Proc. Internat. School of Physics "E. Fermi," 1983, Varenna, Italy, M. Ghil, R. Benzi, and G. Parisi, eds., North-Holland, Amsterdam, 1985, pp. 84–87. [131]

[224] V. PELLER, *A description of Hankel operators of class $\mathfrak{S}_p$ for $p > 0$, an investigation of the rate of rational approximation, and other applications*, Math. USSR Sbornik, 50 (1985), pp. 465–492. [The Russian version was published in 1983.] [170]

[225] P. PETRUSHEV, *Direct and converse theorems for spline and rational approximation and Besov spaces*, in Function Spaces and Applications, M. Cwikel et al., eds., Lecture Notes in Math. 1302, Springer-Verlag, New York, 1988.

[226] P. PETRUSHEV AND V. POPOV, *Rational Approximation of Real Functions*, Cambridge Univ. Press, Cambridge, U. K., 1988. [169]

[227] W. L. PRESS, *Wavelet-based compression software for FITS images*, in Astronomical Data Analysis Software and Systems I, APS Conference Series, vol. 25, Astronom. Soc. Pacific, San Francisco, 1992. [196]

[228] H. QUEFFELEC, *Dérivabilité de certaines sommes de séries de Fourier lacunaire*, C. R. Acad. Sci. Paris Sér. A, 273 (1971), pp. 291–293.

[229] H. REEVES, *Patience dans l'azur*, Seuil, Paris, 1981. [197]

[230] G. M. RICHTER, *Zur auswertung astronomischer aufnahmen mit dem automatischen flächenphotometer*, Astronom. Nachr., 299 (1978), pp. 283–303. [194]

[231] X. RODET, *Time-domain format-wave-function synthesis*, Comput. Music J., 8 (1985). [69]

[232] S. ROQUES, F. BOURZEIX, AND K. BOUYOUCEF, *Soft-thresholding technique and restoration of 3C273 jet*, Astrophys. Space Sci., 239 (1986), pp. 297–304. [192, 193]

[233] S. ROUX, *Analyse en ondelettes de l'auto-similarité de signaux en turbulence plainement développée*, Ph.D thesis, Univ. of Aix–Marseille, Marseille, France, 1996. [136]

[234] J. SCHAUDER, *Zur Theorie stetiger Abbildungen in Funktionalräumen*, Math. Z., 26 (1927), pp. 47–65. [18]

[235] E. SÉRÉ, *Localisation fréquentielle des paquets d'ondelettes*, Rev. Mat. Iberoamericana, 11 (1995), pp. 334–354. [106]

[236] E. SLEZAK, A. BIJAOUI, AND G. MARS, *Identification of structures from galaxy counts: Use of the wavelet transform*, Astronom. Astrophys., 227 (1990), pp. 301–316. [199]

[237] E. SLEZAK, V. DE LAPPARENT, AND A. BIJAOUI, *Objective detection of voids and high-density structures in the first CfA redshift survey slice*, Astronom. J., 409 (1993), pp. 517–529. [198, 199, 200]

[238] M. J. T. SMITH AND T. P. BARNWELL III, *A procedure for designing exact reconstruction filter banks for tree structured subband coders*, in Proc. IEEE Internat. Conf. Acoust. Speech Signal Process., IEEE Press, Piscataway, NJ, 1984. [207]

[239] ———, *Exact reconstruction techniques for tree structured coders*, IEEE Trans. Acoust. Speech Signal Process., 34 (1986), pp. 434–441. [207]

[240] J.-L. STARK, F. MURTAGH, AND A. BIJAOUI, *Image Processing and Data Analysis. The Multiscale Approach*, Cambridge Univ. Press, Cambridge, U. K., 1998. [197]

[241] J.-L. STARK, F. MURTAGH, B. PIRENNE, AND M. ALBRECHT, *Astronomical image compression based on noise suppression*, Pub. Astronom. Soc. Pacific, 108 (1996), pp. 446–455. [197]

[242] E. M. STEIN, *Singular Integrals and Differentiability Properties of Functions*, Princeton Univ. Press, Princeton, NJ, 1970. [23]

[243] ———, *Topics in Harmonic Analysis Related to the Littlewood–Paley Theory*, Princeton Univ. Press, Princeton, NJ, 1970. [23]

[244] G. STRANG AND G. FIX, *An Analysis of the Finite Element Method*, Prentice-Hall, Englewood Cliffs, NJ, 1973. [51]

[245] J.-O. STRÖMBERG, *A modified Franklin system and higher-order spline systems on $\mathbb{R}$ as unconditional bases for Hardy spaces*, in Conference on Harmonic Analysis in Honor of Antoni Zygmund, vol. II, W. Beckner et al., eds., Wadsworth, Belmont, CA, 1983, pp. 475–494. [15, 28]

[246] P. TCHAMITCHIAN, *Biorthogonalité et théorie des opérateurs*, Rev. Mat. Iberoamericana, 3 (1987), pp. 163–189. [63]

[247] P. TCHAMITCHIAN AND B. TORRÉSANI, *Ridge and skeleton extraction from the wavelet transform*, in Wavelets and Their Applications, M. B. Ruskai et al., eds., Jones and Bartlett, Boston, MA, 1992, pp. 123–151. [104]

[248] A. N. TIKHONOV, *Regularization of incorrectly posed problems*, Soviet Math. Dokl., 4 (1963), pp. 1624–1627. [189]

[249] B. TORRÉSANI, *Analyse continue par ondelettes*, InterÉditions/CNRS Éditions, Paris, 1995.

[250] R. VAUTARD AND M. GHIL, *Singular spectrum analysis in nonlinear dynamics, with applications to paleoclimatic time series*, Phys. D, 35 (1989), pp. 359–424. [5]

[251] J. P. VÉRAN AND J. R. WRIGHT, *Compression software for astronomical images*, in Astronomical Data Analysis Software and Systems III, ASP Conference Series, vol. 61, Astronom. Soc. Pacific, San Francisco, 1994. [194]

[252] M. VERGASSOLA, B. DUBRULLE, U. FRISCH, AND A. NOULLEZ, *Burgers' equation, devil's staircases and the mass distribution for large-scale structures*, Astronom. Astrophys., 289 (1994), pp. 325–356. [165]

[253] M. VETTERLI AND J. KOVAČEVIĆ, *Wavelet and Subband Coding*, Prentice–Hall, Englewood Cliffs, NJ, 1995. [207]

[254] J. VILLE, *Théorie et applications de la notion de signal analytique*, Câbles et Transmissions, Laboratoire de Télécommunications de la Société Alsacienne de Construction Mécanique, 2A (1948), pp. 61–74. [25, 67, 72, 87, 89, 90]

[255] C. F. VON WEIZSÄCKER, *Das Spektrum der Turbulenz bei großen Reynoldschen Zahlen*, Z. Phys., 124 (1948), pp. 614–627. [128]

[256] K. WEIERSTRASS, *Über continuierliche Functionen eines reellen Arguments, die für kienen Werth des letzeren einen bestimmten Differentialquotienten besitzen*, in Matematische Werke II, Abhandlung 2, Georg Olms, Verlagsbuchhandlung, Mildesheim; Johnson Reprint Corp., New York, 1967, pp. 71–74. [150]

[257] E. WESFREID, *Vocal command signal segmentation and phonemes classification*, in Proc. Second Symposium on Artificial Intelligence, Havana, Cuba, A. Ochoa, M. Ortiz, and R. Santana, eds., Éditorial Academia Cuba, Havana, 1999, pp. 45–50. [97]

[258] E. WESFREID AND V. WICKERHAUSER, *Adapted local trigonometric transform and speech processing*, IEEE Trans. Signal Process., 41 (1993), pp. 3596–3600. [97]

[259] R. L. WHITE, *High-Performance Compression of Astronomical Images*, tech. report, Space Telescope Science Institute, Baltimore, MD, 1992. [196]

[260] E. P. WIGNER, *On the quantum correction for thermodynamic equilibrium*, Phys. Rev., 40 (1932), pp. 749–759. [86]

[261] K. G. WILSON, *Renormalization group and critical phenomena* II: *Phase-space cell analysis of critical behavior*, Phys. Rev. B, 4 (1971), pp. 3184–3205. [32, 33, 90]

[262] P. WOJTASZCZYK, *The Franklin system is an unconditional basis in H_1*, Ark. Mat., 20 (1982), pp. 293–300. [28]

[263] J. W. WOODS AND S. O'NEIL, *Subband coding of images*, IEEE Trans. Acoust. Speech Signal Process., 34 (1986), pp. 1278–1288. [58]

[264] N. ZABUSKY, *Computational synergetics*, Phys. Today, July (1984), pp. 36–46. [138]

Author Index

Abry, Patrice, 140
Adelson, E. H., ix, 31, 36, 49, 50, 52, 55, 57
Arneodo, Alain, 16, 20, 125, 127, 130, 134–137, 139, 140, 142, 143, 145, 200
Aubry, J.-M., 143

Bacry, Emmanuel, 142
Balian, Roger, 9, 15, 67, 90
Baraniuk, Richard, xi, 80, 100
Barnwell, T. P., 207
Barthes, Roland, 1
Batchelor, G. K., 130
Battle, Guy, ix, xi, 32, 145
Benassi, Albert, 21
Benveniste, Albert, 132
Bernard, Claude, 118
Bertoin, J., 165
Bijaoui, Albert, xi, 187, 194–197, 199, 201
Bobichon, Yves, xi, 194, 196
Boulez, Pierre, 69
Brillouin, Léon, ix, 9
Brislawn, Christopher, 6
Burt, P. J., ix, 31, 50, 52, 55, 57

Calderón, Alberto, 13, 15, 27, 32
Candès, E., 184
Castaing, B., 131
Charlier, Charles, 198, 200
Ciesielski, Zbigniew, 21, 24
Cohen, Albert, xi, 45, 62, 63, 147, 177, 180, 183
Coifman, Ronald, 26, 86, 92
Cordoba, A., 87
Couder, Yves, 139

Croisier, A., 31, 35

Dahmen, W., 147
Daubechies, Ingrid, ix, 9, 11, 16, 31, 46, 63, 91, 180
De Bruijn, N. G., 87
DeVore, Ronald, 71, 147, 167, 169, 170, 172–175, 177, 182, 183
Donoho, David, 140, 167, 173, 177, 179, 180, 183, 184, 236
Du Bois-Reymond, Paul, 17
Duistermaat, J., 158, 159

Einstein, Albert, 128
Esteban, D., 31, 35, 58, 207

Faber, G., 18
Falconer, Kenneth, 147
Fang, X., 97
Farge, Marie, 90, 104, 140, 200
Fauve, S., 140
Feauveau, Jean-Christophe, 63
Federbush, Paul, ix, 32, 145
Fefferman, C., 87
Flandrin, Patrick, 85, 167
Fourier, Joseph, 13, 16, 31
Fournier d'Albe, Edward, 198, 200
Franklin, Philip, 23
Freidman, Alexander, 198
Freud, Geza, 150
Frick, P., 145
Frisch, Uriel, 127, 130, 131, 133, 134, 136, 149
Froment, Jacques, 115

Gabor, Dennis, ix, 9, 15, 67, 90
Gagne, Y., 130

Galand, Claude, 31, 35, 36, 41, 58, 207
Gerver, Joseph L., xi, 158, 163, 164
Glimm, James, ix, 77
Goldstine, Herman H., 138
Gribonval, Rémi, 71
Grossmann, Alex, ix, 8, 15, 27

Haar, Alfred, 9, 17, 31
Hardy, G. H., 2, 157, 158, 164
Harrison, Edward, 198
Hausdorff, F., 148
Haykin, Simon, 100
Heisenberg, Werner, 128
Herschel, William, 198
Hingorani, R., 36, 49
Holschneider, Matthias, 3, 15, 19, 215, 218
Hopfinger, E. J., 130
Hubble, Edwin, 198
Hunt, J. C. R., 104

Innocent, J. M., 103, 104
Itatsu, Seiichi, 158, 164

Jaffard, Stéphane, ix, x, 16, 20, 21, 91, 142, 143, 176
Jaffe, Arthur, ix, 78
Jawerth, Björn, 167, 169, 170, 172–175
Johnstone, Iain, 167
Jones, Douglas, 100
Jones, L. K., 71
Journé, Jean-Lin, ix, 91
Julesz, Bela, 118

Kahane, Jean-Pierre, 16, 131
Kant, Immanuel, 198
Kerkyacharian, Gérard, 21, 168
Kevlahan, N. K.-R, 104
Kolmogorov, A. N., 128, 129
Krim, Hamid, xi
Kruskal, M. D., 138

Lambert, Johann, 198
Lang, Serge, 158
Lannes, André, 187, 189
Laroch, C., 140
Lebesgue, Henri, 17
Lemarié-Rieusset, Pierre Gilles, 16, 65, 145

Leray, J., 128
Lévy, Paul, 20, 21, 179
Liénard, Jean-Sylvain, 68, 92, 112
Lions, Jacques-Louis, 5
Littlewood, J. E., 22, 157, 164
Low, Francis, 9, 15, 90
Lucier, B., 175, 182
Lusin, N., 25, 158
Lyche, T., 175

Magnen, Jacques, ix
Mallat, Stéphane, 8, 23, 31, 41, 43, 57, 122, 135
Malvar, Henrique, ix, 9, 90, 91
Mandelbrot, Benoît, 10, 16, 21, 127, 130, 131, 200
Mann, Steve, 80, 100
Marcinkiewicz, J., 26
Marr, David, ix, 6, 11, 12, 23, 32, 49, 117–120, 122, 168, 181
Mars, G., 199
Meyer, François, 114, 115
Meyer, Yves, x, 15, 31, 57, 92, 94, 142, 176, 177, 219
Minsky, Marvin, 117
Morel, Jean-Michel, 182
Mørken, K., 175
Morlet, Jean, ix, 8, 15, 27
Moyal, J. E., 86
Mumford, David, 5, 13, 182
Muzy, Jean-François, 142

Newman, D. J., 169
Nicolleau, F., 137

Obukhov, A. M., 128, 129
O'Neil, S., 58
Onsager, L., 128
Osher, S., 182, 183

Paley, R. E. A. C., 22
Parisi, Giorgio, 127, 130, 131, 133, 134, 136, 149
Peetre, J., 169
Pekarskii, A., 169, 171
Peller, V., 169, 170
Penzias, Arno, 198
Petrushev, P., 167, 169, 172, 177, 183
Peyrière, Jacques, 131
Picard, Dominique, 168

Popov, V. A., 167, 169, 170, 172–174

Rayner, John, xi
Reeves, Hubert, 197
Richter, G. M., 194
Riemann, Bernhard, 3, 19
Rodet, X., 69
Roques, Sylvie, xi, 192, 193
Roux, Daniel, 21
Rudin, L. I., 182, 183
Ryan, Robert, x

Schauder, J., 18
Sellan, Fabrice, 21, 65, 181
Seneor, Roland, ix
Séré, Eric, 106
Shah, J., 182
Shannon, Claude, ix, 43
Simoncelli, E., 36, 49
Slezak, E., 198, 199
Smith, M. J. T., 207
Stark, Jean-Luc, 197
Strömberg, J.-O., 13, 15, 24, 31
Swedenborg, Emanuel, 198

Tajchman, Marc, xi
Tchamitchian, Philippe, 3, 19, 63, 104
Torrésani, Bruno, xi, 103, 104
Townsend, A. A., 130

van Ness, J. W., 21
Vassilicos, J. C., 104, 137
Vial, P., 180
Ville, Jean, 25, 67, 68, 72, 73, 79, 87, 89, 90, 95, 112, 184
Vjacheslavov, N. S., 169
von Neumann, John, ix, 9, 15, 138
von Weizsäcker, C. F., 128

Weierstrass, Karl, 150
Weiss, Guido, 26
Wesfreid, Eva, xi
Weyl, Hermann, 73
Wickerhauser, Victor, 86, 140
Wiener, Norbert, ix
Wigner, Eugene, ix, 73, 76, 86
Willsky, Alan S., 132
Wilson, Kenneth, ix, 9, 15, 32, 90, 91
Wilson, Robert, 198
Wöhler, Friedrich, 2, 119
Wojtaszczyk, P., 28
Woods, J., 58

Xu, H., 177, 183

Zabusky, Norman, 138
Zimin, V., 145
Zygmund, Antoni, 22

Subject Index

admissibility condition, 209, 215
aliasing, 206
ambiguity function, 74
analytic signal, 11, 25, 79
    associated with an asymptotic signal, 81
approximation of irrationals by continued fractions, 161
astronomical data, 194
asymptotic signals, 81
atomic decomposition, 3, 7, 8, 26
atoms, 18, 26, 67

Balian–Low theorem, 9, 37, 91
bases
    chirplet, 101
    local Fourier, 13
    wavelet, 13
Bernoulli measures, 130, 131
Bernstein's theorem, 40
Besov spaces
    characterized by wavelet coefficients, 234
    homogeneous, 234
    nonhomogeneous, 234
best-basis algorithm, 72, 97, 101, 114
big bang, 198
biorthogonal wavelets, 63, 70
    divergence-free, 65
Bobichon–Bijaoui algorithm, *See* ht_compress
Brownian motion, 20
    fractional (fBm), 21, 65, 129, 181
    realization of, 20
    regularity of, 20, 21
Burgers's equation, 165

Burt and Adelson's algorithms, *See* pyramid algorithms

Calderón's identity, 15, 26, 27, 29
cartography, an illustration of scale, 49–50
cartoon image, 168
chirplets, 80, 100
chirps, 80
    first definition, 142
    hyperbolic, 86
    linear, 83
    second definition, 142
    three-dimensional, 143
    in turbulence, 141
coding
    textures, 114
coherent structures, 127, 137
conjugate quadrature filters, 207
Couder's experiment, 139

Daubechies's wavelets, 31, 79
    construction, 46
decimation operator, 37, 38, 52
devil's staircase, 130
DeVore–Lucier model, 183
discrete cosine transform, 91
discrete sine transform, 91
DNA, 2, 125, 136
dyadic blocks, 22, 30

entropy criterion, 89, 112
entropy of a vector, 96
estimator, 168, 178, 179
    optimal, 168, 178
    suboptimal, 178
extension operator, 51, 60

fast Fourier transform, 47
fast wavelet transform, 47
filter bank, general two-channel, 205
filter, definition of, 205
fingerprints, storage by FBI, 6, 70
FIR, *See* impulse response
Fix and Strang condition, 51
fluctuation, 40, 41, 55, 56
    continuous, 42
Fourier analysis, 2
Fourier–Bros–Iagolnitzer
    transform, 87
Fourier series, 2, 16, 19
Franklin system, 23, 24, 26
Freud's method, 151
functions of bounded variation (BV),
    176, 178, 184

Gabor wavelets, 9, 33, 69, 105
    optimal localization of, 106
geometric images, 181
Gerver's theorem, 19
global warming, 5
gravitational waves, 86, 102, 103
Grossmann–Morlet wavelets, 7, 9, 11,
    13, 29, 100

Hölder condition, 19
Hölder exponents, 19, 20
    algorithm for computing, 155
Hölder spaces, 19
    $C^\alpha(\mathbb{R})$, 153
    homogeneous, 233
    nonhomogeneous, 233
Haar system, 9, 15, 18, 31, 46, 62, 79
Haar wavelets, *See* Haar system
Hardy spaces, 25
    real version, 28
Hausdorff dimension, 19, 147, 148
Hausdorff measure, 148
hcompress, 194, 195
Heisenberg boxes, 72, 78
    associated with level sets of the
        Wigner–Ville transform, 89,
        90
Heisenberg uncertainty principle, 105
Hilbert basis, 18
ht_compress, 194, 196
Hubble Space Telescope, 187

IDEA, a deconvolution algorithm,
    189–193
IIR, *See* impulse response
image, *See* signal, two-dimensional
image processing, *See* signal processing
    fundamental problem, 115
impulse response, 39, 205
    finite (FIR), 205
    infinite (IIR), 205
inertial zone, 129
instantaneous frequency, 67, 79, 85
    of an asymptotic signal, 81
    relation with instantaneous spectrum, 80
    via matching pursuit, 83
    Ville's definition, 80
instantaneous spectrum, 73, 79
intermittency, 129
inversion formulas for the wavelet
    transform, 153, 210
    generalized, 215

Jacobi's Theta function, 159
Jarník's theorem, 162
JPEG-2000, x

$|k|^{-5/3}$ law, 129
knot removal, 175

Legendre inversion formula, 133
Lemarié–Meyer wavelets, 78, 94
Lévy processes, 165
linear chirps, *See* chirps
Littlewood–Paley analysis, 9, 22, 23,
    26, 29, 151
Littlewood–Paley–Stein function, 23
Littlewood–Paley–Stein theory, 29
Lusin's wavelet, 27, 211, 213

Mallat's algorithm, 41–42, 47, 124
    continuous version, 42
Mallat's conjecture, 117, 121, 122,
    125
    a case where it is true, 229
    a counterexample, 219
Mallat's matching pursuit algorithm,
    See pursuit algorithms
Mallat's theorem
    convergence to wavelet analysis,
        44

Malvar–Wilson bases, 100, 101, 115
  optimal, 97
Malvar–Wilson wavelets, 9, 23, 35, 89, 90, 92, 94, 107, 114
mammogram analysis, 175
Marr's conjecture, 117, 120, 122, 123, 125
  a counterexample to, 121
Marr's wavelet, 120, 122
microlocal analysis, 87
models for image processing, 168
Moyal's identity, 75, 84
multifractal analysis, 137
multifractal formalism, 127, 130, 133, 165
  an extension of, 143, 144
  failure of, 136, 141
multifractal objects, 19
multilayered analysis, 104
multiplicative cascade, 130
multiresolution analysis, 8, 9, 42, 57, See also pyramid algorithms
  regularity of, 58
multiscale system theory, 132
Mumford–Shah model, 182, 183

Navier–Stokes equations, 128, 145
nonlinear approximation, 175
Nyquist condition, 51

optimal algorithms, the search for, 10, 11
orthogonal pyramids, See pyramid algorithms
orthonormal basis, 18
oscillation exponent, 142
Osher–Rudin model, 182, 183
Oslo algorithm, 175

paraproduct algorithms, 147
partial isometry, 43, 210
partition functions, 135
perfect reconstruction, 39, 206
point-spread function, 188
pseudodifferential calculus, See Wigner–Ville transform
pursuit algorithms, 71
  Mallat's, 71
pyramid algorithms, 8, 31, 36
  Burt and Adelson's, 50–53, 55, 56, 59
  coding scheme, 57
  examples, 54–55
  image compression, 55
  orthogonal pyramids, 58–60
  relation with multiresolution analyses, 58

q-sparse, 174
quadrature mirror filters, 8, 31, 36, 38, 41, 59, 111, 207
  examples, 39–40, 43
quantization, See signal processing
quantization noise, 4, 36

rational approximation, 168–171
  versus spline approximation, 172
representation, Marr's ideas, 12
restriction operator, 50, 51, 60
ridgelets, 168, 184, 185
Riemann's function, 3, 13, 15, 132, 142, 150, 165, 211, 218
  belongs to $C^{1/2}(\mathbb{R})$, 155
  spectrum of singularities of, 132, 163
Riesz basis, 58

Schauder basis, 18, 20, 23
  to represent Brownian motion, 20
segmentation, 10
Shannon's theorem, 30, 36, 112
Shannon's wavelets, 43
signal, 1–2
  frequency modulated, 100–102
  nonstationary, 7
  stationary, 7
  two-dimensional, 2
signal processing, 2
  analysis, 2
  coding
    entropy, 4
    linear prediction, 35
    transform, 3, 35
    by zero-crossings, 3
  compression, 3, 31
  diagnostics, 4
  quantization, 4, 62
  restoration, 5
  storage, 3
  transmission, 3

sparse wavelet expansion, 167, 170, 171, 173, 176
sparsity, *See* sparse wavelet expansion
spectrum of oscillating singularities, 143
spectrum of singularities, 131, 133, 165
spline approximation, 171
spline function, 51
    basic cubic, 51, 219
split-and-merge algorithm, 96
splitting algorithms, 112
statistical modeling, 5, 6, 70, 128
Strömberg's wavelets, 24, 28
structure functions, 129, 133, 134
    of fractional Brownian motion, 129
subband coding
    ideal filters, 36–37
    two channels, 38
subsampling, *See* decimation operator

Taylor hypothesis, 129, 130
textures, 181, 182
theta modular group, 159
thresholding, 173, 174, 179, 181
    soft, 181, 192
Tikhonov regularization, 189
time-frequency algorithms, 13
time-frequency analysis, 67, 86, 87, 89, 105
time-frequency atoms, 68, 72, 89, 105
    a collection Ω, 69
    Gabor's, 68, 69
    Liénard's, 69
    precise definition, 79
time-frequency plane, 67, 72
time-frequency wavelets, 7, 9, 15
time-scale algorithms, 13
time-scale wavelets, 7, 15
transfer function, 205
transition operator, 60

trend, 40, 41, 55, 56
    continuous, 42
Tukey's window, 124, 229

unconditional basis, 28

vortex filaments, 137

Walsh system, 110, 111
wavelet analysis, 3
wavelet coefficients, 27
wavelet methods for PDEs, 147
wavelet packets, 35, 36, 38, 89, 90, 107, 114, 115
    basic, 108, 109
    general, 111
wavelet shrinkage, 168, 180, 181, 183
wavelet thresholding, *See* thresholding
wavelet transform modulus maxima algorithm, 135, 137
wavelets
    divergence-free, 145
wave packets of Cordoba–Fefferman, 87
weak l^p, 176
Weierstrass's function, 13, 132, 149
    belongs to $C^{-\frac{\log B}{\log A}}(\mathbb{R})$, 157
Weyl–Heisenberg group, 33
Weyl symbol, 76
Wigner–Ville transform, 72, 73
    of an asymptotic signal, 81–83
    cross terms, 85
    properties, 74, 75
    pseudodifferential calculus, 76
    quantum mechanics, 74
    relation with ambiguity function, 74
    relation with Weyl symbol, 77
WTMM algorithm, *See* wavelet transform modulus maxima algorithm

zero-crossings, 117, 120